VOLUME ONE HUNDRED AND TWENTY SIX

PROGRESS IN MOLECULAR BIOLOGY AND TRANSLATIONAL SCIENCE

Mechanotransduction

VOLUME ONE HUNDRED AND TWENTY SIX

Progress in Molecular Biology and Translational Science

Mechanotransduction

Edited by

ADAM J. ENGLER
Department of Bioengineering, University of California, San Diego, La Jolla, California, USA

SANJAY KUMAR
Department of Bioengineering, University of California, Berkeley, Berkeley, California, USA

ELSEVIER

AMSTERDAM • BOSTON • HEIDELBERG • LONDON
NEW YORK • OXFORD • PARIS • SAN DIEGO
SAN FRANCISCO • SINGAPORE • SYDNEY • TOKYO
Academic Press is an imprint of Elsevier

Academic Press is an imprint of Elsevier
225 Wyman Street, Waltham, MA 02451, USA
525 B Street, Suite 1800, San Diego, CA 92101-4495, USA
32 Jamestown Road, London NW1 7BY, UK
The Boulevard, Langford Lane, Kidlington, Oxford OX5 1GB, UK

First edition 2014

ISBN: 978-0-12-394624-9
ISSN: 1877-1173

For information on all Academic Press publications visit our website at store.elsevier.com

CONTENTS

CONTRIBUTORS

Zeinab Al-Rekabi
Centre for Interdisciplinary NanoPhysics, Department of Physics, University of Ottawa, Ottawa, Ontario, Canada, and Department of Mechanical Engineering, University of Washington, Seattle, Washington, USA

Samer Alam
Department of Chemical Engineering, University of Florida, Gainesville, Florida, USA

Soline Chanet
Department of Biology, Massachusetts Institute of Technology, Cambridge, Massachusetts, USA

Erin J. Cram
Department of Biology, Northeastern University, Boston, Massachusetts, USA

Richard B. Dickinson
Department of Chemical Engineering, University of Florida, Gainesville, Florida, USA

Adam J. Engler
Department of Bioengineering, University of California, San Diego, La Jolla, California, USA

Gregory R. Fedorchak
Department of Biomedical Engineering & Weill Institute for Cell and Molecular Biology, Cornell University, Ithaca, New York, USA

Nikolche Gjorevski
Institute of Bioengineering, Ecole Polytechnique Fédérale de Lausanne, Lausanne, Switzerland

Wolfgang H. Goldmann
Center for Medical Physics and Technology, Biophysics Group, Friedrich-Alexander-University of Erlangen-Nuremberg, Erlangen, Germany

Kristina Haase
Centre for Interdisciplinary NanoPhysics, Department of Physics, University of Ottawa, Ottawa, Ontario, Canada

Hiroaki Hirata
Mechanobiology Institute, National University of Singapore, Singapore, Singapore

Brenton D. Hoffman
Department of Biomedical Engineering, Duke University, Durham, North Carolina, USA

Ashley Kaminski
Department of Biomedical Engineering & Weill Institute for Cell and Molecular Biology, Cornell University, Ithaca, New York, USA

Gaurav Kaushik
Department of Bioengineering, University of California, San Diego, La Jolla, California, USA

Sanjay Kumar
Department of Bioengineering, University of California, Berkeley, Berkeley, California, USA

Jean-Cheng Kuo
Institute of Biochemistry and Molecular Biology, National Yang-Ming University, Taipei, Taiwan

Jan Lammerding
Department of Biomedical Engineering & Weill Institute for Cell and Molecular Biology, Cornell University, Ithaca, New York, USA

Tanmay P. Lele
Department of Chemical Engineering, University of Florida, Gainesville, Florida, USA

Chwee Teck Lim
Mechanobiology Institute, and Department of Biomedical Engineering, National University of Singapore, Singapore, Singapore

David B. Lovett
Department of Chemical Engineering, University of Florida, Gainesville, Florida, USA

Shaoying Lu
Department of Bioengineering, Institute of Engineering in Medicine, University of California, San Diego, La Jolla, California, USA

Matthias Lutolf
Institute of Bioengineering, Ecole Polytechnique Fédérale de Lausanne, Lausanne, Switzerland

Adam C. Martin
Department of Biology, Massachusetts Institute of Technology, Cambridge, Massachusetts, USA

Andrew E. Pelling
Centre for Interdisciplinary NanoPhysics, Department of Physics; Department of Biology, and Institute for Science Society and Policy, University of Ottawa, Ottawa, Ontario, Canada

Kyle J. Roux
Sanford Children's Health Research Center, University of South Dakota, Sioux Falls, South Dakota, USA

Masahiro Sokabe
Mechanobiology Institute, National University of Singapore, Singapore, Singapore, and Mechanobiology Laboratory, Nagoya University Graduate School of Medicine, Nagoya, Japan

Yingxiao Wang
Department of Bioengineering, Institute of Engineering in Medicine, University of California, San Diego, La Jolla, California, USA

Sophie Y. Wong
Department of Bioengineering, University of California, Berkeley, Berkeley, California, USA

PREFACE

The past few decades have observed an explosion of genomic and proteomic technologies that have fueled dramatic advances in our understanding of how specific genes and gene products influence cell behavior. While these efforts have been instrumental to our understanding of cell-intrinsic regulation of cell biology, it has also become increasingly evident that cell-extrinsic signals from the tissue microenvironment can direct and/or facilitate many critical cell functions. Physical changes in the niche, including alterations to tissue structure and mechanics, strongly activate and modulate a wide variety of biochemical signals, and the growing recognition of this fact in the 1970s and 1980s spawned the field of mechanotransduction. The origins of this field are rooted at the interface of cell biology and engineering, where significant technological advances in optical and nonoptical imaging, the ability to mechanically manipulate cells, and computational modeling have enabled researchers to identify cell signaling pathways in which application of force produces biochemical changes in specific molecules. These changes include but are not limited to cells pulling on individual molecules to directly change conformation and activity, expose kinase phosphorylation sites, and uncover cryptic binding sites. These molecular-scale events can have enormous phenotypic consequences, influencing growth, apoptosis, differentiation, motility, and many other functions. Given the fundamental, life-defining nature of these events, mechanotransduction—which may be defined to include mechanical regulation of cell behavior, force sensing, and force-dependent signaling, among other topics—is becoming a central focus not just for bioengineers but for biomedical scientists more broadly.

The chapters in this edition of the *Progress in Molecular Biology and Translational Science* series highlight a range of topics leading from the development and use of tools in molecular biology (Chapters 1 and 2) to commonly studied cellular structures, including focal adhesions (Chapters 3–6) and the nucleus (Chapters 7 and 8). One emerging and particularly exciting notion is the idea that application of force can serve as a master switch that can direct quiescent or uncommitted cells toward specific phenotypic endpoints. An example of this is a stem cell that generates relatively low force in its undifferentiated state. As differentiation is initiated, cellular stresses and strains increase, and these changes directly contribute to lineage commitment and differentiation. Chapters 9–11 focus on various aspects of mechanotransduction in these uncommitted cells: in cardiac cells during their maturation (Chapter 9), in the malignant behavior of tumor-initiating cells (Chapter 10), and in stem cells (Chapter 11). Finally, as the field scales lessons learned from cell culture studies up to complex tissues, it is important to realize that mechanotransduction in tissue is likely not a single-cell phenomenon. Instead, mechanical cues and forces can be transmitted through cells and tissues and felt over hundred of microns to millimeters, much like growth factors can diffuse over long distances through a tissue. As such, our final chapters focus on mechanotransduction in tissues (Chapters 12 and 13). In model organisms, it is also possible to study mechanotransduction not just in space but in time during development. These chapters provide insight into those aspects of mechanotransduction as well as provide the reader with a more complete understanding of the role played by mechanically induced regulation, sensing, and signaling in intact tissues.

At the conclusion of this edition of the series, we hope that the insights offered by these chapters will give the novice reader an introductory appreciation of the field and a more experienced reader a snapshot of this field's state of the art and major open questions. We also hope that this work provokes significant dialogue among authors and other scientists within the field to tackle problems described by these chapters, notably the integration of signaling across length scales, the development of new tools in non-transparent tissues, and even basic definitions of where mechanical inputs end and biochemical signals begin. Thus with much anticipation, we encourage the reader to use the force and appreciate mechanotransduction from the perspectives of tools, cells, and tissues.

Adam J. Engler
Department of Bioengineering, University of California, San Diego
La Jolla, CA

Sanjay Kumar
Department of Bioengineering, University of California, Berkeley
Berkeley, CA

PART ONE

Subcellular Tools for Activating and Measuring Mechanotransductive Signaling

CHAPTER ONE

The Detection and Role of Molecular Tension in Focal Adhesion Dynamics

Brenton D. Hoffman
Department of Biomedical Engineering, Duke University, Durham, North Carolina, USA

Contents

Abstract

Cells are exquisitely sensitive to the mechanical nature of their environment, including applied force and the stiffness of the extracellular matrix (ECM). Recent evidence has shown that these variables are critical regulators of diverse processes mediating embryonic development, adult tissue physiology, and many disease states, including cancer, atherosclerosis, and myopathies. Often, detection of mechanical stimuli is mediated by the structures that link cells that surround ECM, the focal adhesions (FAs). FAs are intrinsically force sensitive and display altered dynamics, structure, and composition in response to applied load. While much progress has been made in determining the proteins that localize to and regulate the formation of these structures, less is known about the role of tension across specific proteins in this process. A recently developed class of force-sensitive biosensors is enabling a greater understanding of the molecular bases of cellular mechanosensitivity and cell migration.

Progress in Molecular Biology and Translational Science, Volume 126
ISSN 1877-1173
http://dx.doi.org/10.1016/B978-0-12-394624-9.00001-4

1. BRIEF INTRODUCTION TO MECHANOBIOLOGY

The mechanical nature of the cellular microenvironment is increasingly recognized as a key determinant of many developmental, physiological, and pathophysiological processes[1–6] as well as an important variable in tissue engineering and regenerative medicine.[7–9] *In vivo* cells adhere to a deformable extracellular matrix (ECM) that is both a source of applied forces and a means of mechanical support.[10–12] Cells detect, interpret, and respond to these mechanical signals through a poorly understood process called mechanotransduction.[13,14] Mechanosensitive signaling affects several fundamental cellular processes, including cell contraction,[15] migration,[16] differentiation,[17] and growth.[18] For instance, during gastrulation in *Drosophila melanogaster*, germband extension causes compression of the stomodeal cells required for subsequent tissue invagination.[19] Similarly, mechanical effects also mediate many physiological principles. These include Wolfe's law, which describes how mechanical loading leads to enhanced bone formation to enable adaptation of the skeleton,[20] as well as the Bayliss effect, which describes the reduction in the diameter of arterioles after a pressure increase to maintain constant flow in downstream capillaries.[21]

Pathological mechanotransduction, often due to alterations in the mechanical nature of the microenvironment, is critically important in many prevalent and poorly understood human diseases.[4,6,22] Inside the vasculature, cyclic blood flow leads to dynamic shear stresses on endothelial cells and alterations in blood pressure, stretching the vessel wall. Atherosclerotic lesions preferentially form in areas with perturbed hemodynamics and reduced forces applied to cells. Tumors are characterized by enhanced cell growth and perturbed ECM structure, leading to changes in the local tissue stiffness.[15] These alterations in mechanical properties enable palpation exams for "lumps" as a common method of tumor detection. Recently, these rigidity changes have been shown to have a causative role in tumor progression.[23]

Traditionally, biological regulation has been understood through the principles of solution chemistry; reaction rates, diffusion, and binding affinities have been considered the dominant molecular scale variables. A central premise of biochemistry is that protein structure dictates function, suggesting that the diverse properties exhibited by proteins are due to their intricate three-dimensional shapes. A significant challenge in the field of mechanobiology is determining how macroscale mechanical variables alter molecular scale biochemical processes.[24] A major advance was the demonstration that proteins are deformable at forces that can be generated by cells.[25–27] Thus, a simple mechanism for mechanotransduction is that applied load alters protein structure, leading to novel functions. While many ground-breaking studies have enumerated the changes in cell signaling or protein expression after mechanical stimulation,[28,29] relatively few have focused on determining the relationship between protein deformation and alterations in biochemical properties, such as differential enzymatic activity or binding lifetimes, in living cells. Recently, a new class of biosensor has been developed that reports the deformation within or the tension across specific proteins in living cells.[30–32] These have started to reveal some of the molecular mechanisms mediating mechanotransduction, particularly in the context of adhesion biology.

In this article, I first review mechanosensitive behavior in the subcellular structure most associated with mechanosensitive phenomena, the focal adhesion (FA).[33] We divide this process into mechanotransmission, mechanosensing, and mechanoresponse and briefly highlight critical molecular determinants of each step. Then, I discuss issues associated with the development and use of force-sensitive biosensors. I also describe recent advances in the understanding of the force-sensitive regulation of FA dynamics made possible by the development of these new sensors. I end by highlighting several future experiments that would further our understanding of the biophysical and biochemical processes mediating mechanotransduction.

2. FOCAL ADHESIONS IN MECHANOSENSING

2.1. Focal adhesion structure

While the signaling pathways mediating mechanotransduction are still being elucidated, important subcellular structures have been identified.[34] These include the structures that cells utilize to interact with the ECM, FAs.[35] FAs are dynamic mechanosensitive scaffolds containing >150 proteins that mechanically link the ECM and actin cytoskeleton.[36,37] Connections to the ECM are mediated by integrins, heterodimeric transmembrane proteins which mediate cell adhesion through conformational regulation.[38] Integrins primarily mediate changes in FA structure through the direct and indirect recruitment of proteins. Recent work using super-resolution microscopy with enhanced resolution in the vertical direction has shown that the proteins within FAs are arranged in a weakly stratified structure, comprising an integrin signaling layer, a force transduction layer, an actin regulatory layer, and finally actin-based stress fibers. The stress fibers simultaneously load the FA with forces generated by the cytoskeleton and resist deformation due to the application of external loads[39] (Fig. 1.1). The cytoplasmic domains of integrins mediate interactions with numerous adaptor proteins (e.g., talin, paxillin, kindlin) and recruit, directly or indirectly, a host of signaling proteins (e.g., FA kinase, Src family kinases) to comprise the integrin signaling layer. The force transmission layer comprises many additional adaptor proteins (e.g., vinculin, talin, zyxin) that enable the dynamic and biochemically regulated transmission of force between the other layers. These proteins bind to a host of actin regulatory proteins (e.g., VASP) and actin cross-linking proteins (e.g, actinin) that mediate formation and reinforcement of the actin stress fibers. Note that some proteins exist in multiple layers. Talin, for instance, is thought to be oriented at an angle to link multiple layers.[39]

2.2. Molecular mechanisms of mechanotransduction

Descriptions of mechanotransduction typically involve three distinct steps: transmission of the applied load to specialized structures, transduction of the force into a biochemically detectable signal, and the subsequent response of the cell[13,40] (Fig. 1.2). These are commonly referred to as mechanotransmission, mechanosensing, and mechanoresponse, respectively. Here, I focus on the first two steps. Mechanoresponses, including the long-term adaptation of cellular adhesion structures and the actin cytoskeleton,

Figure 1.1 The stratified structure of a focal adhesion (FA). Integrins are transmembrane proteins that bind to the ECM and several intracellular proteins, including paxillin and FAK. Many of the proteins found near the plasma membrane are signaling proteins, which determine the activation state and binding partners of integrins. Slightly higher in the FA reside many adaptor proteins whose main functions are to transmit force between the layers and transduce these forces into biochemically detectable signals. Key elements include talin and vinculin. Higher still is a layer of proteins associated with actin regulation. This class is typified by VASP. At the highest point, the FA links to actin stress fibers. These layers work in concert to regulating the mechanosensitive behavior of the FA. (See the color plate.)

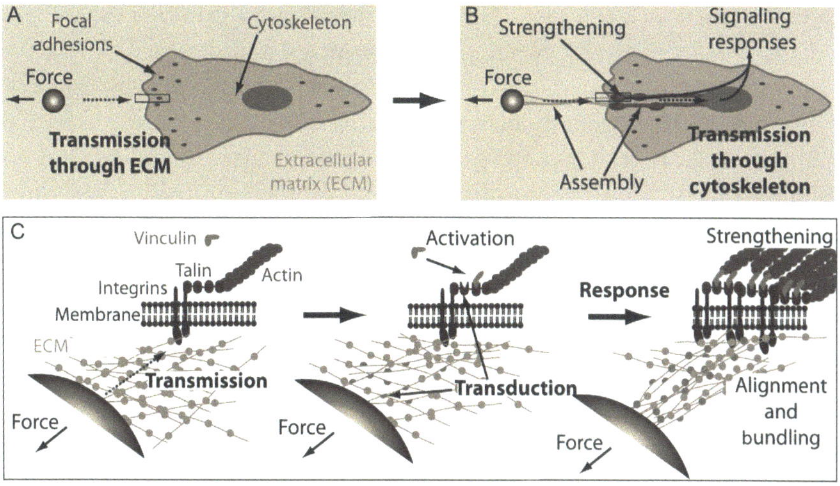

Figure 1.2 Steps of mechanotransduction. (A) A depiction of a scenario common in mechanosensitive phenomena. Cells are adhered to a deformable ECM, which a magnetic bead is used to deform. Force is transmitted through the ECM to the cell surface, where it is detected by FAs. (B) The applied load results in the enhanced assembly of the ECM, strengthening of the focal adhesion, and formation of actin-based stress fibers. Forces are then transmitted through the cytoskeleton and to the nucleus to mediate changes in signaling and gene expression. (C) An enlargement of the box shown in (A) and (B) with a simplified molecular scale representation of a FA. Transmission of applied load results in force-induced conformation changes in proteins within the ECM and FA. Conformation changes in the ECM lead to enhanced alignment and bundling. In FAs, larger loads are associated with the exposure of a cryptic binding site in talin that enables binding of vinculin. This is a critical step in FA strengthening.

activation of transcription factors leading differential protein expression, as well as many physiological processes mediating tissue homeostasis in response to mechanical perturbations, are not necessarily force-dependent and have been extensively reviewed elsewhere.[5,21,28,34,41]

While mechanotransmission typically refers to the ability of mechanical forces to be transmitted over long distances, like on the length scale of the whole cell, it also must occur on smaller scales, like within a FA. Recent evidence has suggested that this transmission is dynamic and/or load-dependent. From a purely physical basis, the noncovalent bonds that mediate protein–protein interactions are force sensitive and will rupture if pulled on long enough or strong enough,[42,43] ceasing mechanotransmission. While most bonds are slip bonds[44] and their lifetime decreases with load, many biological important linkages, such as integrin–fibronectin[45] and actin–myosin,[46] show an increase in lifetime over certain ranges of force. These are referred to as catch bonds.[47] These interactions set a critical timescale for cells as mechanotransduction must be completed and detected during the characteristic lifetime of the load-bearing bonds.[13] Consistent with this concept, studies simultaneously tracking the flow of the actin cytoskeleton and FA proteins have shown that there is an apparent slip plane within the FA whose strength is dependent on membrane protrusion.[48] Together, these data suggest the interesting possibility that dynamically regulated mechanotransmission within the FA is a critical determinant of mechanosensitive signaling.

The step that enables the conversion of mechanical force into a biochemically detectable signal, mechanotransduction, is the least understood. Figure 1.2C depicts a schematic of some of the most well-understood mechanisms. Force-based activation has only been demonstrated for several proteins, including fibronectin,[49] p130CAS,[50] and talin.[27] Talin, which is the most well understood, is a dimeric scaffolding protein that binds integrins and some intracellular components of the FA, including the force-generating actin cytoskeleton.[51] When inactive, it is held in the cytosol in autoinhibited conformation, where the binding sites for actin and integrin are hidden from the protein surface.[52] Upon activation, talin binds to integrins, actin, and vinculin, forming some of the initial linkages in FAs. The application of larger forces exposes cryptic binding sites within talin, which are typically buried within its protein structure, for vinculin.[27]

Vinculin is a FA adaptor protein that binds to talin and actin, as well as several FA adaptor proteins.[53] It comprises a head domain, which binds to talin, and a tail domain, which binds to actin, that are joined by a flexible linker.[54] When inactive, vinculin assumes an autoinhibited conformation, mediated by a strong interaction between head and tail domains.[55,56] Importantly, vinculin is recruited to FAs in response to increased load,[33] and the enhanced talin–vinculin–actin linkages are thought to be a critical aspect of adhesion strengthening.[27] Broadly, consistent with this idea, cells lacking talin do not form mature FAs, typically associated with load-bearing, and display disturbed interactions (i.e., loss of long-term adhesion stability) with adhesive surfaces.[57] Cells lacking vinculin exert lower traction forces than wild-type cells and have migration and adhesion defects.[58,59] Furthermore, reports have suggested that talin can be extended up to seven times its original length in load-bearing FAs.[60] This massive extension is thought to be mediated by the partial unfolding of repeated helical domains to mediate vinculin binding along the talin rod. While the exact role of force in vinculin activation is unknown, its recruitment is particularly effective in

FA strengthening, as it binds to actin as well as a number of other proteins known to be important in FA regulation.[61] Together these data suggest that talin–vinculin interactions are one of several key regulators of the mechanosensitive nature of FAs.

3. DESIGN AND USE OF OPTICALLY BASED MOLECULAR TENSION SENSORS

The increased interest in the response of cells to mechanical forces has led to many studies regarding the forces that proteins experience and the resulting effects on biochemical signaling. While there are many techniques for detecting the force generation of cells,[10] until recently there have been few tools for measuring the forces experienced by specific proteins in living cells. Now there are several designs, with various advantages and disadvantages, but most are based on sensors that lead to force-induced changes in the emission of fluorescent proteins or other fluorescent molecules.[30,31] In this section, I review the physical principles enabling the development of such sensors, as well as important controls and aspects of the experimental protocols necessary for the proper use of this new class of biosensor.

3.1. Basics of Forster Resonance Energy Transfer (FRET)

Technologies capable of reporting molecular scale deformation (e.g., nanometer displacements) within living cells are somewhat limited and are currently being developed.[62] One approach is Forster Resonance Energy Transfer, which is based on the nonradiative transfer of energy between an excited donor fluorophore and an adjacent acceptor fluorophore or a quenching group. Acceptor fluorophores will then emit a photon, while quenching groups dissipate absorbed energy as shown in Fig. 1.3.[63] While analogous arguments hold for quencher-based systems, here I mainly focus on FRET pairs comprising two fluorophores. In this case, FRET results in the quenching of donor fluorescence and the sensitized emission of the acceptor fluorophore in response to donor excitation light. Importantly, direct excitation and detection of emission of the acceptor fluorophore can be used as an indication of local concentration, as this signal is independent of FRET. FRET is quantified through either measurements of the percentage of photons that undergo FRET, referred to as the FRET efficiency, or relative measurements, typically based on enhanced emission of the acceptor fluorophore.[64] The amount of FRET is dictated by the optical properties, separation distance, relative orientation, and mobility of the

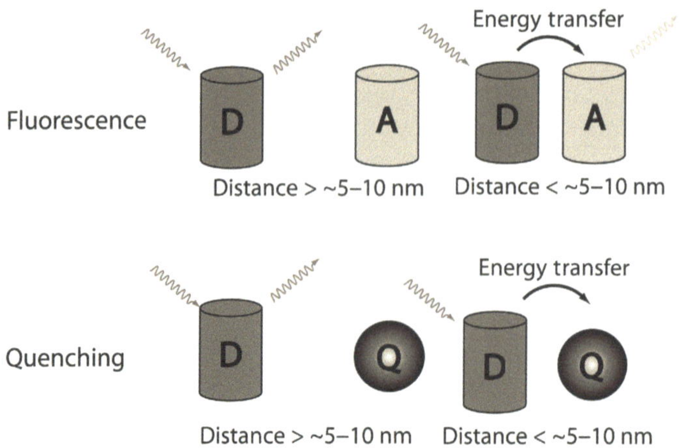

Figure 1.3 Schematic showing the basic principle of FRET: FRET is due to non-radiative energy transfer between two molecules with overlapping emission and absorption spectra. If two fluorophores are present, enhanced emission of the acceptor occurs when the separation distance between the dyes is less than approximately 5–10 nm. If a quencher is used, then donor excitation will be suppressed at small separation distances and no fluorescence will be observed.

fluorophores. It is commonly assumed that the fluorophores diffuse randomly and have no preferred orientation.[65] In this limit, FRET efficiency (E) can be well described with the simple equation:

$$E = \frac{R_o^6}{R_o^6 + r}$$

where R_o is the Forster distance and is defined as the fluorophore separation distance where 50% FRET efficiency is achieved and r is the separation distance of the two fluorophores. The effects of varying fluorophores are captured within the R_o parameter. Two key properties are the degree of overlap between the donor emission wavelengths and the acceptor excitation wavelengths and the wavelengths at which this overlap occurs, as overlap at longer wavelengths is more favorable for FRET.[63] The value of R_o is an indication of the ability of a pair of fluorophores to undergo FRET. The most commonly used organic dye pair is Cy3 and Cy5 ($R_o = 6.0$ nm);[66] however, these are often difficult to use in living cells due to the technical challenges associated with cellular incorporation. Given the recent advances in the discovery and engineering of fluorescent proteins, many FRET pairs based on genetically encodable fluorescent proteins have emerged.[67] Sensors are created with standard, DNA-based molecular cloning techniques and readily expressed in cells using common transfection reagents or viral methods. Typically used pairs include mCerulean and Venus ($R_o = 5.4$) and mTFP1 and Venus ($R_o = 6.1$ nm).[68] Instead of an acceptor fluorophore, nonfluorescent quenchers can also be used. One common pair is Alexa Fluor 647 and QSY 21 with ($R_o = 6.9$).[69] These sensors have the advantage of imaging in a single channel and a simplified analysis, but comparisons of concentration between samples can be difficult, as no acceptor fluorophore is present.

3.2. Designs of FRET-based force-sensitive biosensors

A wide array of fluorescent and FRET-based biosensors for many biologically relevant phenomena, including pH, ion concentration, and the activity state of kinases and small GTPases,[67,70] exist. Given the strong distance dependence of FRET efficiency, many of the sensors for biochemical activity are approximated as binary detection systems. Here, we focus on FRET biosensors which explicitly use the distance,[30,31] or in some cases the orientation,[71] sensitivity of FRET to report the forces across specific proteins (Fig. 1.4).

3.2.1 Extensible domain

In the most common design,[30,31] an extensible domain is placed between two fluorophores, and, while the actual situation is more complex, the general principle can be understood from Hooke's law, $F = kx$. As force (F) is applied to the sensor, the domain will extend, increasing the separation distance of the fluorophores (x) and reducing the amount of FRET. If the mechanical properties of the extensible domain are known, represented by the spring constant k in this simplified example, then the magnitude of the force, also referred to as the tension, can be inferred.[31] Note, that no vector or directional information is reported by these sensors. Several genetically encodable extensible linkers have been used. Most have involved well-understood biopolymers, including single-stranded DNA,[72] alpha-helical domains,[30] spectrin–repeat domains,[73] and domains based on flagelliform,[31] a main component of spider silk. The flagelliform-based tension sensor enables particularly quantitative measurements as single-molecule fluorescent force spectroscopy[66] was used to calibrate its force sensitivity. This allows direct calculation of the force experienced by the tension senor as a quantifiable analog readout, if FRET efficiency is measured.[31,74] For most other extensible sensors, it is only known that FRET changes occur at an approximate load (e.g., ~6 pN). Importantly, these genetically encoded

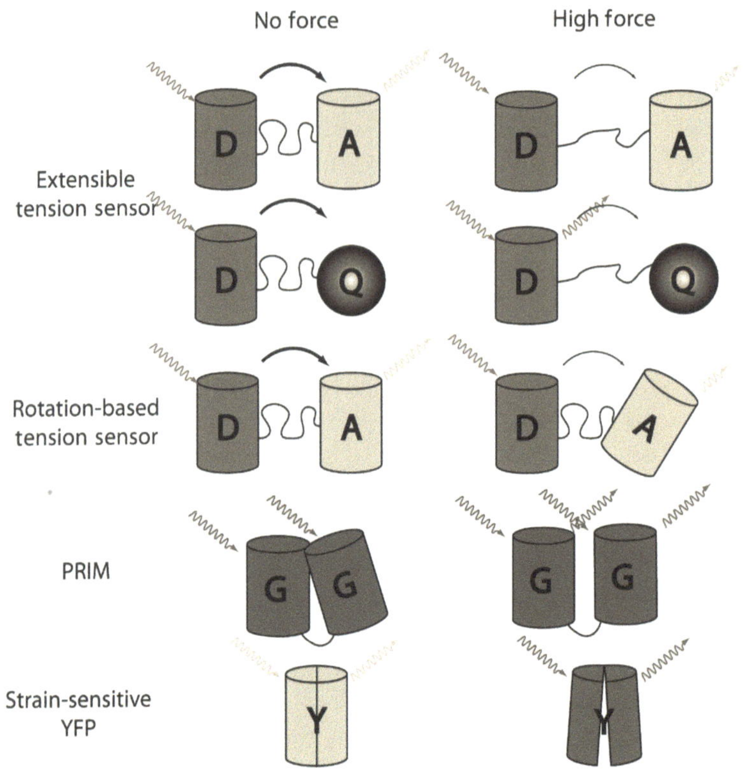

Figure 1.4 Schematics of the most common designs used to create force or strain-sensitive biosensors: extension-based tension sensors report reduced donor quenching as a calibrated spring is extended. This can be detected through the reduced emission of an acceptor fluorophore (A) or enhanced emission of the donor fluorophore (D). The later measurements typically involved a quenching group (Q) that dissipates absorbed energy as heat. Rotation-based sensors utilize the orientation dependence of FRET to report increased strain. Proximity imaging (PRIM) is based on the fact that two green fluorescent proteins (GFP, G in the figure) in contact have different optical properties. Load application could separate the fluorophores, restoring fluorescent emission. Recently, a version of yellow fluorescent protein was engineered to have strain-dependent photophysical properties. (See the color plate.)

tension sensors should be compatible with advanced three-dimensional cell culture systems as well as *in vivo* studies.

Recently, new designs using organic components have been reported. To enable single-molecule imaging and to avoid technical challenges associated with the introduction of protein or synthetic constructs into cells, efforts have focused on the cell–substrate interface. These sensors often involve the use of organic dyes,[74] such as Cy3 and Cy5, or quenchers,[69,75] to enable measurements ranging from ensemble averages to single-molecule resolution due to the greater brightness and photostability of organic dyes. Nongenetically encoded extensible linkers are typically comprised of extremely well-understood synthetic polymers, such as polyethylene glycol (PEG).[69,75] This allows the creation of tension sensors with rationally designed force sensitivities and bypasses the necessity of single-molecule measurements for calibration. Additionally, sensors based on other forms of fluorescent quenching can be used. Specifically, gold nanoparticles have recently been used to create a sensor with $1/r^4$ distance dependence,[75] enabling accurate measurements of forces over a larger range (1–25 pN) with a single construct. Also, a simplified technique based on the well-understood dissociation kinetics of DNA hairpins has also been developed.[76] This technique is digital, as opposed to fully analog tension sensors described thus far, and reports whether the forces are above or below a tunable threshold.

Force-sensitive biosensors based on extensible domains have been created for actinin,[30] spectrin,[30] filamin,[30] vinculin,[31] E-cadherin,[77] VE-cadherin,[78] as well as the tension between integrins and an engineered ECM surface[74,75] and the tension generated in early stages of epidermal growth factor receptor endocytosis.[69]

3.2.2 Rotatable domain

Changes in orientation of fluorophores can be used to modulate FRET.[63] This dependence is captured by the orientation factor, κ^2. κ^2 varies between 0 and 1 when fluorophores are oriented side by side and aligned perpendicularly or in parallel. Recently, based on the premise that an applied torsion would alter the orientation of two aligned fluorophores between these limits, a new class of load-sensitive FRET probes was reported.[71] Sensors respond to loads of approximately 5–10 pN. This sensor has the advantage of having a hypothetically larger dynamic range than the extension-based sensors, but the relationships between force, continuous changes in κ^2, and FRET are not well understood. However, whether this sensor only responds to torsion or has sensitivity to extension, as the fluorophores are linked by a short extensible linker, is unknown. Additionally, the role of torsion in mechanosensitive cellular processes has rarely been studied. This design has been used to create torsion sensors for spectrin.[71]

3.2.3 Other designs

Several new strain-sensitive optical probes have recently been reported, but the forces that the relevant transitions occur at are unknown. Another force-sensitive imaging modality is proximity imaging (PRIM).[79] PRIM is based on the direct contact of two GFPs, which can lead to structural perturbations and detectable shifts in spectral properties. Using this technique, a sensor that reports the binding of myosin heads to actin filaments, which should be indicative of cellular force generation, due to the strains induced within the sensor upon binding has been created. Recently, a variant of YFP exhibiting a strain-sensitive shift in emission spectrum was created using circular permutation and an extensible bridge peptide.[80] Once calibrated, this strain-sensitive element will have the advantage of being approximately one-half to one-third the size of the existing tension sensor modules. Suggesting a new design modalities, it was recently proposed that actin filaments are naturally force sensitive, as tension would induce structural changes affecting interactions with actin-binding proteins.[81]

3.3. Use of FRET-based tension sensors: Relative versus absolute measurements

Given the prominence of FRET-based biosensors in cell signaling studies, which do not necessarily rely on the absolute distance between the fluorophores, many methods for imaging and analyzing FRET constructs have emerged.[63,64] FRET efficiency refers to the percentage of photons that undergo FRET, and is directly related to the separation distance of the fluorophores, and thus force in the case of a calibrated tension sensor. FRET efficiency can be measured in cells with advanced single-molecule techniques,[74] fluorescence lifetime imaging (FLIM)–FRET,[31,82] and quenching-based methods.[75] With proper calibration, as has been conducted for the flagelliform-based tension sensor[31] and some of the quenching methods,[69,75] these type of data can be converted into the average force exerted on molecules within an imaging voxel. However, these techniques do have some drawbacks, as they involve relatively specialized equipment or require long-exposure times that may miss critical dynamics. As the equipment required for single-molecule techniques and FLIM is rapidly evolving, these constraints may be nonlimiting soon.

Several forms of FRET analysis have been developed that are quantitative, but only lead to a relative measure of FRET.[64] These are referred to as FRET index measurements. The most popular of these is sensitized emission and involves the detection of acceptor emission due to donor excitation. This can be readily detected with standard epi-fluorescent and line-scanning confocal microscopes equipped with the proper excitation sources, filter sets, and camera/photomultiplier tube.[31] However, it is challenging to convert these FRET index measurements into FRET efficiency measurements, although methods utilizing constructs with known FRET efficiencies have been developed.[83] In most cases, FRET index measurements should be interpreted as a relative indication of tension across the force-sensitive module. The main advantage of these techniques is that data can be obtained at low-light densities and a high rate (as fast as tens to hundreds of milliseconds).[84]

3.4. Critical control experiments and assumptions involved in the creation and use of FRET-based biosensors

Creating a FRET-based tension sensor, especially those designed to function as endogenous proteins, is a significant undertaking. This is because the tension-sensing modules, typically comprising two fluorescent proteins and a force-sensitive element, must be incorporated into the middle of the protein, and there is a significant chance that this will perturb protein function. Often times, it is necessary to try several insertion sites to create a new sensor. To establish functionality, controls for proper protein localization, protein–protein interactions, dynamic properties, and the ability of the sensor to rescue a cellular function (e.g., migration, division, or adhesion) in cells where endogenous protein is lacking, at a minimum, are required.[31]

For the sensors to accurately report force-sensitive changes, all FRET must occur intramolecularly. Intermolecular FRET will lead to large background signals, potentially masking small changes. This is particularly problematic in adhesion structures, which are quite dense, and in small dimeric proteins. The existence of intermolecular FRET can be determined from cotransfection of constructs containing individual donor and acceptor fluorophores.[31] Alternatively, control constructs, based on the tension sensors, where one of the fluorophores has been rendered nonfluorescent through mutation can also be used.[78]

The use of the calibrated, extensible sensors to report absolute forces across a specific protein in a living cell requires several additional assumptions. First, force application must not lead to changes in κ^2. In the extensible sensors, these effects have been shown to be small, likely due to vibrational modes within the proteins.[31,65] Technically, the modules report the forces across the two fluorophores. For this to be applicable to the entire protein of interest, it must be assumed that the portions of the protein and the tension module act like springs in series. For most cases, where proteins are thought to be relatively linear and under large extensional forces, as is the case for vinculin, this is a natural assumption. Additionally, the extensible domains are subject to thermal fluctuations, causing them to have a slightly extended conformation at rest. Thus, it is possible that compressive forces could reduce fluorophore separation distance, increasing FRET. This possibility can be evaluated through the use of a force-insensitive FRET construct that would report the anticipated FRET levels at no force. Force-insensitive control constructs can be made by removing a portion of the protein required for loading (e.g., an actin-binding domain)[11,31] or by inserting the tension sensor module at the end of the protein of interest.[71]

3.5. Conformation sensors versus tension sensors

A common form of regulation among many mechanically relevant proteins is based on conformation. In an "inactive form," the protein assumes a conformation that blocks binding to other proteins or subcellular structures to maintain a cytosolic distribution, as in vinculin[56] and talin.[52] Often, this is mediated by the ends of the protein and referred to as head–tail inhibition.[56] As adhesion strengthening involves the recruitment of proteins from the cytosol, a conformation change is often required to mediate incorporation into subcellular structures. In biochemical studies, these are commonly referred to as force-sensitive changes, as they are often associated with signaling pathways known to manipulate myosin force generation. However, it is often unclear if the changes are due to the force-based activation of biochemical pathways or loading of the protein. Several FRET-based biosensors have been created to detect large, often biochemically regulated conformation changes in proteins (Fig. 1.5). Pertinent examples include fibronectin[49] and vinculin.[56] Tension sensors and conformation sensors cannot necessarily be used interchangeably. If conformation sensors are sensitive to biochemically induced changes, the resulting signal will not be indicative of the loads that proteins are experiencing and calibrations between force and FRET efficiency will not be valid. Established solution-based biochemical techniques, which are somewhat protein specific, can typically be used to initiate conformation changes to determine if tension sensors are sensitive to these changes.[55] The comparison of tension sensors and conformation sensors in dynamic FAs, as is possible for vinculin, will likely lead to an enhanced molecular scale understanding of mechanotransduction.

4. THE ROLE OF MOLECULAR TENSION IN FOCAL ADHESION DYNAMICS

The response of FAs to force is complex. The most commonly discussed aspect is that FAs assemble under applied load, referred to as adhesion strengthening.[33,85] This is often due to cellular force generation following

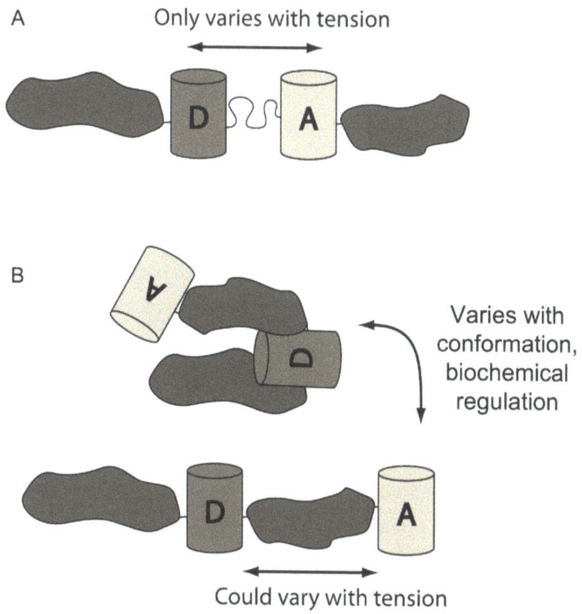

Figure 1.5 Schematics of a tension sensor versus a conformation sensor. (A) A tension sensor only shows changes in FRET due to application of tension. This can be verified by conducting biochemically based control experiments (see text for details). (B) Conformation sensors typically show large changes in FRET due to changes in the protein structure. These can be induced by biochemical regulation or binding of other proteins. Once opened, further changes in FRET are possible due to the extension of the protein of interest or flexible regions of the linkers to the fluorescent proteins.

the activation of the molecular motor myosin. Common initiators of adhesion strengthening include the direct activation of RhoA,[86] endogenous activation of contraction at the leading edge of a migrating cell,[87] and the application of forces to the outside of the cell.[33,88] Interestingly, all three processes seem to be governed by similar mechanosensitive signaling pathways.[89]

Much of the initial work regarding the molecular scale physical basis of this process was based on mathematical models.[24] These can be sorted into two classes: those based on thermodynamically unfavorable protein extension inducing protein recruitment[90,91] and those that explicitly considered force-activated biochemical signaling.[92] While the bases of these models are quite distinct, their central predictions are similar; force application leads to tension on FA constituent proteins, initiating assembly to alleviate tension as the FA grows. Recently, the creation of the vinculin tension sensor allowed direct testing and confirmation of this hypothesis through the tracking of assembling FAs through time.[31] Consistent with this idea, several studies have shown that large cellular stresses are often associated with small and assembling FAs.[93,94]

However, force is also associated with the disassembly of FAs.[85] This is particularly important in enabling efficient cell migration, as cells must release FAs at the trailing edge.[95] FA disassembly or sliding, which is a controlled form of disassembly, preferentially occurs at the trailing edge of migrating cells.[96] Given the fact that traction force microscopy has shown that forces are also exerted at the trailing edge of cells,[97] this behavior is in stark contrast to the observations in the leading edge of cells discussed above. Recently, the vinculin tension sensor was used to settle this apparent paradox.[31] When cellular forces are directed to FAs and across vinculin, FAs assemble. Interestingly, when vinculin was found to not be under tension, as the trailing edge of migrating cells, FAs were found to disassemble, even if the FAs were located in regions associated with cellular force exertion. Thus, vinculin tension appears to be a switch that dictates the response of a FA to applied load (Fig. 1.6). Consistent with the notion that leading and trailing edge FAs are distinct, local dissociation of these structures, through the use of soluble ECM-based peptide, has been shown to lead to drastically different responses in terms of cell shortening and alterations in overall cellular force generations.[98] The dynamic responses of other FA protein-based tension sensors have not been analyzed in individual FAs, but a gradient of actinin tension was observed in migrating cells.[30] Similarly, disruption of a single

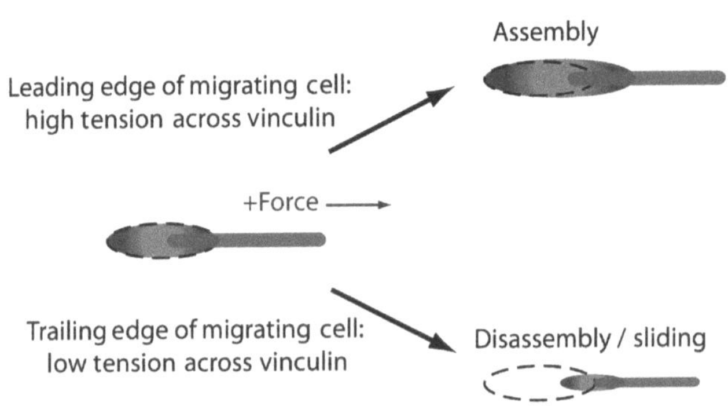

Figure 1.6 Vinculin as a force-sensitive switch controlling FA dynamics. Cellular force generation is associated with both FA assembly and disassembly. The vinculin tension sensor has been used to show that applied loads leading to tension across vinculin are associated with FA growth, as is typically observed in the leading edge of migrating cells. Lack of vinculin tension in response to applied loads is associated with FA disassembly, which is typically seen in the trailing edge of migrating cells.

stress fiber leads to complex changes in the vinculin tension throughout the cell, consistent with a rich interplay between the actin cytoskeleton, myosin force generation, and protein-specific tension.[99]

5. FUTURE OUTLOOK

The analysis of the spatiotemporal dynamics of the tension across a specific FA protein has advanced our understanding of mechanotransduction. However, the FA is a complex structure comprising a large number of mechanosensitive proteins. The creation of novel tension sensors and the examination of their spatiotemporal dynamics will likely lead to a much greater understanding of the molecular mechanisms mediating mechanosensitivity. Critical points requiring clarification include the relationship between molecular tension and biochemical activity and force-induced protein–protein interactions. A recent study has shown that the exposure of cryptic binding sites for MAPK1 within vinculin may play a critical role in the differentiation of human stem cells.[100]

Additionally, several studies have suggested that adherens junctions, the subcellular structures that mediate mechanical linkages between cells, are also mechanosensitive.[101,102] These structures are mediated by transmembrane proteins called cadherins. Recently, tension sensors for E-cadherin[77] and VE-cadherin,[78] which support cell–cell adhesion in epithelial and endothelial cells, respectively, have been created. Future studies of the spatiotemporal dynamics of these sensors will likely lead to a much greater understanding of the mechanosensitivity of adherens junctions and tissue dynamics. The successful creation and use of a wide range of molecular tension sensors may be a critical step toward a more complete understanding of the mechanical basis of cell migration as well as the mechanosensitive signaling pathways that mediate tissue formation and mechanosensitive diseases.

ACKNOWLEDGMENTS

This work was supported by a Searle Scholar Award and a Basil O'Conor Starter Scholar Award from the March of Dimes (5-FY13-221).

REFERENCES

1. DuFort CC, Paszek MJ, Weaver VM. Balancing forces: architectural control of mechanotransduction. *Nat Rev Mol Cell Biol*. 2011;12(5):308–319.
2. Wozniak MA, Chen CS. Mechanotransduction in development: a growing role for contractility. *Nat Rev Mol Cell Biol*. 2009;10(1):34–43.

3. Polacheck WJ, Li R, Uzel SG, Kamm RD. Microfluidic platforms for mechanobiology. *Lab Chip*. 2013;13(12):2252–2267.
4. Jaalouk DE, Lammerding J. Mechanotransduction gone awry. *Nat Rev Mol Cell Biol*. 2009;10(1):63–73.
5. Mammoto T, Mammoto A, Ingber DE. Mechanobiology and developmental control. *Annu Rev Cell Dev Biol*. 2013;29:27–61.
6. Hahn C, Schwartz MA. Mechanotransduction in vascular physiology and atherogenesis. *Nat Rev Mol Cell Biol*. 2009;10(1):53–62.
7. Nemir S, West JL. Synthetic materials in the study of cell response to substrate rigidity. *Ann Biomed Eng*. 2010;38(1):2–20.
8. Sun Y, Fu J. Mechanobiology: a new frontier for human pluripotent stem cells. *Integr Biol (Camb)*. 2013;5(3):450–457.
9. Takahashi K, Kakimoto Y, Toda K, Naruse K. Mechanobiology in cardiac physiology and diseases. *J Cell Mol Med*. 2013;17(2):225–232.
10. Schoen I, Pruitt BL, Vogel V. The Yin-Yang of rigidity sensing: how forces and mechanical properties regulate the cellular response to materials. *Annu Rev Mater Res*. 2013;43:589–618.
11. Conway DE, Schwartz MA. Flow-dependent cellular mechanotransduction in atherosclerosis. *J Cell Sci*. 2013;126(Pt 22):5101–5109.
12. Ricca BL, Venugopalan G, Fletcher DA. To pull or be pulled: parsing the multiple modes of mechanotransduction. *Curr Opin Cell Biol*. 2013;25(5):558–564.
13. Hoffman BD, Grashoff C, Schwartz MA. Dynamic molecular processes mediate cellular mechanotransduction. *Nature*. 2011;475(7356):316–323.
14. Geiger B, Spatz JP, Bershadsky AD. Environmental sensing through focal adhesions. *Nat Rev Mol Cell Biol*. 2009;10(1):21–33.
15. Paszek MJ, Zahir N, Johnson KR, et al. Tensional homeostasis and the malignant phenotype. *Cancer cell*. 2005;8(3):241–254.
16. Lo CM, Wang HB, Dembo M, Wang YL. Cell movement is guided by the rigidity of the substrate. *Biophys J*. 2000;79(1):144–152.
17. Engler AJ, Sen S, Sweeney HL, Discher DE. Matrix elasticity directs stem cell lineage specification. *Cell*. 2006;126(4):677–689.
18. Nelson CM, Jean RP, Tan JL, et al. Emergent patterns of growth controlled by multicellular form and mechanics. *Proc Natl Acad Sci USA*. 2005;102(33):11594–11599.
19. Desprat N, Supatto W, Pouille PA, Beaurepaire E, Farge E. Tissue deformation modulates twist expression to determine anterior midgut differentiation in Drosophila embryos. *Dev Cell*. 2008;15(3):470–477.
20. Duncan RL, Turner CH. Mechanotransduction and the functional response of bone to mechanical strain. *Calcif Tissue Int*. 1995;57(5):344–358.
21. Osol G. Mechanotransduction by vascular smooth muscle. *J Vasc Res*. 1995;32(5): 275–292.
22. Ingber DE. Mechanobiology and diseases of mechanotransduction. *Ann Med*. 2003;35(8):564–577.
23. Levental KR, Yu H, Kass L, et al. Matrix crosslinking forces tumor progression by enhancing integrin signaling. *Cell*. 2009;139(5):891–906.
24. Bershadsky A, Kozlov M, Geiger B. Adhesion-mediated mechanosensitivity: a time to experiment, and a time to theorize. *Curr Opin Cell Biol*. 2006;18(5):472–481.
25. Erickson HP. Reversible unfolding of fibronectin type III and immunoglobulin domains provides the structural basis for stretch and elasticity of titin and fibronectin. *Proc Natl Acad Sci USA*. 1994;91(21):10114–10118.
26. Rief M, Gautel M, Oesterhelt F, Fernandez JM, Gaub HE. Reversible unfolding of individual titin immunoglobulin domains by AFM. *Science*. 1997;276(5315): 1109–1112.

27. del Rio A, Perez-Jimenez R, Liu R, Roca-Cusachs P, Fernandez JM, Sheetz MP. Stretching single talin rod molecules activates vinculin binding. *Science.* 2009;323(5914):638–641.
28. Davies PF. Flow-mediated endothelial mechanotransduction. *Physiol Rev.* 1995;75(3):519–560.
29. Chien S. Mechanotransduction and endothelial cell homeostasis: the wisdom of the cell. *Am J Physiol Heart Circ Physiol.* 2007;292(3):H1209–1224.
30. Meng F, Suchyna TM, Sachs F. A fluorescence energy transfer-based mechanical stress sensor for specific proteins in situ. *FEBS J.* 2008;275(12):3072–3087.
31. Grashoff C, Hoffman BD, Brenner MD, et al. Measuring mechanical tension across vinculin reveals regulation of focal adhesion dynamics. *Nature.* 2010;466(7303): 263–266.
32. Wang Y, Meng F, Sachs F. Genetically encoded force sensors for measuring mechanical forces in proteins. *Commun Integr Biol.* 2011;4(4):385–390.
33. Riveline D, Zamir E, Balaban NQ, et al. Focal contacts as mechanosensors: externally applied local mechanical force induces growth of focal contacts by an mDia1-dependent and ROCK-independent mechanism. *J Cell Biol.* 2001;153(6):1175–1186.
34. Schwartz MA, DeSimone DW. Cell adhesion receptors in mechanotransduction. *Curr Opin Cell Biol.* 2008;20(5):551–556.
35. Geiger B, Bershadsky A. Assembly and mechanosensory function of focal contacts. *Curr Opin Cell Biol.* 2001;13(5):584–592.
36. Zaidel-Bar R, Itzkovitz S, Ma'ayan A, Iyengar R, Geiger B. Functional atlas of the integrin adhesome. *Nat Cell Biol.* 2007;9(8):858–867.
37. Zaidel-Bar R, Geiger B. The switchable integrin adhesome. *J Cell Sci.* 2010;123(Pt 9):1385–1388.
38. Hynes RO. Integrins: bidirectional, allosteric signaling machines. *Cell.* 2002;110(6): 673–687.
39. Kanchanawong P, Shtengel G, Pasapera AM, et al. Nanoscale architecture of integrin-based cell adhesions. *Nature.* 2010;468(7323):580–584.
40. Orr AW, Helmke BP, Blackman BR, Schwartz MA. Mechanisms of mechanotransduction. *Dev Cell.* 2006;10(1):11–20.
41. De R, Zemel A, Safran SA. Theoretical concepts and models of cellular mechanosensing. *Methods Cell Biol.* 2010;98:143–175.
42. Bell GI. Models for the specific adhesion of cells to cells. *Science.* 1978;200(4342): 618–627.
43. Merkel R, Nassoy P, Leung A, Ritchie K, Evans E. Energy landscapes of receptor-ligand bonds explored with dynamic force spectroscopy. *Nature.* 1999;397(6714): 50–53.
44. Evans EA, Calderwood DA. Forces and bond dynamics in cell adhesion. *Science.* 2007;316(5828):1148–1153.
45. Kong F, Garcia AJ, Mould AP, Humphries MJ, Zhu C. Demonstration of catch bonds between an integrin and its ligand. *J Cell Biol.* 2009;185(7):1275–1284.
46. Guo B, Guilford WH. Mechanics of actomyosin bonds in different nucleotide states are tuned to muscle contraction. *Proc Natl Acad Sci USA.* 2006;103(26):9844–9849.
47. Thomas WE, Vogel V, Sokurenko E. Biophysics of catch bonds. *Annu Rev Biophys.* 2008;37:399–416.
48. Hu K, Ji L, Applegate KT, Danuser G, Waterman-Storer CM. Differential transmission of actin motion within focal adhesions. *Science.* 2007;315(5808):111–115.
49. Smith ML, Gourdon D, Little WC, et al. Force-induced unfolding of fibronectin in the extracellular matrix of living cells. *PLoS Biol.* 2007;5(10):e268.
50. Sawada Y, Tamada M, Dubin-Thaler BJ, et al. Force sensing by mechanical extension of the Src family kinase substrate p130Cas. *Cell.* 2006;127(5):1015–1026.

51. Critchley DR. Biochemical and structural properties of the integrin-associated cytoskeletal protein talin. *Annu Rev Biophys*. 2009;38:235–254.
52. Goult BT, Xu XP, Gingras AR, et al. Structural studies on full-length talin1 reveal a compact auto-inhibited dimer: implications for talin activation. *J Struct Biol*. 2013;184(1):21–32.
53. Peng X, Nelson ES, Maiers JL, DeMali KA. New insights into vinculin function and regulation. *Int Rev Cell Mol Biol*. 2011;287:191–231.
54. Bakolitsa C, Cohen DM, Bankston LA, et al. Structural basis for vinculin activation at sites of cell adhesion. *Nature*. 2004;430(6999):583–586.
55. Chen H, Cohen DM, Choudhury DM, Kioka N, Craig SW. Spatial distribution and functional significance of activated vinculin in living cells. *J Cell Biol*. 2005;169(3): 459–470.
56. Cohen DM, Kutscher B, Chen H, Murphy DB, Craig SW. A conformational switch in vinculin drives formation and dynamics of a talin-vinculin complex at focal adhesions. *J Biol Chem*. 2006;281(23):16006–16015.
57. Zhang X, Jiang G, Cai Y, Monkley SJ, Critchley DR, Sheetz MP. Talin depletion reveals independence of initial cell spreading from integrin activation and traction. *Nat Cell Biol*. 2008;10(9):1062–1068.
58. Mierke CT, Kollmannsberger P, Zitterbart DP, Smith J, Fabry B, Goldmann WH. Mechano-coupling and regulation of contractility by the vinculin tail domain. *Biophys J*. 2008;94(2):661–670.
59. Xu W, Baribault H, Adamson ED. Vinculin knockout results in heart and brain defects during embryonic development. *Development*. 1998;125(2):327–337.
60. Margadant F, Chew LL, Hu X, et al. Mechanotransduction in vivo by repeated talin stretch-relaxation events depends upon vinculin. *PLoS Biol*. 2011;9(12):e1001223.
61. Carisey A, Tsang R, Greiner AM, et al. Vinculin regulates the recruitment and release of core focal adhesion proteins in a force-dependent manner. *Curr Biol*. 2013;23(4): 271–281.
62. Oddershede LB. Force probing of individual molecules inside the living cell is now a reality. *Nat Chem Biol*. 2012;8(11):879–886.
63. Lakowicz JR. *Principles of Fluorescence Spectroscopy*. 3rd ed. New York: Springer; 2006.
64. Zeug A, Woehler A, Neher E, Ponimaskin EG. Quantitative intensity-based FRET approaches—a comparative snapshot. *Biophys J*. 2012;103(9):1821–1827.
65. Evers TH, van Dongen EM, Faesen AC, Meijer EW, Merkx M. Quantitative understanding of the energy transfer between fluorescent proteins connected via flexible peptide linkers. *Biochemistry*. 2006;45(44):13183–13192.
66. Hohng S, Zhou R, Nahas MK, et al. Fluorescence-force spectroscopy maps two-dimensional reaction landscape of the holliday junction. *Science*. 2007;318(5848):279–283.
67. Okumoto S, Jones A, Frommer WB. Quantitative imaging with fluorescent biosensors. *Annu Rev Plant Biol*. 2012;63:663–706.
68. Day RN, Booker CF, Periasamy A. Characterization of an improved donor fluorescent protein for Forster resonance energy transfer microscopy. *J Biomed Opt*. 2008;13(3): 031203.
69. Stabley DR, Jurchenko C, Marshall SS, Salaita KS. Visualizing mechanical tension across membrane receptors with a fluorescent sensor. *Nat Methods*. 2012;9(1):64–67.
70. Wang Y, Shyy JY, Chien S. Fluorescence proteins, live-cell imaging, and mechanobiology: seeing is believing. *Annu Rev Biomed Eng*. 2008;10:1–38.
71. Meng F, Sachs F. Orientation-based FRET sensor for real-time imaging of cellular forces. *J Cell Sci*. 2012;125(Pt 3):743–750.
72. Shroff H, Reinhard BM, Siu M, Agarwal H, Spakowitz A, Liphardt J. Biocompatible force sensor with optical readout and dimensions of 6 nm3. *Nano Lett*. 2005;5(7): 1509–1514.

73. Meng F, Sachs F. Visualizing dynamic cytoplasmic forces with a compliance-matched FRET sensor. *J Cell Sci*. 2011;124(Pt 2):261–269.
74. Morimatsu M, Mekhdjian AH, Adhikari AS, Dunn AR. Molecular tension sensors report forces generated by single integrin molecules in living cells. *Nano Lett*. 2013;13(9):3985–3989.
75. Liu Y, Yehl K, Narui Y, Salaita K. Tension sensing nanoparticles for mechano-imaging at the living/nonliving interface. *J Am Chem Soc*. 2013;135(14):5320–5323.
76. Wang X, Ha T. Defining single molecular forces required to activate integrin and notch signaling. *Science*. 2013;340(6135):991–994.
77. Borghi N, Sorokina M, Shcherbakova OG, et al. E-cadherin is under constitutive actomyosin-generated tension that is increased at cell-cell contacts upon externally applied stretch. *Proc Natl Acad Sci USA*. 2012;109(31):12568–12573.
78. Conway DE, Breckenridge MT, Hinde E, Gratton E, Chen CS, Schwartz MA. Fluid shear stress on endothelial cells modulates mechanical tension across VE-cadherin and PECAM-1. *Curr Biol*. 2013;23(11):1024–1030.
79. Iwai S, Uyeda TQ. Visualizing myosin-actin interaction with a genetically-encoded fluorescent strain sensor. *Proc Natl Acad Sci USA*. 2008;105(44):16882–16887.
80. Ichimura T, Fujita H, Yoshizawa K, Watanabe TM. Engineering strain-sensitive yellow fluorescent protein. *Chem Commun (Camb)*. 2012;48(63):7871–7873.
81. Galkin VE, Orlova A, Egelman EH. Actin filaments as tension sensors. *Curr Biol*. 2012;22(3):R96–101.
82. Austen K, Kluger C, Freikamp A, Chrostek-Grashoff A, Grashoff C. Generation and analysis of biosensors to measure mechanical forces within cells. *Methods Mol Biol*. 2013;1066:169–184.
83. Chen H, Puhl 3rd HL, Koushik SV, Vogel SS, Ikeda SR. Measurement of FRET efficiency and ratio of donor to acceptor concentration in living cells. *Biophys J*. 2006;91(5):L39–41.
84. Na S, Collin O, Chowdhury F, et al. Rapid signal transduction in living cells is a unique feature of mechanotransduction. *Proc Natl Acad Sci USA*. 2008;105(18):6626–6631.
85. Parsons JT, Horwitz AR, Schwartz MA. Cell adhesion: integrating cytoskeletal dynamics and cellular tension. *Nat Rev Mol Cell Biol*. 2010;11(9):633–643.
86. Chrzanowska-Wodnicka M, Burridge K. Rho-stimulated contractility drives the formation of stress fibers and focal adhesions. *J Cell Biol*. 1996;133(6):1403–1415.
87. Galbraith CG, Yamada KM, Sheetz MP. The relationship between force and focal complex development. *J Cell Biol*. 2002;159(4):695–705.
88. Choquet D, Felsenfeld DP, Sheetz MP. Extracellular matrix rigidity causes strengthening of integrin-cytoskeleton linkages. *Cell*. 1997;88(1):39–48.
89. Chen CS. Mechanotransduction—a field pulling together? *J Cell Sci*. 2008;121(Pt 20):3285–3292.
90. Shemesh T, Geiger B, Bershadsky AD, Kozlov MM. Focal adhesions as mechanosensors: a physical mechanism. *Proc Natl Acad Sci USA*. 2005;102(35): 12383–12388.
91. Nicolas A, Geiger B, Safran SA. Cell mechanosensitivity controls the anisotropy of focal adhesions. *Proc Natl Acad Sci USA*. 2004;101(34):12520–12525.
92. Besser A, Safran SA. Force-induced adsorption and anisotropic growth of focal adhesions. *Biophys J*. 2006;90(10):3469–3484.
93. Beningo KA, Dembo M, Kaverina I, Small JV, Wang YL. Nascent focal adhesions are responsible for the generation of strong propulsive forces in migrating fibroblasts. *J Cell Biol*. 2001;153(4):881–888.
94. Tan JL, Tien J, Pirone DM, Gray DS, Bhadriraju K, Chen CS. Cells lying on a bed of microneedles: an approach to isolate mechanical force. *Proc Natl Acad Sci USA*. 2003;100(4):1484–1489.

95. Lauffenburger DA, Horwitz AF. Cell migration: a physically integrated molecular process. *Cell.* 1996;84(3):359–369.
96. Ballestrem C, Hinz B, Imhof BA, Wehrle-Haller B. Marching at the front and dragging behind: differential alphaVbeta3-integrin turnover regulates focal adhesion behavior. *J Cell Biol.* 2001;155(7):1319–1332.
97. Munevar S, Wang Y, Dembo M. Traction force microscopy of migrating normal and H-ras transformed 3T3 fibroblasts. *Biophys J.* 2001;80(4):1744–1757.
98. Munevar S, Wang YL, Dembo M. Distinct roles of frontal and rear cell-substrate adhesions in fibroblast migration. *Mol Biol Cell.* 2001;12(12):3947–3954.
99. Chang CW, Kumar S. Vinculin tension distributions of individual stress fibers within cell-matrix adhesions. *J Cell Sci.* 2013;126(Pt 14):3021–3030.
100. Holle AW, Tang X, Vijayraghavan D, et al. In situ mechanotransduction via vinculin regulates stem cell differentiation. *Stem Cells.* 2013;31(11):2467–2477.
101. Maruthamuthu V, Sabass B, Schwarz US, Gardel ML. Cell-ECM traction force modulates endogenous tension at cell-cell contacts. *Proc Natl Acad Sci USA.* 2011;108(12):4708–4713.
102. Liu Z, Tan JL, Cohen DM, et al. Mechanical tugging force regulates the size of cell-cell junctions. *Proc Natl Acad Sci USA.* 2010;107(22):9944–9949.

CHAPTER TWO

Single-Cell Imaging of Mechanotransduction in Endothelial Cells

Shaoying Lu, Yingxiao Wang
Department of Bioengineering, Institute of Engineering in Medicine, University of California, San Diego, La Jolla, California, USA

Contents

Abstract

Endothelial cells (ECs) are constantly exposed to chemical and mechanical microenvironment *in vivo*. In mechanotransduction, cells can sense and translate the extracellular mechanical cues into intracellular biochemical signals, to regulate cellular processes. This regulation is crucial for many physiological functions, such as cell adhesion, migration, proliferation, and survival, as well as the progression of disease such as atherosclerosis. Here, we overview the current molecular understanding of mechanotransduction in ECs associated with atherosclerosis, especially those in response to physiological shear stress. The enabling technology of live-cell imaging has allowed the study of spatiotemporal molecular events and unprecedented understanding of intracellular signaling responses in mechanotransduction. Hence, we also introduce recent studies on mechanotransduction using single-cell imaging technologies.

Progress in Molecular Biology and Translational Science, Volume 126
ISSN 1877-1173
http://dx.doi.org/10.1016/B978-0-12-394624-9.00002-6

1. INTRODUCTION

Mechanical forces play crucial roles in regulating pathophysiological processes, e.g., atherosclerosis, the leading cause of death in the USA and most developed countries.[1] Shear stress without a clear direction can lead to endothelial cell (EC) dysfunction and atherogenesis,[2–4] but there is a lack of understanding on how ECs perceive the spatiotemporal cues and transduce them into biochemical activities to regulate cellular functions. The underlying pathology of atherosclerosis involves a chronic inflammatory process of vessel wall[5] due to endothelial dysfunction with increased permeability and recruitment of immune cells, including monocytes, resulting from upregulation of adhesion molecules and cytokine secretion by ECs.[5] Atherosclerosis occurs preferentially at vascular curvature and branch sites where the vessel walls and ECs are exposed to disturbed flow, which has been reported to facilitate atherogenesis.[6–8] It is possible that the greater spatiotemporal heterogeneity of shear-stress distribution under atheroprone disturbed flows than that of atheroprotective laminar flows contributes to the pathophysiological modulation of EC responses in the subsequent mechanotransduction.[2,9–15]

Since the plasma membrane provides an interface between the cell and environment, it is expected to be an important subcellular structure in mechanotransduction. Transmembrane receptors such as integrins, G-protein-coupled receptor (GPCR), platelet endothelial cell adhesion molecule (PECAM-1), and transient receptor potential channels (TRPC6) can be activated by shear to regulate downstream signals, i.e., membrane-associated tyrosine kinases, Ca^{2+}, and small GTPases.[16–18] Shear stress has been shown to activate Src, which leads to ERK activation and translocation between nucleus and cytosol, and subsequent atheroprone genetic changes.[19–21] TRPC6 can also sense shear and trigger Ca^{2+} influx to regulate EC permeability.[22,23] The plasma membrane also consists of different microdomains,[24] including lipid rafts,[25] which play important roles for mechanotransduction.[26] Various mechanosensing elements can localize at or proximal to the plasma membrane to regulate downstream intracellular functions.[27]

The wide applications of fluorescence proteins (FPs) in single live-cell imaging have revolutionized the whole research field of cell biology, including mechanotransduction in ECs. Fluorescence resonance energy transfer (FRET)-based biosensors have been engineered and applied to molecular activities in live cells.[28–37] The ratiometric measurement of the biosensors utilizes the ratio of the donor to acceptor fluorescence intensity to represent the target molecular activity. This measurement is self-normalizing and independent of the heterogeneous biosensor expression level among various cells.[38] The genetically engineered FRET biosensors also allow subcellular localization to cytosol, plasma membrane, or organelles, which can provide versatile measurement of local molecular activities. As a result, the FRET-based biosensors have been widely used in live cells to monitor molecular signals in real time.[39,40]

Utilizing fluorescence imaging technologies including FPs and FRET biosensors, a large amount of video imaging data of live cells have been collected. In order to precisely and efficiently interpret the underlying biological mechanism, automated, intelligent, and objective image analysis tools are in high demands.[39] Automatic methods for the accurate detection of cell and subcellular features are crucial for the high-throughput image analysis and quantification of subcellular molecular interactions. For example, the water algorithm has been widely used to detect FAs in fluorescent images.[41–45] Quantitative image-based analysis can also reveal the hidden spatial pattern and temporal sequence of signaling events.[46,47] Integrated computational tools are needed to quantify the activity and localization of intracellular molecules, which can provide the basis for the quantitative analysis elucidating subcellular molecular interactions and therefore contribution to EC migration and atherosclerosis.[39,47–49]

2. ATHEROSCLEROSIS, EC WOUND HEALING, AND MECHANOTRANSDUCTION

Atherosclerosis is a cardiovascular disease characterized by the patchy deposit of fatty materials in the arterial walls and reduced/blocked blood flows.[50] It occurs preferentially at vascular curvature and branch sites where the vessel walls are exposed to disturbed flows, but not at the straight parts of vessels where laminar flows dominate.[50] Vascular ECs, which form a monolayer of endothelium lining along and protecting the vessel wall from the circulating blood,[51] are continuously exposed to shear stress resulted from these different flows. It has been shown that ECs subjected to disturbed flows, but not to laminar flows, tend to have a high and sustained permeability, which facilitates the formation of atherosclerosis.[52,53] Recent evidence also indicates that the effect of disturbed flows on ECs is pro-inflammatory, whereas that of laminar flows is anti-inflammatory.[50,54–57] For example, disturbed flows induced the expression of pro-inflammatory BMP-4 and cytokines and adhesion receptors such as intercellular adhesion molecule-1 (ICAM-1) and vascular cell adhesion molecule (VCAM-1).[58–60] In contrast, laminar flows can inhibit the inflammatory signaling cascades induced by TNFα.[61–63]

Following the atherosclerosis, the injury of endothelium and loss of ECs after bypass surgery or balloon angioplasty is the main cause of restenosis. The EC wound healing process, which involves EC migration at the wounding edge followed by EC proliferation, serves as the critical step to restore the endothelium integrity and prevent restenosis.[64] Shear stress has been shown to affect the EC wound healing process by modulating EC junctions. EC junctions in a monolayer can be disrupted by shear stress and recovered if continuously exposed to laminar flows for long term as the cells adapt to the new environment.[52,53] The transient disruption of cell junctions under laminar flows will last long during the wound healing process because it takes time for cells to repair the wound, restore a stable monolayer, and adapt to the new environment. This junction disruption would hence keep the ECs disengaged from each other and free to move. As such, the laminar flow-directed protrusion and migration of ECs at the upstream side of wounded area can be promoted and maintained before the wound closure, thus facilitating the EC migration toward the wounded area. At the downstream side, the pushing effect of laminar flows on cells along the flow direction negates the junction disruption-induced motility enhancement. Although a wound *per se* can also induce a migration of cells at both wounding edges to close the wound, results have revealed that laminar flows promote cells at the upstream side to migrate into the wounded area faster, whereas the rate of cell migration downstream of flow is comparable to that without flow.[65] This asymmetric effect of laminar flows results in an increased net speed of wound healing comparing to that without flow. In fact, both *in vitro* and *in vivo* experiments indicate that laminar flows enhance EC migration and consequently wound healing.[66–68] Laminar flows have been shown to enhance the wound healing by modulating cell–cell and cell–extracellular matrix (ECM) adhesions, in particular, vein endothelial (VE)-cadherin-mediated adherens junction (AJ)[69,70] and β1 integrins.[71] Membrane fluidity, cytoskeleton, and tyrosine kinases also appear to be important for the laminar flow-induced EC wound healing.[65,72,73] Interestingly, disturbed flows induce sustained disruption of AJ.[52,53] However, the EC migration speed toward the wounded area under disturbed flows is comparable to that without flow.[65] It was revealed that ECs under disturbed flows have strong staining of focal adhesion proteins due to the unstable mechanical environment,[65] which may impair the cell detachment from the substrate necessary for EC migration and hence negate the motility enhancement due to the junction disruption.[74] It remains an intriguing question how ECs coordinate the multiple signaling events in space and time under different flows to regulate AJ, protrusion, motility, migration, and wound healing.

It has been well documented that shear stress can activate a variety of signaling cascades and gene expressions to regulate EC functions and pathophysiological processes.[7,8,75] For example, a wide range of signaling molecules and structures, including the plasma membrane,[76,77] membrane proteins/receptors (e.g., integrins,[16,19] GPCR,[17] cadherin,[18,53] PECAM-1,[18,78,79] VCAM-1,[80,81] ICAM-1,[80] and ion channels[75]), actin filaments,[82] microtubules,[83,84] and intermediate filaments[85–87] were identified to play important roles in transmitting shear stress into biochemical signaling cascades, i.e., mechanotransduction. Atheroprone disturbed fluid shear stress has been reported to induce the expression of fibronectin (FN) gene in human ECs and enhance its assembly into fibril matrix.[88] In particular, the PECAM-1/NF-κB pathway was found to be essential for FN accumulation in atheroprone regions of the aortic arch, which positively feedbacks and promotes the activation of NF-κB and atherogenesis.[88] Evidence also indicates that both the temporal and spatial gradient of shear stress can affect the cellular functions.[2,10–12,14,15,76,89] In fact, it was realized that the subcellular characteristics of shear stress are heterogeneous in a single cell and impact significantly on cellular responses.[12,19,90,91] Calcium flux in response to shear stress showed significant subcellular directionality.[92,93] The lateral diffusion coefficient of lipids in the plasma membrane also increased at regions upstream of the nucleus while decreased at the downstream regions upon shear-stress application.[94]

3. SIGNALING MOLECULES INVOLVED IN MECHANOSENSING AND MECHANOTRANSDUCTION

TRPCs at the plasma membrane can regulate Ca^{2+} entry and play crucial roles in regulating EC actomyosin contractility and permeability[22,23]; increased TRPC6 expression has been revealed as a hallmark of atherosclerosis[95] and atheroprone flow shear stress has been shown to induce TRPC6 expression in ECs.[96] TRPC6 can also interact with caveolin,[97] which is localized at VE-cadherin-mediated AJs.[98] In fact, TRPC4, a close family member of TRPC6, has been shown to colocalize with VE-cadherin at

AJs.[99] Therefore, the Ca^{2+} influx mediated by TRPC6 may affect the local $[Ca^{2+}]$ in the intercellular junction, at least transiently, and modulate VE-cadherin functions to regulate EC permeability. While TRPC6 can mediate the effect of shear stress on triggering the Ca^{2+} influx under different stimuli,[100] it remains controversial whether TRPC6 can be directly activated by mechanical tension[101] or indirectly via GPCRs and their effector phospholipase C (PLC).[102,103] Since shear stress has been reported to stimulate GPCRs and their coupled G-proteins,[17,104,105] it is possible that shear stress utilizes GPCR/PLC pathway to activate TRPC6 besides imposing direct mechanical impact.

Src kinase is recruited to the plasma membrane and activated differentially at the different membrane microdomains under various conditions (Fig. 2.1).[34,106] Shear stress can activate Src[19,107,108] through plasma membrane mechanosomes containing PKG/Shp2.[109] This shear-induced Src can then lead to the phosphorylation and activation of ERK[21,110,111] via Shc/Grb2/SOS/Ras/MEK pathway.[109] The phosphorylated ERK can translocate into the nucleus[112] and undergo an oscillatory translocation between nucleus and cytosol[20] to regulate downstream gene expression of MCP-1 for monocyte recruitment.[113] It has been hypothesized that the spatiotemporal patterns of these oscillatory activities of transcription factors can regulate the strength and time course of the target gene expression.[20] Recent evidence indicates that molecular functions are dependent on its subcellular localization. For example, Src inhibits Rho GTPase at the focal adhesion sites,[114] whereas it activates Rho GTPase at podosomes.[115] Furthermore, junction remodeling takes place at a molecularly and phenotypically distinct subset of VE-cadherin adhesions, focal AJs, and vinculin associates with focal AJs and stabilizes the junctions from opening during their force-dependent remodeling.[116] Therefore, the different spatiotemporal characteristics of disturbed and laminar flows may cause differential TRPC6 activation and/or Src localization/function at membrane microdomains to regulate AJs and MCP-1, which modulate endothelial permeability and monocyte recruitment, respectively.[5]

4. THE EFFECT OF SUBCELLULAR STRUCTURE ON MECHANOTRANSDUCTION

The plasma membrane is not uniform in structure[24] and has different compartments, e.g., lipid rafts, which are rich in cholesterol, sphingomyelin, and saturated fatty acids.[25] These compartment structures and their

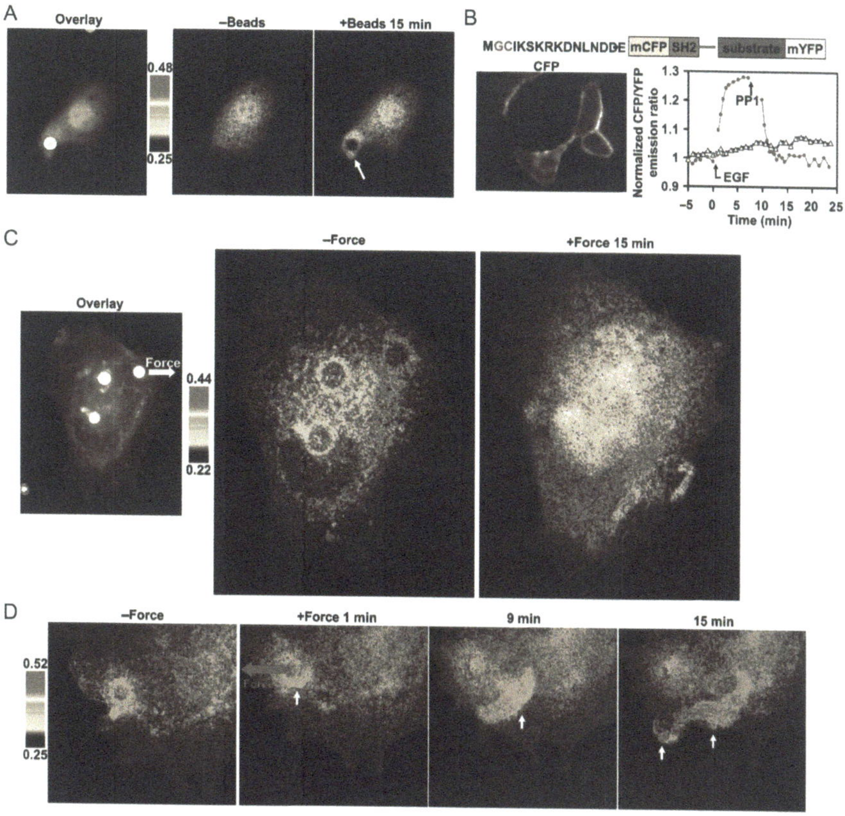

Figure 2.1 Directional and long-range propagation of Src induced by mechanical force. (A) A fibronectin-coated bead (white spot from phase contrast image overlaid on CFP cell image) induced FRET responses around the bead. White arrow points to the spot with activated Src. Color bar represents CFP/YFP emission ratio values. (B) The schematic diagram in the upper panel shows the design strategy of membrane targeting. The CFP-only image on the left shows the effective tethering of the reporter on the plasma membrane. The EGF-induced FRET responses of the reporter is reversed by PP1 (red line) and prevented by pretreatment with PP1 (blue line). (C) Laser-tweezer traction on the bead at the upper right corner of the cell (shown on the left) caused FRET responses. White arrow represents force direction. (D) FRET responses of a cell with clear directional wave propagation away from the site of mechanical stimulation. *This research was originally published in the* Nature *Journal.*[34] (See the color plate.)

interaction with cytoskeleton are involved in the regulation of intracellular signaling.[26] Indeed, Src family kinase (SFK) members, including Fyn and Lyn, are anchored at lipid rafts to become activated following N-terminal palmitoylation and myristoylation,[117] whereas H-Ras depends on the non-rafts anchoring at the plasma membrane to be functional.[117,118] Evidence also indicates that the plasma membrane and its compartments are involved in mechanotransduction. For example, different subtypes of G-proteins are partitioned into membrane compartments to regulate the mechanical-activated signaling molecules, e.g., nitric oxide production[119] and MAPK activity.[120] The modulation of the plasma membrane fluidity also altered the shear stress-induced MAPK signaling pathway,[94] further underscoring the importance of the plasma membrane in mechanotransduction.

Src kinase is a nonreceptor tyrosine kinase critical for a variety of cellular processes.[114] At resting state, Src localizes at the microtubule-associated perinuclear regions[34,121–126] and/or at the nonrafts regions on the plasma membrane.[127–129] Recent evidence indicates that Rho small GTPases and the associated actin network can facilitate the transportation of Src from perinuclear area to actin-associated cell periphery,[130–132] possibly through the Src SH3 domain, but not the catalytic domain.[133,134] The inhibition of RhoA or actin stress fibers but not microtubules resulted in the blockage of Src translocation/activation in response to various stimuli.[130–132] The compartmental structures at the plasma membrane are also involved in the localization and regulation of Src kinase.[26,135] Indeed, SFKs can be transported to distinct compartments of plasma membrane through different types of endosomes.[136] SFK members such as Lyn and Fyn can reside in lipid rafts of the plasma membrane,[137] via their N-terminal myristoylation and palmitoylation sites.[118] Src kinase itself has only myristoylation motif and it is not clear whether Src kinase localizes within the lipid rafts at the plasma membrane.[127–129,138–140] In mouse fibroblasts, Src was shown to be excluded from the detergent-resistant membrane (DRM) fractions in one study, while another publication suggested that Src resides in DRM fraction.[127,140] Different groups also reported different Src localizations in PC12 cells.[129,139] This inconsistency is likely attributed to the controversial effects of nonionic detergents and the detergent extraction method used in these reports for isolating DRMs.[141,142] Advanced methods are hence needed to visualize the Src translocation and activation at different compartments on the plasma membrane in live cells.

Src contributes to cell protrusion and migration in many ways. Src, mediated by focal adhesion kinase (FAK), can phosphorylate p130cas, which recruits Crk and DOCK180 through the interaction of SH3 domain on Crk and PXXP motif on DOCK180. DOCK180 subsequently binds to ELMO and activates Rac, which leads to the activation of Wave1/Scar1.[143] Recent results indicate that Src can also directly phosphorylate

Scar1.[144] Activated Scar1 can bind to and activate Arp2/3, which causes the branching growth of actin filaments and the formation of actin arcs adjacent to the plasma membrane.[143] The polymerized actin filaments bend beneath membrane, and the subsequently accumulated thermodynamic energy *in situ* may eventually promote the protrusion of the lipid layer at the leading edge along migration direction.[145] It has been reported that Src can be activated in live human umbilical VE cells by applying laser-tweezer traction on FN-coated beads adhering to the cells.[34] A rapid distal Src activation and a slow directional wave propagation of Src activation along the plasma membrane have been observed. The wave propagated away from the stimulation site with a relatively constant speed of about 18.1 nm/s. This force-induced directional and long-range activation of Src can be abolished by the disruption of actin filaments or microtubules. Therefore, the transmission of mechanically induced Src activation is considered a dynamic process that directs signals via the cytoskeleton to spatial destinations (Fig. 2.2).[34]

The permeability of endothelium and consequently atherosclerosis involves EC junctions.[69,147] Among the three major types of intercellular connections, viz., AJ, gap junction, and tight junction,[148] AJ is the most ubiquitous.[149] In ECs, AJ is mainly comprised of a membrane receptor VE-cadherin, with its intracellular domain separated into the juxtamembrane domain (JMD) and the catenin-binding domain (CBD). JMD provides putative docking sites for p120^{ctn}, which is a substrate molecule for Src. CBD binds directly to β-catenin and γ-catenin, which possibly bridge the VE-cadherin complex to actin-based cytoskeleton.[148,150,151]

Active Src perturbs the cadherin-mediated cell–cell adhesion. For example, AJ was severely deteriorated in v-Src transformed fibroblasts.[152,153] Constitutively active Src protein also caused the tyrosine phosphorylation of E-cadherin and a concurrent loss of cell–cell contact.[154] Further, ERK is constitutively activated in Src-transformed cells.[110,111] The SH2/SH3 domains of Src can recruit[155] and activate ERK, resulting in the phosphorylation of an ERK substrate molecule myosin light-chain kinase (MLCK).[117] The phosphorylation of MLCK ultimately leads to the phosphorylation of myosin light chain and activation of actomyosin machinery to cause the *in situ* contractility and the breakage of AJ.[155] Shear stress has been shown to activate both Src and ERK in bovine aortic endothelial cells (BAECs).[19,107,156] It is, however, not clear how different flows activate Src, ERK, and MLCK in space and time to regulate AJ.

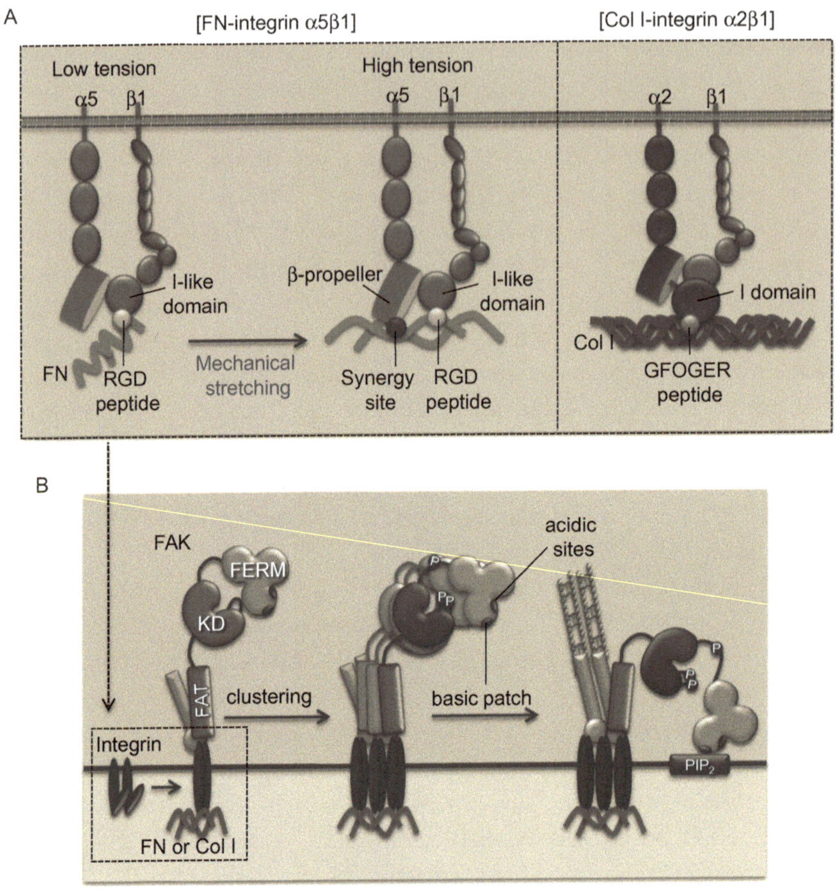

Figure 2.2 Proposed model of FAK mechanoactivation mechanisms via different ECM and integrin subtypes during cell adhesion process. (A) Integrin α5β1 can be fully activated in the tensioned state where both RGD peptide (yellow circle [white in print version]) and synergy site (red circle [dark gray in print version]) bind to α5 and β1 subunits, respectively. Because FN synergy site is exposed only in the high-tensional state, the FAK activation via integrin α5β1 is dependent on the mechanical environment. In contrast, integrin α2β1 can directly bind to the constitutively exposed GFOGER motif (orange circle [light gray in print version]) in Col I, thus causing the activation of integrin α2β1 and FAK independent of mechanical tension. (B) Integrin activation can recruit and induce the transphosphorylation of FAK. This leads to the FAK activation, which is maintained by the interaction between the FERM basic patch (blue oval [light gray in print version]) and PIP2 to prevent the inhibitory interaction of myosin II with FERM acidic sites (red oval [dark gray in print version]). *This research was originally published in* Nature Communication.[146]

5. FOCAL ADHESION AND FAK

Focal adhesions are the contact sites of cells to outside ECM through transmembrane proteins integrins. Integrins are heterodimeric receptors containing α and β subunits, and so far 24 different subtypes of integrins have been identified in vertebrates with the combination of 18α and 8β subunits. These integrin subtypes allow the diverse and specific recognition of various ECM proteins, e.g., FN, fibrinogen, collagen, vitronectin, and laminin.[157]

After the ligation of integrins with ECM proteins, many structural and signaling proteins are recruited to focal adhesions. Signaling proteins at focal adhesions include kinases, e.g., Src, FAK, integrin-linked kinase, and phosphatase, e.g., receptor-like tyrosine phosphatase α. Since integrin itself lacks the enzymatic activity, these signaling proteins at focal adhesions are crucial to transfer extracellular mechanical information inside the cells. For example, integrin-mediated activation of Src and FAK can regulate Rho GTPases, which then regulate the organization of actin cytoskeleton.[158,159] Src and FAK coordinate and regulate downstream signals in focal adhesions (Fig. 2.2).[146,160] It has been reported that growth factor-induced FAK activation is mediated and maintained by Src activity, while FAK activation on cell adhesion is independent of and in fact essential for the Src activation.[146] FAK also interacts with integrin receptors and ECM proteins to sense the mechano-environment (Fig. 2.2).[160] Seong *et al.* found that matrix protein FN-mediated FAK activation is dependent on mechanical tension, which may expose the otherwise hidden FN synergy site to integrin α5. In contrast, the binding motif of type I collagen to its receptor integrin α2 is constitutively exposed. Hence, FAK can be sufficiently activated on type I collagen independent of tension. Therefore, different ECM proteins can differentially transmit or shield mechanical forces from the environment to the functional molecules in the cell and regulate cellular functions (Fig. 2.2).[160]

Structural proteins, for example, talin, paxillin, vinculin, and zyxin, link other focal adhesion proteins and actin cytoskeleton. For example, ECM-bound integrins can recruit an adaptor molecule talin to focal adhesions. Talin has the binding sites for another molecule vinculin, which then bind to actin cytoskeleton.[161,162] As such, outside ECM can be physically connected to intracellular cytoskeleton through integrin and focal adhesion proteins. Therefore, due to their physical location at the interface between extracellular microenvironments and intracellular space, focal adhesions play crucial roles in sensing mechanical signals from both outside and inside.[163,164]

6. TOOLS TO MONITOR SIGNAL TRANSDUCTION IN LIVE CELLS

6.1. FPs, FRET, and fluorescence lifetime imaging microscopy

Molecular-tagged FPs and specific FRET biosensors are capable of monitoring cellular events in live cells.[28–37] YPet, a variant of yellow FP, was paired with enhanced cyan fluorescence protein (ECFP) and severalfolds increase in sensitivity was observed for various biosensors.[165,166] These ECFP/YPet-based biosensors, however, only allow the visualization of one active molecular event in a single live cell. Several studies have used one FP as the common donor or acceptor for two FRET biosensors to visualize different molecular signals in the same cell.[167–169] Other pairs, including Ametrine paired with tdTomato,[170] mOrange with mCherry,[171–174] TagFP with mPlum,[175] and T-Sapphire with DsRed,[176] have been explored as a second FRET pair.[171–174] These approaches either require sophisticated means to quantify the signals or are difficult for multi-FRET imaging in a single cell. The potential of mOrange2/mCherry as a new FRET pair, together with ECFP and YPet, was also demonstrated for dual-FRET imaging.[177]

Although mOrange2 and mCherry can act as a new FRET pair to monitor active signaling events,[166] there is some overlap between their excitation spectra to cause nonspecific cross talk,[171,172] and this may have contributed to the lower sensitivity of a MT1-MMP FRET biosensor *in vitro* using the mOrange2/mCherry pair.[166] Because fluorescence lifetime imaging microscopy (FLIM) only monitors the donor lifetime to measure FRET signals without the need to measure the acceptor lifetime, it can avoid the nonspecific contamination of acceptor excitation/emission. FLIM is also independent of the local concentrations of fluorescent molecules and can separate the population of "FRETing" donors from those of noninteracting ones based on the lifetime distribution, thus enhancing the accuracy of FRET detection.[178–181] Hence, FLIM is ideal for the visualization of multiple FRET biosensors simultaneously in the same live cell.

Recently, FRET techniques have been applied to visualize signal transduction in response to shear stress. For example, GFP-fused Rac and Alexa568-p21-binding domain of PAK1 (PBD) were used to monitor the Rac activation in live cell by measuring FRET between GFP to Alexa568.[182] With this FRET-based biosensor, shear stress was shown to induce a directional activation of Rac concentrated at the leading edge of the cell along flow direction.[183] Shear stress has also been shown to induce a polarized Cdc42 activation along flow direction visualized by the FRET between a GFP-Cdc42 and an Alexa568-PBD.[184] In another study, a separated pair of ECFP-fused relA and EYFP-fused IκBα was used to monitor the interaction of relA and IκBα. The FRET efficiency between ECFP-relA and EYFP-IκBα decreased upon shear-stress application, indicating a mechanical force-induced dissociation of relA and IκBα.[185] CFP and YFP have also been fused to human B_2 bradykinin receptor, a GPCR, to detect the activation of GPCR. Shear stress was shown to activate B_2 bradykinin GPCR within 2 min, which can be inhibited by B_2-selective antagonist.[17] These results suggest that B_2 bradykinin GPCR may serve as a mechanosensing molecule in response to shear stress. In the mechanosensing and mechanotransducing elements included focal adhesion and the cytoskeleton network, FRET biosensors have been developed for many molecules, including Src, FAK, α-actinin, vinculin, and talin.[34,146,186]

6.2. Quantitative image-based analysis for live cells

The development and application of FPs and related biosensors have greatly advanced our knowledge of signaling transduction in live cells. The vast amount of imaging data produced by these FPs and biosensors demands the development of automated, intelligent, and objective image analysis tools to allow precise and efficient interpretation of complex biological information.[39,44] Automated algorithms integrated with image-based analysis have been developed and utilized to analyze the kinetics of FPs, discretize the intracellular space, track the cell movement, and detect the localization of the FPs. Here, we briefly review the widely used fluorescence recovery after photobleach (FRAP) analysis methods for quantifying the diffusion kinetics of molecules and provide an example of quantitative polarity analysis based on images.

6.3. The FRAP analysis and finite-element-based diffusion analysis

The diffusion kinetics of a FRET biosensor, or a fluorescent molecule, can be examined by FRAP analysis, a widely used technique in estimating the apparent diffusion coefficient of molecules in live cells.[187–190] In classical FRAP analysis, the fluorescence recovery curve is obtained by monitoring the recovery process of the average fluorescence intensity in the small region, which underwent photobleaching and the subsequent recovery.

The apparent diffusion coefficient of fluorescent molecules can be estimated by performing parameter fitting on the recovery curve.[187] Results from the FRAP analysis has been revealed the characteristics of transport kinetics for many important molecules.[190–192] However, there are limitations with this approach on the geometry of the cell, the geometry of the photobleached spot, and the actual photobleaching process that can be analyzed.[187–189,193] Most recently, FRAP analysis using computational approaches, such as the model-based finite-element (FE) and finite-difference methods and the simulation-based computational particle and Monte Carlo methods, has been developed to address these limitations.[192,194–200]

In particular, the FE method is a computational approach for defining a set of linear equations, which approximate the partial differential equations in the value of solutions. It is well known for the flexibility in resolving complex geometry of tissue and cellular structures for approximating the elastic and diffusion equations.[201,202] This method has been used to estimate the apparent diffusion constant in inhomogeneous tissues[200] and model protein transport in single cells.[203] Newly developed FE-based image analysis and simulation tool for FRAP enforce no specific requirement on the geometry of cell, the bleaching light beam, or the photobleaching process. This tool provided a general approach for evaluating the accuracy of the diffusion model. With this FE-based tool integrated with diffusion model, the diffusion of FRET-based Src biosensors has been evaluated, simulated, and subsequently subtracted from the biosensor FRET signals in live cells.[46]

6.4. Automatic tracking of moving cells and subcellular features

Image registration is the process of transforming different sets of images into one coordinate system. It is widely used in engineering and science for automatically finding the pixel-wise transformation from the data to reference images by detecting and matching their feature points and the interpolation of values among these feature points.[204,205] For example, it was used in live-cell FRET imaging to physically align the fluorescent images obtained at different wavelengths simultaneously.[31,182,206] However, the registration of the video images of moving cells is a considerably more challenging problem, due to the relatively nonlinear movement of cell body during migration or adhesion. To circumvent the difficulty involved in the image registration of moving cells, some quantification methods have been developed to divide the whole cell into small wedges in the polar coordinated system and quantify the cell motility, protein distribution, or molecular activity in these wedges with different fixed angles.[206–208] These methods are very useful to align and quantify the molecular localization in the angular direction among different cells or in a cell which changes the shape in time. However, a robust and accurate whole-cell image registration method capable of tracking all intracellular pixels is needed for the precise image analysis and model simulation of molecular activities in migrating cells.

In addition to whole-cell analysis, it is very important to detect and track molecular locations and activities within subcellular structures. For example, the image segmentation-based water algorithm has been developed to detect individual FAs based on their intensity profiles. It was initially used to characterize the molecular heterogeneity between and within FAs and to study the effect of kinase inhibitors on FA disassembly.[41] It was also applied successfully for analyzing the correlation between the local force and the physical property of FAs in subcellular locations, such as orientation, size, and shape.[42] This algorithm has further helped to discover that the FA dynamics is regulated by the pattern of integrin clusters,[209] and to characterize the effect of different drugs on the structure and organization of FAs in a high-throughput fashion.[43]

Polarized molecular activities play important roles in guiding the cell toward persistent and directional migration as those observed in wound healing and the growth of blood vessels. Molecular polarity has been quantified using automated analysis method, and the related time difference among different molecules can be used to compare the time sequence in the signaling pathways (Fig. 2.3).[46,47] Lu *et al.* showed the polarized distributions of the activities of phosphatidylinositol 3-kinase (PI3K) and the Rac1 small GTPase monitored using chimeric FPs in cells. The cells were constrained on micropatterned strips, with one end connecting to a neighboring cell (junction end) and the other end free of cell–cell contact (free end, see Fig. 2.3). The subcellular distribution of FPs and the edge position and velocity at the free end of the cells were quantified to analyze their correlation and interpret the signaling sequence. The initiation of the edge extension occurred before the activation of PI3K, which led to a stable extension of the free end followed by the Rac1 activation.[47] The results showed the power of quantitative image-based analysis in deciphering coordinated sequential signaling events regulating the lamellipodia extension and migration in live cells.[47]

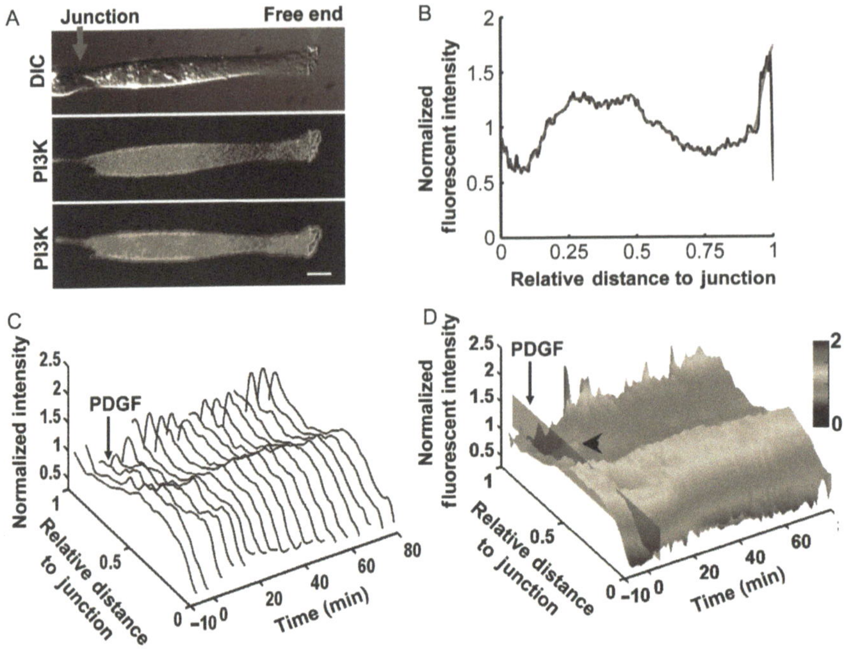

Figure 2.3 The quantification methods and protein localization results. (A) The alignment and outline of a cell before quantification. Top panel: The DIC image of a cell with a junction connecting to a neighboring cell on a patterned FN strip and a free end capable of lamellipodial protrusion. Middle panel: The fluorescence image shows the intensity distribution of PH-Akt-GFP expressed in the cell of interest. Bottom panel: The cell was rotated and aligned along a horizontal direction, and the boundary of the cell calculated and overlaid in red with the fluorescence intensity image. Scale bar: 10 μm. (B) The normalized fluorescence intensity of PH-Akt-GFP plotted against the relative distance to the junction end with the blue and red lines representing the raw and smoothed data, respectively. (C) The sequence of the normalized fluorescence intensity curves arranged along the time axis with a 3D view. The platelet-derived growth factor (PDGF) simulation was applied at 0 min. (D) The 3D intensity surface of PI3K activity landscape was color-coded by the fluorescence intensity values and visualized as a function of time and distance to the junction end, based on the data in (C). The time of PDGF stimulation is indicated by the red plane. A transient secondary peak can be observed between the cell body and the free end (arrowhead). *This research was originally published in* PLoS One.[210] (See the color plate.)

7. CONCLUSION

ECs can sense the mechanical force from the environment and convert it into molecular signals regulating cellular function. Studies on the regulation and function, especially mechanotransduction, of ECs provide important insights into the prevention and treatment of cardiovascular diseases such as atherosclerosis. Many molecules function at the plasma membrane or focal adhesions at the interface between the cell and the environment, acting as mechanosensors. Transmembrane receptors and focal adhesion-associated molecules, such as integrin, CGPR, PECAM-1, TRPC6, Src, FAK, p130Cas, and vinculin, are examples of mechanotransducing molecules sensitive to shear stress, stretch, or the stiffness of the microenvironment. Novel technologies such as fluorescence imaging and FRET biosensors provide powerful tools to study the cellular and molecular functions with high spatiotemporal resolution at the single-cell level. Quantitative image-based analysis tools allow the systematic quantification and inference of single-cell imaging results with rigor and accuracy. Therefore, the future in single-cell imaging of mechanotransduction will be illuminated by genetic engineered fluorescence biosensors, which enables the systematic and quantitative exploration of molecular networks in live cells.

REFERENCES

1. Makowski L, Hotamisligil GS. Fatty acid binding proteins—the evolutionary crossroads of inflammatory and metabolic responses. *J Nutr Sep*. 2004;134(9):2464S–2468S.
2. Bao X, Clark CB, Frangos JA. Temporal gradient in shear-induced signaling pathway: involvement of MAP kinase, c-fos, and connexin43. *Am J Physiol Heart Circ Physiol*. 2000;278(5):H1598–H1605.
3. Bao X, Lu C, Frangos JA. Temporal gradient in shear but not steady shear stress induces PDGF-A and MCP-1 expression in endothelial cells: role of NO, NF kappa B, and egr-1. *Arterioscler Thromb Vasc Biol*. 1999;19(4):996–1003.
4. Gimbrone Jr MA, Garcia-Cardena G. Vascular endothelium, hemodynamics, and the pathobiology of atherosclerosis. *Cardiovasc Pathol*. 2013;22(1):9–15.
5. Weber C, Noels H. Atherosclerosis: current pathogenesis and therapeutic options. *Nat Med*. 2011;17(11):1410–1422.
6. Tartaglia M, Gelb BD. Noonan syndrome and related disorders: genetics and pathogenesis. *Annu Rev Genomics Hum Genet*. 2005;6:45–68.
7. Davies PF, Spaan JA, Krams R. Shear stress biology of the endothelium. *Ann Biomed Eng*. 2005;33(12):1714–1718.
8. Boo YC, Jo H. Flow-dependent regulation of endothelial nitric oxide synthase: role of protein kinases. *Am J Physiol Cell Physiol*. 2003;285(3):C499–C508.
9. Li S, Butler P, Wang Y, et al. The role of the dynamics of focal adhesion kinase in the mechanotaxis of endothelial cells. *Proc Natl Acad Sci USA*. 2002;99(6):3546–3551.
10. Butler PJ, Weinbaum S, Chien S, Lemons DE. Endothelium-dependent, shear-induced vasodilation is rate-sensitive. *Microcirculation*. 2000;7(1):53–65.
11. Davies PF. Overview: temporal and spatial relationships in shear stress-mediated endothelial signalling. *J Vasc Res*. 1997;34(3):208–211.
12. DePaola N, Gimbrone Jr MA, Davies PF, Dewey Jr CF. Vascular endothelium responds to fluid shear stress gradients. *Arterioscler Thromb*. 1992;12(11):1254–1257.
13. Frangos JA, Huang TY, Clark CB. Steady shear and step changes in shear stimulate endothelium via independent mechanisms—superposition of transient and sustained nitric oxide production. *Biochem Biophys Res Commun*. 1996;224(3):660–665.

14. Nagel T, Resnick N, Dewey Jr CF, Gimbrone Jr MA. Vascular endothelial cells respond to spatial gradients in fluid shear stress by enhanced activation of transcription factors. *Arterioscler Thromb Vasc Biol.* 1999;19(8):1825–1834.
15. Tardy Y, Resnick N, Nagel T, Gimbrone Jr MA, Dewey Jr CF. Shear stress gradients remodel endothelial monolayers in vitro via a cell proliferation-migration-loss cycle. *Arterioscler Thromb Vasc Biol.* 1997;17(11):3102–3106.
16. Tzima E, del Pozo MA, Shattil SJ, Chien S, Schwartz MA. Activation of integrins in endothelial cells by fluid shear stress mediates Rho-dependent cytoskeletal alignment. *EMBO J.* 2001;20(17):4639–4647.
17. Chachisvilis M, Zhang YL, Frangos JA. G protein-coupled receptors sense fluid shear stress in endothelial cells. *Proc Natl Acad Sci USA.* 2006;103(42):15463–15468.
18. Tzima E, Irani-Tehrani M, Kiosses WB, et al. A mechanosensory complex that mediates the endothelial cell response to fluid shear stress. *Nature.* 2005;437(7057):426–431.
19. Jalali S, Li YS, Sotoudeh M, et al. Shear stress activates p60src-Ras-MAPK signaling pathways in vascular endothelial cells. *Arterioscler Thromb Vasc Biol.* 1998;18(2):227–234.
20. Shankaran H, Ippolito DL, Chrisler WB, et al. Rapid and sustained nuclear-cytoplasmic ERK oscillations induced by epidermal growth factor. *Mol Syst Biol.* 2009;5:332.
21. Brunton VG, Avizienyte E, Fincham VJ, et al. Identification of Src-specific phosphorylation site on focal adhesion kinase: dissection of the role of Src SH2 and catalytic functions and their consequences for tumor cell behavior. *Cancer Res.* 2005;65(4):1335–1342.
22. Singh I, Knezevic N, Ahmmed GU, Kini V, Malik AB, Mehta D. Galphaq-TRPC6-mediated Ca^{2+} entry induces RhoA activation and resultant endothelial cell shape change in response to thrombin. *J Biol Chem.* 2007;282(11):7833–7843.
23. Tiruppathi C, Ahmmed GU, Vogel SM, Malik AB. Ca^{2+} signaling, TRP channels, and endothelial permeability. *Microcirculation.* 2006;13(8):693–708.
24. Simons K, Vaz WL. Model systems, lipid rafts, and cell membranes. *Annu Rev Biophys Biomol Struct.* 2004;33:269–295.
25. Bickett DM, Green MD, Berman J, et al. A high throughput fluorogenic substrate for interstitial collagenase (MMP-1) and gelatinase (MMP-9). *Anal Biochem.* 1993; 212(1):58–64.
26. Jacobson K, Mouritsen OG, Anderson RG. Lipid rafts: at a crossroad between cell biology and physics. *Nat Cell Biol.* 2007;9(1):7–14.
27. Chiu JJ, Chien S. Effects of disturbed flow on vascular endothelium: pathophysiological basis and clinical perspectives. *Physiol Rev.* 2011;91(1):327–387.
28. Kunkel MT, Ni Q, Tsien RY, Zhang J, Newton AC. Spatio-temporal dynamics of protein kinase B/Akt signaling revealed by a genetically encoded fluorescent reporter. *J Biol Chem.* 2005;280(7):5581–5587.
29. Miyawaki A, Llopis J, Heim R, et al. Fluorescent indicators for Ca^{2+} based on green fluorescent proteins and calmodulin. *Nature.* 1997;388(6645):882–887.
30. Mochizuki N, Yamashita S, Kurokawa K, et al. Spatio-temporal images of growth-factor-induced activation of Ras and Rap1. *Nature.* 2001;411(6841):1065–1068.
31. Pertz O, Hodgson L, Klemke RL, Hahn KM. Spatiotemporal dynamics of RhoA activity in migrating cells. *Nature.* 2006;440(7087):1069–1072.
32. Ting AY, Kain KH, Klemke RL, Tsien RY. Genetically encoded fluorescent reporters of protein tyrosine kinase activities in living cells. *Proc Natl Acad Sci USA.* 2001;98(26):15003–15008.
33. Violin JD, Zhang J, Tsien RY, Newton AC. A genetically encoded fluorescent reporter reveals oscillatory phosphorylation by protein kinase C. *J Cell Biol.* 2003;161(5):899–909.
34. Wang Y, Botvinick EL, Zhao Y, et al. Visualizing the mechanical activation of Src. *Nature.* 2005;434(7036):1040–1045.

35. Zhang J, Hupfeld CJ, Taylor SS, Olefsky JM, Tsien RY. Insulin disrupts beta-adrenergic signalling to protein kinase A in adipocytes. *Nature*. 2005;437(7058):569–573.
36. Zhang J, Ma Y, Taylor SS, Tsien RY. Genetically encoded reporters of protein kinase A activity reveal impact of substrate tethering. *Proc Natl Acad Sci USA*. 2001;98(26):14997–15002.
37. Buranachai C, Kamiyama D, Chiba A, Williams BD, Clegg RM. Rapid frequency-domain FLIM spinning disk confocal microscope: lifetime resolution, image improvement and wavelet analysis. *J Fluoresc*. 2008;18(5):929–942.
38. Liao X, Lu S, Zhuo Y, Winter C, Xu W, Wang Y. Visualization of Src and FAK activity during the differentiation process from HMSCs to osteoblasts. *PLoS One*. 2012;7(8):e42709.
39. Wang Y, Shyy JY, Chien S. Fluorescence proteins, live-cell imaging, and mechanobiology: seeing is believing. *Annu Rev Biomed Eng*. 2008;10:1–38.
40. Song Y, Madahar V, Liao J. Development of FRET assay into quantitative and high-throughput screening technology platforms for protein-protein interactions. *Ann Biomed Eng*. 2011;39(4):1224–1234.
41. Zamir E, Katz BZ, Aota S, Yamada KM, Geiger B, Kam Z. Molecular diversity of cell-matrix adhesions. *J Cell Sci*. 1999;112(Pt 11):1655–1669.
42. Balaban NQ, Schwarz US, Riveline D, et al. Force and focal adhesion assembly: a close relationship studied using elastic micropatterned substrates. *Nat Cell Biol*. 2001;3(5):466–472.
43. Paran Y, Ilan M, Kashman Y, et al. High-throughput screening of cellular features using high-resolution light-microscopy; application for profiling drug effects on cell adhesion. *J Struct Biol*. 2007;158(2):233–243.
44. Peng H. Bioimage informatics: a new area of engineering biology. *Bioinformatics*. 2008;24(17):1827–1836.
45. Jaqaman K, Loerke D, Mettlen M, et al. Robust single-particle tracking in live-cell time-lapse sequences. *Nat Methods*. 2008;5(8):695–702.
46. Lu S, Ouyang M, Seong J, Zhang J, Chien S, Wang Y. The spatiotemporal pattern of Src activation at lipid rafts revealed by diffusion-corrected FRET imaging. *PLoS Comput Biol*. 2008;4(7):e1000127.
47. Lu S, Wang Y, Huang H, et al. Quantitative FRET imaging to visualize the invasiveness of live breast cancer cells. *PLoS One*. 2013;8(3):e58569.
48. Sabouri-Ghomi M, Wu Y, Hahn K, Danuser G. Visualizing and quantifying adhesive signals. *Curr Opin Cell Biol*. 2008;20(5):541–550.
49. Machacek M, Hodgson L, Welch C, et al. Coordination of Rho GTPase activities during cell protrusion. *Nature*. 2009;461(7260):99–103.
50. Chien S. Effects of disturbed flow on endothelial cells. *Ann Biomed Eng*. 2008;36(4):554–562.
51. Chien S, Li S, Shyy YJ. Effects of mechanical forces on signal transduction and gene expression in endothelial cells. *Hypertension*. 1998;31(1 Pt 2):162–169.
52. Noria S, Cowan DB, Gotlieb AI, Langille BL. Transient and steady-state effects of shear stress on endothelial cell adherens junctions. *Circ Res*. 1999;85(6):504–514.
53. Miao H, Hu YL, Shiu YT, et al. Effects of flow patterns on the localization and expression of VE-cadherin at vascular endothelial cell junctions: in vivo and in vitro investigations. *J Vasc Res*. 2005;42(1):77–89.
54. Hahn C, Schwartz MA. The role of cellular adaptation to mechanical forces in atherosclerosis. *Arterioscler Thromb Vasc Biol*. 2008;28(12):2101–2107.
55. Berk BC, Abe JI, Min W, Surapisitchat J, Yan C. Endothelial atheroprotective and anti-inflammatory mechanisms. *Ann N Y Acad Sci*. 2001;947:93–109, discussion 109–111.
56. Jo H, Song H, Mowbray A. Role of NADPH oxidases in disturbed flow- and BMP4-induced inflammation and atherosclerosis. *Antioxid Redox Signal*. 2006;8(9–10):1609–1619.

57. World CJ, Garin G, Berk B. Vascular shear stress and activation of inflammatory genes. *Curr Atheroscler Rep*. 2006;8(3):240–244.
58. Chang K, Weiss D, Suo J, et al. Bone morphogenic protein antagonists are coexpressed with bone morphogenic protein 4 in endothelial cells exposed to unstable flow in vitro in mouse aortas and in human coronary arteries: role of bone morphogenic protein antagonists in inflammation and atherosclerosis. *Circulation*. 2007;116(11):1258–1266.
59. Sorescu GP, Song H, Tressel SL, et al. Bone morphogenic protein 4 produced in endothelial cells by oscillatory shear stress induces monocyte adhesion by stimulating reactive oxygen species production from a nox1-based NADPH oxidase. *Circ Res*. 2004;95(8):773–779.
60. Brooks AR, Lelkes PI, Rubanyi GM. Gene expression profiling of vascular endothelial cells exposed to fluid mechanical forces: relevance for focal susceptibility to atherosclerosis. *Endothelium*. 2004;11(1):45–57.
61. Ouyang M, Lu S, Li XY, et al. Visualization of polarized membrane type 1 matrix metalloproteinase activity in live cells by fluorescence resonance energy transfer imaging. *J Biol Chem*. 2008;283(25):17740–17748.
62. Yamawaki H, Lehoux S, Berk BC. Chronic physiological shear stress inhibits tumor necrosis factor-induced proinflammatory responses in rabbit aorta perfused ex vivo. *Circulation*. 2003;108(13):1619–1625.
63. Yamawaki H, Pan S, Lee RT, Berk BC. Fluid shear stress inhibits vascular inflammation by decreasing thioredoxin-interacting protein in endothelial cells. *J Clin Invest*. 2005;115(3):733–738.
64. Schwartz SM, Gajdusek CM, Reidy MA, Selden 3rd SC, Haudenschild CC. Maintenance of integrity in aortic endothelium. *Fed Proc*. 1980;39(9):2618–2625.
65. Hsu PP, Li S, Li YS, et al. Effects of flow patterns on endothelial cell migration into a zone of mechanical denudation. *Biochem Biophys Res Commun*. 2001;285(3):751–759.
66. Albuquerque ML, Waters CM, Savla U, Schnaper HW, Flozak AS. Shear stress enhances human endothelial cell wound closure in vitro. *Am J Physiol Heart Circ Physiol*. 2000;279(1):H293–H302.
67. Sprague EA, Luo J, Palmaz JC. Human aortic endothelial cell migration onto stent surfaces under static and flow conditions. *J Vasc Interv Radiol*. 1997;8(1 Pt 1):83–92.
68. Wu MH, Kouchi Y, Onuki Y, et al. Effect of differential shear stress on platelet aggregation, surface thrombosis, and endothelialization of bilateral carotid-femoral grafts in the dog. *J Vasc Surg*. 1995;22(4):382–390, discussion 390–382.
69. Albuquerque ML, Flozak AS. Wound closure in sheared endothelial cells is enhanced by modulation of vascular endothelial-cadherin expression and localization. *Exp Biol Med (Maywood)*. 2002;227(11):1006–1016.
70. Kondapalli J, Flozak AS, Albuquerque ML. Laminar shear stress differentially modulates gene expression of p120 catenin, Kaiso transcription factor, and vascular endothelial cadherin in human coronary artery endothelial cells. *J Biol Chem*. 2004;279(12):11417–11424.
71. Albuquerque ML, Flozak AS. Lamellipodial motility in wounded endothelial cells exposed to physiologic flow is associated with different patterns of beta1-integrin and vinculin localization. *J Cell Physiol*. 2003;195(1):50–60.
72. Albuquerque ML, Flozak AS. Patterns of living beta-actin movement in wounded human coronary artery endothelial cells exposed to shear stress. *Exp Cell Res*. 2001;270(2):223–234.
73. Gojova A, Barakat AI. Vascular endothelial wound closure under shear stress: role of membrane fluidity and flow-sensitive ion channels. *J Appl Physiol*. 2005;98(6):2355–2362.
74. Lauffenburger DA, Horwitz AF. Cell migration: a physically integrated molecular process. *Cell*. 1996;84(3):359–369.

75. Li YS, Haga JH, Chien S. Molecular basis of the effects of shear stress on vascular endothelial cells. *J Biomech.* 2005;38(10):1949–1971.
76. Butler PJ, Norwich G, Weinbaum S, Chien S. Shear stress induces a time- and position-dependent increase in endothelial cell membrane fluidity. *Am J Physiol Cell Physiol.* 2001;280(4):C962–C969.
77. McCabe JB, Berthiaume LG. Functional roles for fatty acylated amino-terminal domains in subcellular localization. *Mol Biol Cell.* 1999;10(11):3771–3786.
78. Bagi Z, Frangos JA, Yeh JC, White CR, Kaley G, Koller A. PECAM-1 mediates NO-dependent dilation of arterioles to high temporal gradients of shear stress. *Arterioscler Thromb Vasc Biol.* 2005;25(8):1590–1595.
79. Fleming I, Fisslthaler B, Dixit M, Busse R. Role of PECAM-1 in the shear-stress-induced activation of Akt and the endothelial nitric oxide synthase (eNOS) in endothelial cells. *J Cell Sci.* 2005;118(Pt 18):4103–4111.
80. Chen CN, Chang SF, Lee PL, et al. Neutrophils, lymphocytes, and monocytes exhibit diverse behaviors in transendothelial and subendothelial migrations under coculture with smooth muscle cells in disturbed flow. *Blood.* 2006;107(5): 1933–1942.
81. Nagel T, Resnick N, Atkinson WJ, Dewey Jr CF, Gimbrone Jr MA. Shear stress selectively upregulates intercellular adhesion molecule-1 expression in cultured human vascular endothelial cells. *J Clin Invest.* 1994;94(2):885–891.
82. Li S, Kim M, Hu YL, et al. Fluid shear stress activation of focal adhesion kinase. Linking to mitogen-activated protein kinases. *J Biol Chem.* 1997;272(48):30455–30462.
83. Malek AM, Izumo S. Mechanism of endothelial cell shape change and cytoskeletal remodeling in response to fluid shear stress. *J Cell Sci.* 1996;109(Pt 4):713–726.
84. Morita T, Kurihara H, Maemura K, Yoshizumi M, Yazaki Y. Disruption of cytoskeletal structures mediates shear stress-induced endothelin-1 gene expression in cultured porcine aortic endothelial cells. *J Clin Invest.* 1993;92(4):1706–1712.
85. Helmke BP, Davies PF. The cytoskeleton under external fluid mechanical forces: hemodynamic forces acting on the endothelium. *Ann Biomed Eng.* 2002;30(3): 284–296.
86. Helmke BP, Thakker DB, Goldman RD, Davies PF. Spatiotemporal analysis of flow-induced intermediate filament displacement in living endothelial cells. *Biophys J.* 2001;80(1):184–194.
87. Helmke BP, Goldman RD, Davies PF. Rapid displacement of vimentin intermediate filaments in living endothelial cells exposed to flow. *Circ Res.* 2000;86(7): 745–752.
88. Feaver RE, Gelfand BD, Wang C, Schwartz MA, Blackman BR. Atheroprone hemodynamics regulate fibronectin deposition to create positive feedback that sustains endothelial inflammation. *Circ Res.* 2010;106(11):1703–1711.
89. Frangos JA, Eskin SG, McIntire LV, Ives CL. Flow effects on prostacyclin production by cultured human endothelial cells. *Science.* 1985;227(4693):1477–1479.
90. Barbee KA, Davies PF, Lal R. Shear stress-induced reorganization of the surface topography of living endothelial cells imaged by atomic force microscopy. *Circ Res.* 1994;74(1):163–171.
91. Davies PF, Remuzzi A, Gordon EJ, Dewey Jr CF, Gimbrone Jr MA. Turbulent fluid shear stress induces vascular endothelial cell turnover in vitro. *Proc Natl Acad Sci USA.* 1986;83(7):2114–2117.
92. Sun RJ, Muller S, Stoltz JF, Wang X. Shear stress induces caveolin-1 translocation in cultured endothelial cells. *Eur Biophys J.* 2002;30(8):605–611.
93. Yoshikawa N, Ariyoshi H, Ikeda M, Sakon M, Kawasaki T, Monden M. Shear-stress causes polarized change in cytoplasmic calcium concentration in human umbilical vein endothelial cells (HUVECs). *Cell Calcium.* 1997;22(3):189–194.

94. Butler PJ, Tsou TC, Li JY, Usami S, Chien S. Rate sensitivity of shear-induced changes in the lateral diffusion of endothelial cell membrane lipids: a role for membrane perturbation in shear-induced MAPK activation. *FASEB J.* 2002;16(2):216–218.
95. Wuensch T, Thilo F, Krueger K, Scholze A, Ristow M, Tepel M. High glucose-induced oxidative stress increases transient receptor potential channel expression in human monocytes. *Diabetes.* 2010;59(4):844–849.
96. Thilo F, Vorderwulbecke BJ, Marki A, et al. Pulsatile atheroprone shear stress affects the expression of transient receptor potential channels in human endothelial cells. *Hypertension.* 2012;59(6):1232–1240.
97. Bardell TK, Barker EL. Activation of TRPC6 channels promotes endocannabinoid biosynthesis in neuronal CAD cells. *Neurochem Int.* 2010;57(1):76–83.
98. Kronstein R, Seebach J, Grossklaus S, et al. Caveolin-1 opens endothelial cell junctions by targeting catenins. *Cardiovasc Res.* 2012;93(1):130–140.
99. Graziani A, Poteser M, Heupel WM, et al. Cell-cell contact formation governs Ca^{2+} signaling by TRPC4 in the vascular endothelium: evidence for a regulatory TRPC4-beta-catenin interaction. *J Biol Chem.* 2010;285(6):4213–4223.
100. Inoue R, Jensen LJ, Jian Z, et al. Synergistic activation of vascular TRPC6 channel by receptor and mechanical stimulation via phospholipase C/diacylglycerol and phospholipase A2/omega-hydroxylase/20-HETE pathways. *Circ Res.* 2009;104(12):1399–1409.
101. Spassova MA, Hewavitharana T, Xu W, Soboloff J, Gill DL. A common mechanism underlies stretch activation and receptor activation of TRPC6 channels. *Proc Natl Acad Sci USA.* 2006;103(44):16586–16591.
102. Gottlieb P, Folgering J, Maroto R, et al. Revisiting TRPC1 and TRPC6 mechanosensitivity. *Pflugers Arch.* 2008;455(6):1097–1103.
103. Mederos y Schnitzler M, Storch U, Meibers S, et al. Gq-coupled receptors as mechanosensors mediating myogenic vasoconstriction. *EMBO J.* 2008;27(23):3092–3103.
104. Melchior B, Frangos JA. Galphaq/11-mediated intracellular calcium responses to retrograde flow in endothelial cells. *Am J Physiol Cell Physiol.* 2012;303(4):C467–C473.
105. Storch U, Mederos y Schnitzler M, Gudermann T. G protein-mediated stretch reception. *Am J Physiol Heart Circ Physiol.* 2012;302(6):H1241–H1249.
106. Seong J, Lu S, Ouyang M, et al. Visualization of Src activity at different compartments of the plasma membrane by FRET imaging. *Chem Biol.* 2009;16(1):48–57.
107. Okuda M, Takahashi M, Suero J, et al. Shear stress stimulation of p130(cas) tyrosine phosphorylation requires calcium-dependent c-Src activation. *J Biol Chem.* 1999;274(38):26803–26809.
108. Takahashi M, Berk BC. Mitogen-activated protein kinase (ERK1/2) activation by shear stress and adhesion in endothelial cells. Essential role for a herbimycin-sensitive kinase. *J Clin Invest.* 1996;98(11):2623–2631.
109. Rangaswami H, Schwappacher R, Marathe N, et al. Cyclic GMP and protein kinase G control a Src-containing mechanosome in osteoblasts. *Sci Signal.* 2010;3(153):ra91.
110. Gupta SK, Gallego C, Johnson GL, Heasley LE. MAP kinase is constitutively activated in gip2 and src transformed rat 1a fibroblasts. *J Biol Chem.* 1992;267(12):7987–7990.
111. Mansour SJ, Matten WT, Hermann AS, et al. Transformation of mammalian cells by constitutively active MAP kinase kinase. *Science.* 1994;265(5174):966–970.
112. Lenormand P, Sardet C, Pages G, L'Allemain G, Brunet A, Pouyssegur J. Growth factors induce nuclear translocation of MAP kinases (p42mapk and p44mapk) but not of their activator MAP kinase kinase (p45mapkk) in fibroblasts. *J Cell Biol.* 1993;122(5):1079–1088.
113. Ni CW, Wang DL, Lien SC, Cheng JJ, Chao YJ, Hsieh HJ. Activation of PKC-epsilon and ERK1/2 participates in shear-induced endothelial MCP-1 expression that is repressed by nitric oxide. *J Cell Physiol.* 2003;195(3):428–434.

114. Thomas SM, Brugge JS. Cellular functions regulated by Src family kinases. *Annu Rev Cell Dev Biol.* 1997;13:513–609.
115. Berdeaux RL, Diaz B, Kim L, Martin GS. Active Rho is localized to podosomes induced by oncogenic Src and is required for their assembly and function. *J Cell Biol.* 2004;166(3):317–323.
116. Huveneers S, Oldenburg J, Spanjaard E, et al. Vinculin associates with endothelial VE-cadherin junctions to control force-dependent remodeling. *J Cell Biol.* 2012;196(5):641–652.
117. Klemke RL, Cai S, Giannini AL, Gallagher PJ, de Lanerolle P, Cheresh DA. Regulation of cell motility by mitogen-activated protein kinase. *J Cell Biol.* 1997;137(2):481–492.
118. Zacharias DA, Violin JD, Newton AC, Tsien RY. Partitioning of lipid-modified monomeric GFPs into membrane microdomains of live cells. *Science.* 2002;296(5569): 913–916.
119. Ohno M, Gibbons GH, Dzau VJ, Cooke JP. Shear stress elevates endothelial cGMP. Role of a potassium channel and G protein coupling. *Circulation.* 1993;88(1):193–197.
120. Jo H, Sipos K, Go YM, Law R, Rong J, McDonald JM. Differential effect of shear stress on extracellular signal-regulated kinase and N-terminal Jun kinase in endothelial cells. Gi2- and Gbeta/gamma-dependent signaling pathways. *J Biol Chem.* 1997;272(2): 1395–1401.
121. Bard F, Patel U, Levy JB, Jurdic P, Horne WC, Baron R. Molecular complexes that contain both c-Cbl and c-Src associate with Golgi membranes. *Eur J Cell Biol.* 2002;81(1):26–35.
122. Horne WC, Neff L, Chatterjee D, Lomri A, Levy JB, Baron R. Osteoclasts express high levels of pp60c-src in association with intracellular membranes. *J Cell Biol.* 1992;119(4):1003–1013.
123. Krueger JG, Wang E, Garber EA, Goldberg AR. Differences in intracellular location of pp60src in rat and chicken cells transformed by Rous sarcoma virus. *Proc Natl Acad Sci USA.* 1980;77(7):4142–4146.
124. Redmond T, Brott BK, Jove R, Welsh MJ. Localization of the viral and cellular Src kinases to perinuclear vesicles in fibroblasts. *Cell Growth Differ.* 1992;3(9):567–576.
125. Kaplan KB, Swedlow JR, Varmus HE, Morgan DO. Association of p60c-src with endosomal membranes in mammalian fibroblasts. *J Cell Biol.* 1992;118(2):321–333.
126. Kawakatsu H, Sakai T, Takagaki Y, et al. A new monoclonal antibody which selectively recognizes the active form of Src tyrosine kinase. *J Biol Chem.* 1996;271(10):5680–5685.
127. Mukherjee A, Arnaud L, Cooper JA. Lipid-dependent recruitment of neuronal Src to lipid rafts in the brain. *J Biol Chem.* 2003;278(42):40806–40814.
128. Hitosugi T, Sato M, Sasaki K, Umezawa Y. Lipid raft specific knockdown of SRC family kinase activity inhibits cell adhesion and cell cycle progression of breast cancer cells. *Cancer Res.* 2007;67(17):8139–8148.
129. Kasai A, Shima T, Okada M. Role of Src family tyrosine kinases in the down-regulation of epidermal growth factor signaling in PC12 cells. *Genes Cells.* 2005;10(12):1175–1187.
130. Fincham VJ, Unlu M, Brunton VG, Pitts JD, Wyke JA, Frame MC. Translocation of Src kinase to the cell periphery is mediated by the actin cytoskeleton under the control of the Rho family of small G proteins. *J Cell Biol.* 1996;135(6 Pt 1):1551–1564.
131. Timpson P, Jones GE, Frame MC, Brunton VG. Coordination of cell polarization and migration by the Rho family GTPases requires Src tyrosine kinase activity. *Curr Biol.* 2001;11(23):1836–1846.
132. Sandilands E, Cans C, Fincham VJ, et al. RhoB and actin polymerization coordinate Src activation with endosome-mediated delivery to the membrane. *Dev Cell.* 2004;7(6):855–869.
133. Fincham VJ, Brunton VG, Frame MC. The SH3 domain directs acto-myosin-dependent targeting of v-Src to focal adhesions via phosphatidylinositol 3-kinase. *Mol Cell Biol.* 2000;20(17):6518–6536.

134. Fincham VJ, Frame MC. The catalytic activity of Src is dispensable for translocation to focal adhesions but controls the turnover of these structures during cell motility. *EMBO J.* 1998;17(1):81–92.
135. Lasserre R, Guo XJ, Conchonaud F, et al. Raft nanodomains contribute to Akt/PKB plasma membrane recruitment and activation. *Nat Chem Biol.* 2008;4(9):538–547.
136. Sandilands E, Brunton VG, Frame MC. The membrane targeting and spatial activation of Src, Yes and Fyn is influenced by palmitoylation and distinct RhoB/RhoD endosome requirements. *J Cell Sci.* 2007;120(Pt 15):2555–2564.
137. Simons K, Toomre D. Lipid rafts and signal transduction. *Nat Rev Mol Cell Biol.* 2000;1(1):31–39.
138. Arcaro A, Aubert M, Espinosa del Hierro ME, et al. Critical role for lipid raft-associated Src kinases in activation of PI3K-Akt signalling. *Cell Signal.* 2007;19(5): 1081–1092.
139. Na S, Collin O, Chowdhury F, et al. Rapid signal transduction in living cells is a unique feature of mechanotransduction. *Proc Natl Acad Sci USA.* 2008;105(18):6626–6631.
140. Shima T, Nada S, Okada M. Transmembrane phosphoprotein Cbp senses cell adhesion signaling mediated by Src family kinase in lipid rafts. *Proc Natl Acad Sci USA.* 2003;100(25):14897–14902.
141. Lichtenberg D, Goni FM, Heerklotz H. Detergent-resistant membranes should not be identified with membrane rafts. *Trends Biochem Sci.* 2005;30(8):430–436.
142. Bradshaw JM, Mitaxov V, Waksman G. Investigation of phosphotyrosine recognition by the SH2 domain of the Src kinase. *J Mol Biol.* 1999;293(4):971–985.
143. Rodriguez OC, Schaefer AW, Mandato CA, Forscher P, Bement WM, Waterman-Storer CM. Conserved microtubule-actin interactions in cell movement and morphogenesis. *Nat Cell Biol.* 2003;5(7):599–609.
144. Ardern H, Sandilands E, Machesky LM, Timpson P, Frame MC, Brunton VG. Src-dependent phosphorylation of Scar1 promotes its association with the Arp2/3 complex. *Cell Motil Cytoskeleton.* 2006;63(1):6–13.
145. Sheetz MP. Cell control by membrane-cytoskeleton adhesion. *Nat Rev Mol Cell Biol.* 2001;2(5):392–396.
146. Seong J, Ouyang M, Kim T, et al. Detection of focal adhesion kinase activation at membrane microdomains by fluorescence resonance energy transfer. *Nat Commun.* 2011;2:406.
147. Okegawa T, Li Y, Pong RC, Hsieh JT. Cell adhesion proteins as tumor suppressors. *J Urol.* 2002;167(4):1836–1843.
148. Dejana E, Bazzoni G, Lampugnani MG. Vascular endothelial (VE)-cadherin: only an intercellular glue? *Exp Cell Res.* 1999;252(1):13–19.
149. Gumbiner BM. Cell adhesion: the molecular basis of tissue architecture and morphogenesis. *Cell.* 1996;84(3):345–357.
150. Geiger B, Ayalon O. Cadherins. *Annu Rev Cell Biol.* 1992;8:307–332.
151. Yamada KM, Geiger B. Molecular interactions in cell adhesion complexes. *Curr Opin Cell Biol.* 1997;9(1):76–85.
152. Hamaguchi M, Matsuyoshi N, Ohnishi Y, Gotoh B, Takeichi M, Nagai Y. p60v-src causes tyrosine phosphorylation and inactivation of the N-cadherin-catenin cell adhesion system. *EMBO J.* 1993;12(1):307–314.
153. Matsuyoshi N, Hamaguchi M, Taniguchi S, Nagafuchi A, Tsukita S, Takeichi M. Cadherin-mediated cell-cell adhesion is perturbed by v-src tyrosine phosphorylation in metastatic fibroblasts. *J Cell Biol.* 1992;118(3):703–714.
154. Behrens J, Vakaet L, Friis R, et al. Loss of epithelial differentiation and gain of invasiveness correlates with tyrosine phosphorylation of the E-cadherin/beta-catenin complex in cells transformed with a temperature-sensitive v-SRC gene. *J Cell Biol.* 1993;120(3):757–766.

155. Avizienyte E, Fincham VJ, Brunton VG, Frame MC. Src SH3/2 domain-mediated peripheral accumulation of Src and phospho-myosin is linked to deregulation of E-cadherin and the epithelial-mesenchymal transition. *Mol Biol Cell.* 2004;15(6): 2794–2803.
156. Li YS, Shyy JY, Li S, et al. The Ras-JNK pathway is involved in shear-induced gene expression. *Mol Cell Biol.* 1996;16(11):5947–5954.
157. Luo BH, Carman CV, Springer TA. Structural basis of integrin regulation and signaling. *Annu Rev Immunol.* 2007;25:619–647.
158. Mitra SK, Hanson DA, Schlaepfer DD. Focal adhesion kinase: in command and control of cell motility. *Nat Rev Mol Cell Biol.* 2005;6(1):56–68.
159. Mitra SK, Schlaepfer DD. Integrin-regulated FAK-Src signaling in normal and cancer cells. *Curr Opin Cell Biol.* 2006;18(5):516–523.
160. Seong J, Tajik A, Sun J, et al. Distinct biophysical mechanisms of focal adhesion kinase mechanoactivation by different extracellular matrix proteins. *Proc Natl Acad Sci USA.* 2013;110(48):19372–19377.
161. Critchley DR. Cytoskeletal proteins talin and vinculin in integrin-mediated adhesion. *Biochem Soc Trans.* 2004;32(Pt 5):831–836.
162. Critchley DR. Genetic, biochemical and structural approaches to talin function. *Biochem Soc Trans.* 2005;33(Pt 6):1308–1312.
163. Bershadsky AD, Balaban NQ, Geiger B. Adhesion-dependent cell mechanosensitivity. *Annu Rev Cell Dev Biol.* 2003;19:677–695.
164. Geiger B, Spatz JP, Bershadsky AD. Environmental sensing through focal adhesions. *Nat Rev Mol Cell Biol.* 2009;10(1):21–33.
165. Nguyen AW, Daugherty PS. Evolutionary optimization of fluorescent proteins for intracellular FRET. *Nat Biotechnol.* 2005;23(3):355–360.
166. Ouyang M, Sun J, Chien S, Wang Y. Determination of hierarchical relationship of Src and Rac at subcellular locations with FRET biosensors. *Proc Natl Acad Sci USA.* 2008;105(38):14353–14358.
167. Kawai H, Suzuki T, Kobayashi T, et al. Simultaneous real-time detection of initiator- and effector-caspase activation by double fluorescence resonance energy transfer analysis. *J Pharmacol Sci.* 2005;97(3):361–368.
168. Peyker A, Rocks O, Bastiaens PI. Imaging activation of two Ras isoforms simultaneously in a single cell. *Chembiochem.* 2005;6(1):78–85.
169. Wu X, Simone J, Hewgill D, Siegel R, Lipsky PE, He L. Measurement of two caspase activities simultaneously in living cells by a novel dual FRET fluorescent indicator probe. *Cytometry A.* 2006;69(6):477–486.
170. Ai HW, Hazelwood KL, Davidson MW, Campbell RE. Fluorescent protein FRET pairs for ratiometric imaging of dual biosensors. *Nat Methods.* 2008;5(5):401–403.
171. Shaner NC, Campbell RE, Steinbach PA, Giepmans BN, Palmer AE, Tsien RY. Improved monomeric red, orange and yellow fluorescent proteins derived from Discosoma sp. red fluorescent protein. *Nat Biotechnol.* 2004;22(12):1567–1572.
172. Shaner NC, Lin MZ, McKeown MR, et al. Improving the photostability of bright monomeric orange and red fluorescent proteins. *Nat Methods.* 2008;5(6):545–551.
173. Shaner NC, Steinbach PA, Tsien RY. A guide to choosing fluorescent proteins. *Nat Methods.* 2005;2(12):905–909.
174. Goedhart J, Vermeer JE, Adjobo-Hermans MJ, van Weeren L, Gadella Jr TW. Sensitive detection of p65 homodimers using red-shifted and fluorescent protein-based FRET couples. *PLoS One.* 2007;2(10):e1011.
175. Grant DM, Zhang W, McGhee EJ, et al. Multiplexed FRET to image multiple signaling events in live cells. *Biophys J.* 2008;95(10):L69–L71.
176. Niino Y, Hotta K, Oka K. Simultaneous live cell imaging using dual FRET sensors with a single excitation light. *PLoS One.* 2009;4(6):e6036.

177. Ouyang M, Huang H, Shaner NC, et al. Simultaneous visualization of protumorigenic Src and MT1-MMP activities with fluorescence resonance energy transfer. *Cancer Res.* 2010;70(6):2204–2212.
178. Wallrabe H, Periasamy A. Imaging protein molecules using FRET and FLIM microscopy. *Curr Opin Biotechnol.* 2005;16(1):19–27.
179. Chao LK, Clegg RM. *Foerster resonance energy transfer (FRET) for proteins. Wiley Encyclopedia of Chemical Biology Advanced Review.* 2008; 0.
180. Clegg RM. Fluorescence resonance energy transfer. In: Wang XF, Herman B, eds. *Fluorescence Imaging Spectroscopy and Microscopy.* New York: John Wiley & Sons, Inc.; 1996:179–252
181. Clegg RM. Nuts and bolts of excitation energy migration and energy transfer. In: Papageorgiou GC, Govindjee , eds. *Chlorophyll a Fluorescence: A Signature of Photosynthesis.* The Netherlands: Springer; 2005:83–105.
182. Kraynov VS, Chamberlain C, Bokoch GM, Schwartz MA, Slabaugh S, Hahn KM. Localized Rac activation dynamics visualized in living cells. *Science.* 2000;290(5490):333–337.
183. Tzima E, Del Pozo MA, Kiosses WB, et al. Activation of Rac1 by shear stress in endothelial cells mediates both cytoskeletal reorganization and effects on gene expression. *EMBO J.* 2002;21(24):6791–6800.
184. Tzima E, Kiosses WB, del Pozo MA, Schwartz MA. Localized cdc42 activation, detected using a novel assay, mediates microtubule organizing center positioning in endothelial cells in response to fluid shear stress. *J Biol Chem.* 2003;278(33):31020–31023.
185. Ganguli A, Persson L, Palmer IR, et al. Distinct NF-kappaB regulation by shear stress through Ras-dependent IkappaBalpha oscillations: real-time analysis of flow-mediated activation in live cells. *Circ Res.* 2005;96(6):626–634.
186. Eyckmans J, Boudou T, Yu X, Chen CS. A hitchhiker's guide to mechanobiology. *Dev Cell.* 2011;21(1):35–47.
187. Axelrod D, Koppel DE, Schlessinger J, Elson E, Webb WW. Mobility measurement by analysis of fluorescence photobleaching recovery kinetics. *Biophys J.* 1976;16(9):1055–1069.
188. Klonis N, Rug M, Harper I, Wickham M, Cowman A, Tilley L. Fluorescence photobleaching analysis for the study of cellular dynamics. *Eur Biophys J.* 2002;31(1):36–51.
189. Lippincott-Schwartz J, Snapp E, Kenworthy A. Studying protein dynamics in living cells. *Nat Rev Mol Cell Biol.* 2001;2(6):444–456.
190. Lippincott-Schwartz J, Patterson GH. Development and use of fluorescent protein markers in living cells. *Science.* 2003;300(5616):87–91.
191. Phair RD, Misteli T. Kinetic modelling approaches to in vivo imaging. *Nat Rev Mol Cell Biol.* 2001;2(12):898–907.
192. Nehls S, Snapp EL, Cole NB, et al. Dynamics and retention of misfolded proteins in native ER membranes. *Nat Cell Biol.* 2000;2(5):288–295.
193. Carrero G, McDonald D, Crawford E, de Vries G, Hendzel MJ. Using FRAP and mathematical modeling to determine the in vivo kinetics of nuclear proteins. *Methods.* 2003;29(1):14–28.
194. Sbalzarini IF, Mezzacasa A, Helenius A, Koumoutsakos P. Effects of organelle shape on fluorescence recovery after photobleaching. *Biophys J.* 2005;89(3):1482–1492.
195. Sbalzarini IF, Hayer A, Helenius A, Koumoutsakos P. Simulations of (an)isotropic diffusion on curved biological surfaces. *Biophys J.* 2006;90(3):878–885.
196. Ulrich M, Kappel C, Beaudouin J, Hezel S, Ulrich J, Eils R. Tropical–parameter estimation and simulation of reaction–diffusion models based on spatio-temporal microscopy images. *Bioinformatics.* 2006;22(21):2709–2710.

197. Beaudouin J, Mora-Bermudez F, Klee T, Daigle N, Ellenberg J. Dissecting the contribution of diffusion and interactions to the mobility of nuclear proteins. *Biophys J.* 2006;90(6):1878–1894.
198. Houtsmuller AB, Rademakers S, Nigg AL, Hoogstraten D, Hoeijmakers JH, Vermeulen W. Action of DNA repair endonuclease ERCC1/XPF in living cells. *Science.* 1999;284(5416):958–961.
199. Farla P, Hersmus R, Geverts B, et al. The androgen receptor ligand-binding domain stabilizes DNA binding in living cells. *J Struct Biol.* 2004;147(1):50–61.
200. Sniekers YH, van Donkelaar CC. Determining diffusion coefficients in inhomogeneous tissues using fluorescence recovery after photobleaching. *Biophys J.* 2005;89(2): 1302–1307.
201. Barondeau DP, Kassmann CJ, Tainer JA, Getzoff ED. Structural chemistry of a green fluorescent protein Zn biosensor. *J Am Chem Soc.* 2002;124(14):3522–3524.
202. Costa KD, Hunter PJ, Rogers JM, Guccione JM, Waldman LK, McCulloch AD. A three-dimensional finite element method for large elastic deformations of ventricular myocardium: I—cylindrical and spherical polar coordinates. *J Biomech Eng.* 1996;118(4):452–463.
203. Sadegh Zadeh K, Elman HC, Montas HJ, Shirmohammadi A. A finite element model for protein transport in vivo. *Biomed Eng Online.* 2007;6:24.
204. Zitova B, Flusser J. Image registration methods: a survey. *Image Vis Comput.* 2003;21(11):977–1000.
205. Jagalur M, Pal C, Learned-Miller E, Zoeller RT, Kulp D. Analyzing in situ gene expression in the mouse brain with image registration, feature extraction and block clustering. *BMC Bioinformatics.* 2007;8(Suppl 10):S5.
206. Yeo MG, Partridge MA, Ezratty EJ, Shen Q, Gundersen GG, Marcantonio EE. Src SH2 arginine 175 is required for cell motility: specific focal adhesion kinase targeting and focal adhesion assembly function. *Mol Cell Biol.* 2006;26(12):4399–4409.
207. Dormann D, Libotte T, Weijer CJ, Bretschneider T. Simultaneous quantification of cell motility and protein-membrane-association using active contours. *Cell Motil Cytoskeleton.* 2002;52(4):221–230.
208. Dubin-Thaler BJ, Giannone G, Dobereiner HG, Sheetz MP. Nanometer analysis of cell spreading on matrix-coated surfaces reveals two distinct cell states and STEPs. *Biophys J.* 2004;86(3):1794–1806.
209. Cavalcanti-Adam EA, Volberg T, Micoulet A, Kessler H, Geiger B, Spatz JP. Cell spreading and focal adhesion dynamics are regulated by spacing of integrin ligands. *Biophys J.* 2007;92(8):2964–2974.
210. Lu Shaoying, Kim Tae-jin, Chen Chih-En, et al. Computational analysis of the spatiotemporal coordination of polarized PI3K and Rac1 activities in micropatterned live cells. *PLoS One.* 2011;6(6):e21293.

PART TWO

Focal Adhesions as Sensors

CHAPTER THREE

Focal Adhesions Function as a Mechanosensor

Jean-Cheng Kuo
Institute of Biochemistry and Molecular Biology, National Yang-Ming University, Taipei, Taiwan

Contents

Abstract

Focal adhesions (FAs) are complex plasma membrane-associated macromolecular assemblies that engage with the surrounding extracellular matrix (ECM) via integrin receptors and physically connect with the actin cytoskeleton through the recruitment of numerous FA-associated proteins. FAs undergo a maturation process, which is known to be induced by biochemical or physical cues, to grow and change composition. Varying FA size, distribution, dynamics, and compositions during maturation process is required for transducing the specific signaling networks that reflect the requirements of a cell to sense, adapt, and response to a variety of the environments. While advances have been demonstrated in understanding how important FAs are in mediating various biological processes, less is known about how FA composition is regulated and coordinately transduces the specific signals in mediating the distinct biological outcomes, especially cell migration.

1. INTRODUCTION: THE BASIC ORGANIZATION OF FOCAL ADHESIONS

Living cells grow and function while being associated with the extracellular matrix (ECM) in the environment (Fig. 3.1). They sense and respond to a wide range of physical or biochemical signals via the cell–ECM interactions that regulate cellular processes such as growth, survival,

Progress in Molecular Biology and Translational Science, Volume 126
ISSN 1877-1173
http://dx.doi.org/10.1016/B978-0-12-394624-9.00003-8

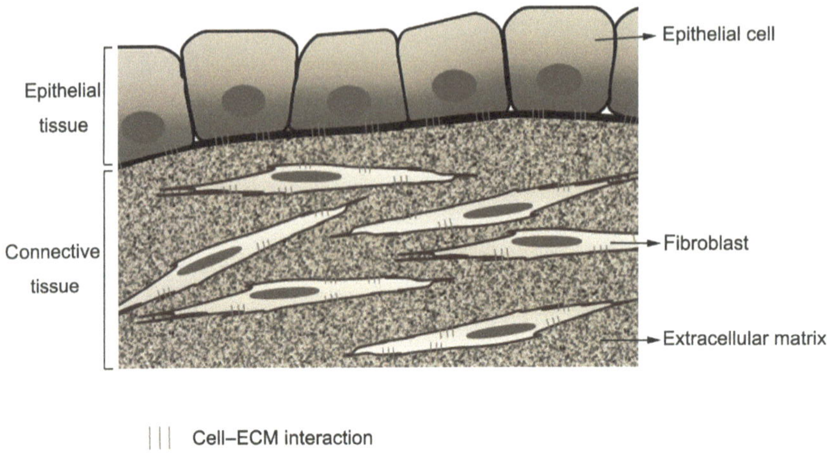

Figure 3.1 Schematic representation of the cell–extracellular matrix (ECM) interactions in living cells. Living cells (e.g., epithelial cells or fibroblasts) associate with the extracellular matrix (ECM) via the cell–ECM interactions. The cell–ECM interactions support cells to grow and function within a wide range of environments.

migration, or differentiation.[1] Focal adhesions are the cell–ECM interactions that form when the central components, integrin receptors, are activated by engagement with the ECM, which then recruit numerous FA-associated proteins to connect with the actin cytoskeleton.[2–4] The macromolecular assembles, FAs, serve as the linkage between the actin cytoskeleton and ECM, thereby integrating environmental signals into the adhesion-mediated signaling networks, which can organize actin cytoskeleton, and change cell morphology, behaviors, and fate. This phenomenon appears in almost every cell type, from prokaryotes to multicellular organisms.

Integrins, the central components of FAs, are type I transmembrane proteins that are heterodimers containing α and β subunits. In mammals, 18 α and 8 β subunits have been characterized to form 24 unique integrin heterodimers (Table 3.1) that recognize specific types of ECM.[3,5,6] For example, the Arg-Gly-Asp (RGD)-binding integrins recognize RGD tripeptide motif containing ECMs such as fibronectin, vitronectin, and fibrinogen; the Leu-Asp-Val (LDV)-binding integrins recognize LDV-related sequences in ECMs such as vascular cell adhesion molecule-1 (MAdCAM-1) and intercellular cell adhesion molecule-1 (ICAM-1); the A domain-containing integrins engage with laminin and collagen; the non-αA domain-containing integrins bind laminin.[6] The interaction between α and β subunits is modulated by ligands from the ECM,[7] thereby

Table 3.1 Four classes of integrin heterodimers

Class	Name
RGD-binding integrins	αVβ1, αVβ3, αVβ5, αVβ6, αVβ8, αIIbβ3, α5β1, α8β1
LDV-binding integrins	α4β1, α4β7, αXβ2, αEβ7, α9β1, αLβ2, αDβ2, αMβ2
A domain-containing integrins	α1β1, α2β1, α10β1, α11β1
Non-αA domain-containing integrins	α3β1, α6β1, α7β1, α6β4

In mammals, 18 α and 8 β are characterized to form 24 unique integrin heterodimers. The lists of 24 integrin heterodimers are classified into 4 classes: Arg-Gly-Asp (RGD)-binding integrins, Leu-Asp-Val (LDV)-binding integrins, A domain-containing integrins and non-αA domain-containing integrins.

changing the conformation of integrin α–β heterodimers, and extends their cytoplasmic tails, which associate with a series of FA proteins.

The association of FA proteins between integrins and the actin cytoskeleton is complex. 180 proteins have been reported to be associated with FAs in order to make up the integrin adhesome[8,9]; these include cytoskeletal proteins, adaptor proteins, and signaling proteins, such as kinases, phosphatases, phospholipases, and regulators of small GTPases. Physical interactions between proteins can depend on the specific scaffolding motifs and can be regulated by signaling events, such as phosphorylation (or dephosphorylation) or the regulation of small GTPases. Therefore, the complex molecular ensembles are composed of scaffolding proteins, signaling enzymes, as well as their substrates.

The proteins with specific scaffolding motifs are classified as the scaffolding proteins that are able to physically connect the actin cytoskeleton to integrin receptors via direct or indirect interactions, while the proteins controlling the connection between integrin receptors and actin filaments through their abilities to modulate the activity, stability, or functionality of the FA components are classified as the regulatory proteins. In detail, the scaffolding proteins include actin-binding proteins and adaptors. Specifically, the actin-binding proteins include proteins that are able to bind directly to the cytoplasmic domains of integrin receptors, such as talin,[10–12] α-actinin,[10,13–15] and filamin A/B/C,[10,16–19] or that are able to connect with integrin receptors via other actin-binding proteins or adaptors, such as vinculin,[10,20] VASP,[10,21,22] and zyxin.[10,23] The adaptors are identified as the FA proteins containing the specific domains, including src homology 2 (SH2), src homology 3 (SH3), pleckstrin homology (PH), FERM (4.1 protein, ezrin, radixin and moesin), calponin homology (CH), and LIM (Lin11, Isl-1 and Mec-3) domains. The SH2 domain typically binds a phosphorylated tyrosine residue present on its target protein,[24,25] while the classic SH3 domain uses proline-rich peptides as its binding partners.[26] PH domains can bind phosphatidylinositol lipid within biological membranes, such as phosphatidylinositol (3,4,5)-trisphosphate and phosphatidylinositol (4,5)-bisphosphate; it thus plays a role in recruiting proteins to specific membranes sites.[27,28] FERM domains consist of three modules (the F1, F2, and F3 subdomains) that are able to form a clover-shaped structure[29]; they play an important role in certain FA proteins that are able to recognize the cytoplasmic tail of β-integrin and mediate integrin activation, such as talin[11,12] and kindlin.[30–33] CH domains are mainly involved in actin binding.[34] LIM domains have highly divergent sequences that are composed of two contiguous zinc finger motifs with a two-amino acid residue hydrophobic linker[35]; these function as a protein-binding interface within many subcellular components such as FAs.[36] On the other hand, the regulatory proteins include the proteins with small guanosine triphosphatase (GTPase) activity, guanine nucleotide exchange factor (GEF) activity, GTPase-activating protein (GAP) activity, proteolytic activity, and activity that regulate protein phosphorylation states. The GTPase activity of the Rho-family proteins, which includes Rac1 and RhoA, is critical for FA maturation and actin cytoskeleton organization.[37–39] The activity of these GTPases is known to be regulated via a switchable cycle that involves GEFs that exchange bound GDP for GTP for activation, and GAPs that promote intrinsic GTP hydrolysis for inactivation.[40,41] Thus, the abundance of GEFs and GAPs regulates the organization of FAs and the actin cytoskeleton through a modulation of GTPase activity. The proteins with proteolytic activity function by cleaving proteins within FAs, thereby disrupting the linkage between integrin and actin, which allows disassembly of FAs. For example, the Ca^{2+}-dependent cysteine-type protease calpain[42] mediates FA disassembly[43–45] via irreversibly cleaving several FA scaffolding proteins, including integrin,[46,47] paxillin,[44] and talin.[44,48] In addition, the proteolytic activity of calpain also regulates the activities of protein tyrosine kinases, such as FAK (focal adhesion kinase)[44,49,50] and SRC,[51] as well as protein tyrosine phosphatases, such as PTP-1B.[51] This feature indicates that FA protein composition is switchable to trigger the specific FA-related signaling networks that regulate the dynamics of FAs, which controls the integrin–actin linkage and organizes the actin cytoskeleton.[2–4,52–56]

The coordination of FAs and the actin cytoskeleton contributes to various cellular behaviors. For example, a cell migration cycle consists of the extension of the leading edge, formation of new FAs, translocation of the cell body, and disassembly of FAs to detach the trailing edge of the cell. When cell migration begins, dendritic actin networks are assembled by polymerizing actin filaments at the leading edge to push the membrane forward.[57] This significant force involved in pushing a cell's leading edge does not involve myosin II motors acting on the actin cytoskeleton.[58–61] Soon after the membrane at the leading edge protrudes, nascent FAs are formed to stabilize the protrusion attached on the substratum. Subsequently, the actomyosin contractile force is generated by myosin II motors sliding on actin filaments, which promotes bundling of filamentous actin (stress fibers) that connect distal points of FAs; this allows the contractile forces to propagate across the cell and applies the force to the substratum through FAs; the result is that the cell body is pulled forward.[62] Finally, the disassembly of FAs at the trailing edge leads to detachment of the cell at the rear. The dynamic responses of the actin cytoskeleton and FAs are fundamental to coordinating the entire process of cell migration. Importantly, the initiative for the integration of FA dynamics and actin cytoskeleton organization depends on the involvement of the specific FA-associated proteins that form functional adhesive organelles and transduce the specific integrin-mediated signaling.

2. MECHANOSENSITIVITY OF FOCAL ADHESIONS

How cells sense, adapt, and respond to a variety of environmental cues to initiate the dynamics of FAs is not well understood. The initial formation of FAs is controlled by the activation of integrin receptors, and then the process of growing FAs is called FA maturation.[63] FAs enlarge during the maturation process, after which they either stabilize or begin to disassemble. Based on their size (~0.1–10 μm^2) and localization, FAs can be classified into nascent adhesions, focal complexes, and mature FAs. Nascent adhesions assemble soon after the integrin receptors engage with the ECM at the edge of lamellipodium, and either are undergoing fast turnover during active protrusions or are evolving into focal complexes within the lamellipodial dendritic actin network. At the lamellipodium–lamellum interface, these adhesions grow and elongate into mature FAs that are connected by bundles of actin filaments (stress fibers), which serve to anchor the cell.[58,64,65] The interplay between all classes of FAs and the organization of actin cytoskeleton is related to the stimuli of the maturation process from the environments.

Mechanisms involved in transducing the maturation stimuli from the environments remain unclear. However, the small G-proteins of the Rho-family are known to regulate assembly and dynamics of FAs,[37–39,66] suggesting that the biochemical signaling cascades could be regulated by the maturation stimuli from the environments. For example, epidermal growth factor, a well-characterized growth factor, promotes FA maturation through activating the small GTPase RhoA.[67] In addition, stimulations of G-coupled receptors, such as thrombin, induce talin binding to integrin αIIbβ3 that activates integrin-mediated signals and triggers cell adhesion and aggregation.[68,69] Previous studies have indicated that the formation of focal complexes is signaled by the activity of small GTPase Rac1,[38] while RhoA signaling promotes the formation of long-lived mature FAs through activating myosin II-driven contractility.[37,66] GTP-bound RhoA activates its target, Rho-associated kinase (ROCK), which increases myosin II-mediated contractility by inhibiting the myosin light chain phosphatase and directly phosphorylating myosin II regulatory light chain (MLC).[70,71] The myosin II-generated contractile force along actin filaments provides the major cellular tension that drives FA maturation.[63] The cellular tension generated directly from actomyosin contractility also can be altered by ECM rigidity, through feedback loops to modulate the pulling forces exerted by the cells,[72–76] or the forces from outside of the cell, such as variation in shear forces. The stimuli from the biochemical or physical cues vary FAs in size, whose maturation state is also indicative of the local balance with respect to the mechanical forces generated from cellular tension or from external forces.[77]

A distinct feature of FA organization in different maturation states is the specific protein composition, which is determined by the local mechanical force.[78,79] FAs can have subtly different compositions in different maturation states to assemble functional adhesive organelles that transduce the specific integrin-mediated signals. However, it is unlikely that all proteins directly sense the mechanical force, rather the recruitment of proteins into FAs is a hierarchical cascade driven by a number of force-sensitive FA proteins.[78,80–83] In response to mechanical force, these force-sensitive FA proteins may undergo structural rearrangement or enzymatic modification that change their binding preferences with respect to other FA-associated proteins (force-responsive FA proteins) and this then further modulates the protein association with FAs. For example, the mechanical stretching forces p130Cas to undergo conformational changes, which expose the site leading to tyrosine phosphorylation by Src family kinases, resulting in the recruitment of phosphorylated p130Cas-binding partners into FAs.[83,84] In addition, application of mechanical force stretches talin rods that expose the binding sites for vinculin.[20] Vinculin controls FA stabilization under force.[81] Thus, the FA components are mutually operative for their associations in FAs that act to strengthen the linkage between integrin and actin filaments and transduce the specific integrin-mediated signals (Fig. 3.2).[10,20,85]

Studies of FA compositional changes in the maturation process provide information on the molecular complexity, diversity, and signals of the integrin-mediated adhesions. Inhibition of myosin II ATPase activity by treatment of blebbistatin results in disrupting FA maturation and keeping FAs in immature states, indicating the contribution of myosin II activity-mediated contractile force in the reorganization of proteins associated with FAs.[78] The proteomic studies of myosin II activity-mediated FA compositional changes reveal a hierarchical cascade of the recruitment of proteins into FAs in the maturation process.[78,79] LIM domain-containing proteins are highly dependent on myosin II activity for FA abundance,[79] suggesting that these proteins may undergo force-dependent unfolding to unmask the binding sites that mediate mechanotransduction.[78–80] The scaffolding proteins, as well as the regulatory proteins, may involve force-triggered unfolding or recruitment that promotes FA association of other components, which are able to produce a physically strengthening of the connection between the integrin receptors and actin filaments.

Despite evidence of the recruitment of scaffolding and regulatory proteins at FAs, proteins that mainly serve as the force-sensitive or force-responsive FA proteins remain to be determined. Theoretically, the force-sensitive or force-responsive FA proteins consist of subsets of scaffolding and regulatory proteins. In Fig. 3.2, the proteins that show increased force-dependent FA abundance could be either positively or negatively regulated by force; these are either force-sensitive or force-responsive FA proteins. To date, many studies based on microscopy and proteomics have revealed that changes in FA components occur in response to mechanical force. In order to understand how FA-related signaling networks modulate the strength of the linkage between integrin and actin, the force-dependent FA abundance of scaffolding and regulatory proteins is organized, as shown in Table 3.2. This provides a broad view of our understanding of how FAs enable cells to respond to their environmental cues via modulation of their composition in a hierarchical cascade.

FA components comprise the linkage between integrin receptors and the actin cytoskeleton, which dictate FAs dynamics (the formation, maturation,

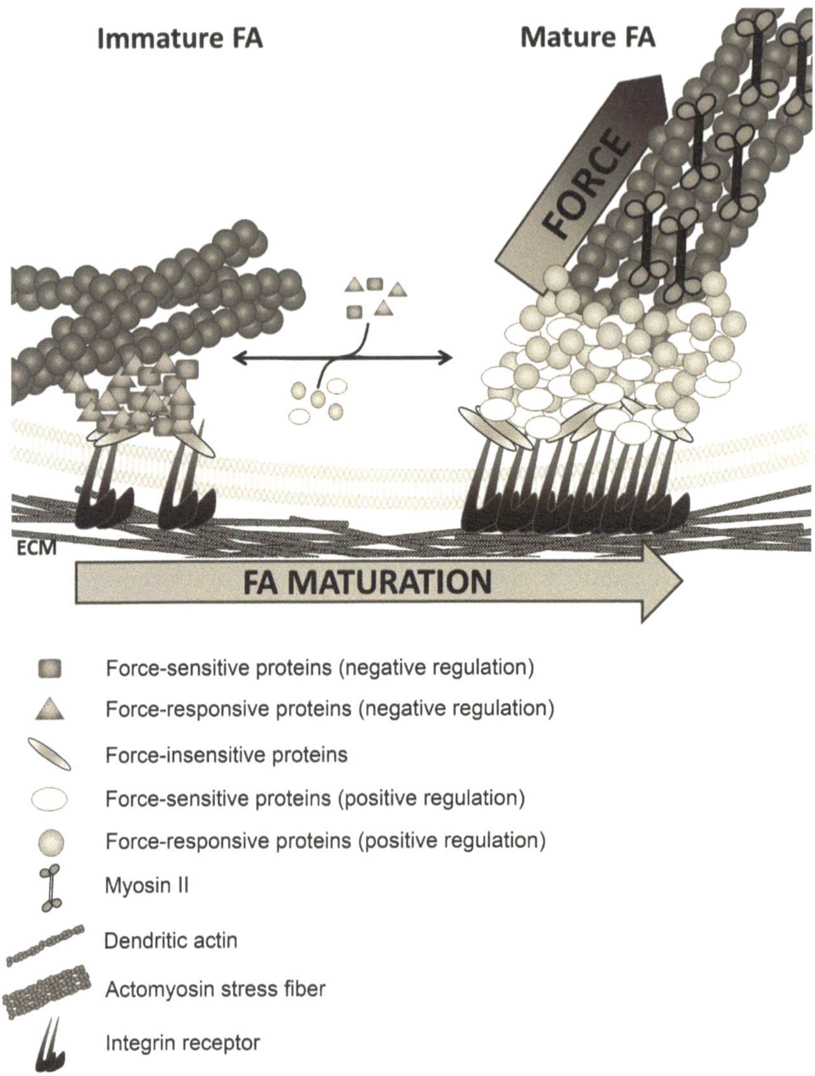

Figure 3.2 Schematic representation of how the protein composition of focal adhesions is reorganized in the maturation process. Focal adhesion protein composition is reorganized during the maturation process. Protein components in immature focal adhesions coordinately transmit the specific integrin-mediated signals that may regulate actin polymerization. During focal adhesion maturation, the force-sensitive proteins in immature focal adhesions respond to the forces generated by myosin II motors and are dissociated from focal adhesions with force-responsive proteins (immature focal adhesion components). Simultaneously, the force-sensitive proteins (mature focal adhesion components) respond to the force and enter focal adhesions, together with the force-responsive proteins (mature focal adhesion components), thereby assembling mature focal adhesions. Protein compositions in mature focal adhesions transmit the specific integrin-mediated signals that may positively regulate actomyosin contractility. The proteins with similar levels of abundance between immature and mature focal adhesions are considered as force-insensitive proteins. (See the color plate.)

Table 3.2 Force-dependent FA abundance of scaffolding and regulatory proteins

Scaffolding protein	
FA abundance positively regulated by force	ABLIM,[79] ACTN1,[56,78,79] ACTN4,[78,79] CNN1,[78,79] CNN2,[78,79] CNN3,[78,79] CORO1C,[78,79] CSRP1,[78,79] CSRP2,[78,79] FBLIM1,[78,79] FHL2,[78,79] FHL3,[78,79] FLNA,[78,79,86] FLNB,[78,79] FLNC,[78,79] DAB2,[78,79] LIMA1,[78,79] LIMCH1,[78] LIMD1,[79] LMO7,[78] LPP,[78,79] MYH9,[78,79] NCK1,[79] PDLIM1,[78,79] PDLIM2,[79] PDLIM4,[78,79] PDLIM5,[78,79] PDLIM7,[78,79] PLEC1,[78,79,87] SH3BP4,[78] SORBS3,[78,79] SPTAN1,[78] SVIL,[78] TES,[55,78,79,88] TGFB1I1,[78,79] TLN1,[78,79] TRIP6,[78,79] VCL,[20,78,79,81] ZYX[55,56,78,79,89]
FA abundance negatively regulated by force	ARP2/3 complex,[78,90] CAPZB,[78] CRIP2,[78] DBNL,[78] EPB41,[78] EPS8,[78,79] FHL1,[78] MICALL1,[78] TNS3[78]
Regulatory protein	
FA abundance positively regulated by force	ARF1,[78] ARF6,[78,79] CAPN1,[78] CAPN2,[78] CAPN5,[78] CSK,[79] DDR2,[78] GIT1,[78] GIT2,[78,79] GNA11,[78] GNA12,[78] GNA13,[78] GNAQ,[78] GNB1,[78] GNB2,[78,79] ILK,[78,79] JAK1,[78] PDGFRB,[78] PTK2,[79] PTPN11,[79] PTPN2,[79] PTPN12,[79] RAB1B,[78,79] RAB14,[78,79] RAB18,[78,79] RAB21,[78,79] RAB23,[78,79] RAB3B,[78] RAB34,[78,79] RAB35,[78,79] RALA,[78,79] RALB,[78,79] RAP1B,[78] RAP2B,[78] RHOA,[78] RHOB,[78] ROR2,[78] RRAS2,[78,79] SRC,[79] YES1[78]
FA abundance negatively regulated by force	ARHGEF7,[78] CSNK2A1,[78] KRAS,[78,79] NRAS,[78] PPP2CB,[78] PTPRF,[78] PTP4A2,[78] PTPRK,[78] RAB11B,[78,79] RAB13,[78] RAB8A,[78] RAN,[78] TENC1[78]

The lists of scaffolding and regulatory proteins are classified into two classes: FA abundance positively regulated by force and FA abundance negatively regulated by force. The proteins in each class could contain force-sensitive or force-responsive proteins.

and disassembly of FAs) as well as cytoskeletal organization. The initial linkage between integrin and actin is built via a FA adaptor, talin, which activates integrin receptor by binding to its cytoplasmic domain (NPXY motif) and also connects to actin filaments.[11,12] Myosin II-mediated contractile force reinforces the linkage by modulating FA composition via a hierarchical cascade. For example, force-dependent talin unfolding reinforces the linkage by binding to the actin-binding protein, vinculin.[20] In addition, myosin II-dependent recruitment of the actin-binding proteins, filamin A/B/C, and the adaptor, migfilin, strengthens the linkage between integrin and actin filaments via a connection that links the integrin receptors indirectly via a FA adaptor, kindlin-2.[91,92]

Mechanical force modulates the integrin-mediated signals transduced from the force-sensitive and force-responsive FA proteins. In response to myosin II activity, the abundance of RhoA enhancers, such as TRIP6 (thyroid hormone receptor interactor 6),[93] testin,[94] and GEF-H1,[95] is increased in FAs. In addition, FA abundance of actin-bundling proteins, such as α-actinin,[96] synaptopodin-2,[97] and supervillin[98,99] as well as several cytoskeletal LIM domain-containing adaptors,[78–80] such as zyxin,[89,100,101] PDLIM1,[102] PDLIM2,[102] PDLIM4,[102] PDLIM5,[102] PDLIM7,[102] and FHL2,[103] is enhanced. This suggests that mechanical force could promote the level of cellular tension in a positive feedback loop through promoting the association of specific FA components that allows the maturation of FAs and creates bundles of filamentous actin (stress fibers).[78]

Cellular tension also contributes to FA turnover,[104] since mature FAs disassembly are blocked by myosin II inhibition.[105] Previous experiments have revealed that the Ca^{2+}-activated protease calpain mediates proteolysis of FA proteins[43,45,48] and endocytosis-mediated pathways are able to recycle FA components; these serve as important mediators in regulating the disassembly of FAs.[106,107] Some disassembly factors are recruited to mature FAs,[78] which may explain how actomyosin contractility mediates FA turnover at the retracting edge of the cells.

Myosin II-mediated contractile force also influences the protein association of immature FAs that transduce signals to promote lamellipodial protrusion.[78,108] In the lamellipodium, actin is arranged as a dendritic network by continuous actin polymerization.[57] This cytoskeletal structure is mainly regulated by the Rho GTPase Rac1, but is also induced by myosin II inhibition.[59] Inhibition of actomyosin contractility enhances the abundance into immature FAs of Rac1 activators, such as RacGEF β-PIX (PAK-interacting exchange factor-β),[109] RacGEF modulator EPS8 (epidermal growth factor receptor pathway substrate 8),[110] MIF (macrophage migration inhibitory factor),[111] and PKA (protein kinase A)[112]; of Rac1 downstream effectors, such as IRSp53 (insulin receptor tyrosine kinase substrate p53)[113] and N-WASP (neuronal Wiskott–Aldrich Syndrome protein)[113,114]; and of Rac1 downstream targets, such as Arp2/3 complex,[115] cofilin,[116] and the actin monomer-binding protein Cap1.[117] Previous studies have shown that the Arp2/3 complex serves as the primary mediator of actin polymerization during lamellipodial protrusion, and Rac1 is sufficient to induce Arp2/3-dependent lamellipodium extension via the Rac1 downstream effectors, IRSP53 and N-WASP. FA association of the actin depolymerization factor cofilin promotes actin polymerization at the lamellipodia through the generation of new barbed ends for binding and this affects the Arp2/3 complex. Therefore, FA association of the Rac1 regulatory modules within the immature FAs explains the negative feedback mechanism of actomyosin contractility on the propagation of continuous membrane protrusions.[78] Taken together, the biochemical signals associated with FAs are adjusted by the local balance of mechanical forces, which dictate FA dynamics, cytoskeletal organization, and the nature of cellular tension.

3. FOCAL ADHESIONS AND THE EFFECTS OF ENVIRONMENTAL PARAMETERS

Various features of a cell's environment have been demonstrated for their ability to support multiple cellular functions. These features include a wide range of chemical molecules[6,118] and the physical properties of the adhesive surfaces, such as the topography,[119] stiffness,[72] and dimensionality.[120–122] FAs serve as the bidirectional linkage that integrates cells with their surrounding ECM through placing spatial restrictions on signaling (outside-in signaling) and ECM remodeling (inside-out signaling). Indeed, the application of adhesive micropattern confines cells within a given geometry that controls cell shape and the organization of traction forces.[123] The spatial distribution of FAs is sufficient to mediate the switch in mesenchymal stem cell (MSC) commitment between adipogenic and osteogenic fates.[124] Furthermore, MSCs sense the rigidity of ECM via their FAs, which are assembled in different maturation states, leading to differentiation into different cell types, such as neuronal, muscle or bone cells.[72] Therefore, the ECM surfaces modulate the spatial distribution, localization, size and maturation states of FAs, which transduce integrin-mediated signals to remodel cytoskeleton and program gene expression for the specific biological outcomes, including cell shape, survival, growth, migration and differentiation.

Mechanisms involved in stimulating FA-mediated signaling for a cell's ECM flexibility and adaptability remain poorly understood. The particular combinations of extracellular domain of integrin receptor determine the binding specificity of the identical ligand on the ECM, reflecting the structural basis of the molecular interaction. The clusters of integrin-ECM interaction orchestrate the FAs-mediated signaling networks via the assembly of functional FA-associated proteins complexes.[125–127] For example, α4β1-VCAM-1 and α5β1-fibronectin exhibit distinct functions in controlling cellular contractile forces and cell migration by the organization of the specific integrin interaction networks. The α5β1-fibronectin forms stronger cytoskeletal interactions, resulting in acceleration of cell spreading, membrane protrusion and cell migration.[128] Therefore, the complexity of the natural ECM controls the cell-specific combinations of integrin receptors with specific FA-mediated signaling networks that enable the attribution of specific cellular responses.

FAs not only regulate cell-ECM adhesion but also organize the functional FA-mediated signaling networks that connect the sensory and operational modules. Thus, disruption of FA dynamics that leads to loss of functions of protein components in the complexes is related to disease.

4. FOCAL ADHESION SIGNALS AND CELL MIGRATION

The importance of cell migration in the maintenance of health and the progress of development raises the point of how FA signals regulate cell migration. Cell migration, a highly dynamic and well regulated process, consists of well-defined steps that include extension of the leading edge and the formation of immature FAs; FA maturation and cell body translocation; and the FA disassembly and rear retraction. Integrin-mediated signals from the FAs steps (assembly, maturation and disassembly), which are modulated by the complex chemosensitive and mechanosensitive environmental cues, regulate actin polymerization and organization. During the migrating cycle, FA dynamics and cytoskeletal organization conjoin to drive this coordinated process.[129]

The initial step of the migration cycle is the extension of the leading edge and formation of nascent adhesions (immature FAs) beneath the lamellipodium. These nascent adhesions not only stabilize the protrusion but also transduce specific signals that continuously promote membrane protrusion. The protein components of nascent adhesions include the Rac1 regulatory module (Rac1 activators, Rac1 downstream effectors and Rac1 downstream targets), which promotes dendritic/branched actin polymerization for continuous protrusion extension and positively enhances the assembly of immature FAs (nascent adhesions and focal complexes).[78,108] Soon after, the immature FAs connect with bundles of actin filaments at the lamellipodia–lamella interface and they undergo a compositional reorganization and enlarge into mature FAs. This compositional reorganization includes force-sensitive and force-responsive FA proteins; these coordinate to reinforce the linkage between integrin and actin, help to form mature FAs and aid bundling of filamentous actin (stress fibers).[78–80]

The RhoA regulatory module associated with mature FAs activates myosin II through the action of downstream effector, ROCK, on upregulating of MLC phosphorylation.[66] Myosin II activation sustains the myosin II-mediated contractile force, and this further enhances the magnitude of the cellular tension. This enhanced cellular tension transmits the pulling force along the actin bundles to the adhesion sites, thereby translocating the cell body forward. The last step of the migration cycle is disassembly of mature FAs at the cell rear, which is also contractile force dependent.[105] Actomyosin contractility promotes FA association with the disassembly factors, including proteases[45,46,48] and the components of endocytosis pathways.[106,107] This disrupts the linkage between integrin and actin by cleaving and recycling the structural proteins that form the mature FAs.[78] Following the action of the disassembly factors, the pulling force supplied by the actomyosin contractility retracts the trailing edge of the cell, completing the migration cycle. Altogether, FAs not only serve as mechanosensors that reorganize their composition in response to the diverse environmental cues but also function as mechanotransducers that mediate specific cellular signaling pathways that regulate FA turnover and cytoskeletal organization, thereby controlling cell behavior and driving cell migration.

FAs enable cells to respond to their various environments, which contain diverse chemical and mechanical properties. They do this by manipulating their protein compositions, which allows the transmission of specific biochemical signals mediating cellular behaviors. Within a range of tissue microenvironments, cells feel and sense the proper matrix elasticity or biochemical gradients, thus displaying their specific biological function in specific tissues. However, this raises the possibility that disorganized FAs may mislead the cells and cause disease progression in some disordered tissues.[130,131]

ACKNOWLEDGMENTS

This work was supported by research grants from the Taiwan Ministry of Science and Technology (NSC 101-2628-B-010-003-MY3), the UST-UCSD International Center of Excellence in Advanced Bioengineering sponsored by the Taiwan Ministry of Science and Technology I-RiCE Program (NSC-100-2911-I-009-101) and the Ministry of Education's "Aim for the Top University Plan". J.C.K is also supported by the Yen Tjing Ling Medical Foundation.

REFERENCES

1. Alberts B, Johnson A, Lewis J, Raff M, Roberts K. *Molecular Biology of the Cell*. 5th ed. New York/Abingdon: Garland Science, Taylor & Francis Group, LLC; 2008.
2. Burridge K, Fath K, Kelly T, Nuckolls G, Turner C. Focal adhesions: transmembrane junctions between the extracellular matrix and the cytoskeleton. *Annu Rev Cell Biol*. 1988;4:487–525.
3. Hynes RO. Integrins: bidirectional, allosteric signaling machines. *Cell*. 2002;110: 673–687.
4. Jockusch BM, Bubeck P, Giehl K, et al. The molecular architecture of focal adhesions. *Annu Rev Cell Dev Biol*. 1995;11:379–416.
5. Arnaout MA, Goodman SL, Xiong JP. Structure and mechanics of integrin-based cell adhesion. *Curr Opin Cell Biol*. 2007;19:495–507.
6. Humphries JD, Byron A, Humphries MJ. Integrin ligands at a glance. *J Cell Sci*. 2006;119:3901–3903.
7. Leisner TM, Wencel-Drake JD, Wang W, Lam SC. Bidirectional transmembrane modulation of integrin alphaIIbbeta3 conformations. *J Biol Chem*. 1999;274: 12945–12949.
8. Zaidel-Bar R, Geiger B. The switchable integrin adhesome. *J Cell Sci*. 2010;123: 1385–1388.
9. Zaidel-Bar R, Itzkovitz S, Ma'ayan A, Iyengar R, Geiger B. Functional atlas of the integrin adhesome. *Nat Cell Biol*. 2007;9:858–867.
10. Calderwood DA, Shattil SJ, Ginsberg MH. Integrins and actin filaments: reciprocal regulation of cell adhesion and signaling. *J Biol Chem*. 2000;275:22607–22610.
11. Moser M, Legate KR, Zent R, Fassler R. The tail of integrins, talin, and kindlins. *Science*. 2009;324:895–899.
12. Tadokoro S, Shattil SJ, Eto K, et al. Talin binding to integrin beta tails: a final common step in integrin activation. *Science*. 2003;302:103–106.
13. Mimura N, Asano A. Isolation and characterization of a conserved actin-binding domain from rat hepatic actinogelin, rat skeletal muscle, and chicken gizzard alpha-actinins. *J Biol Chem*. 1986;261:10680–10687.
14. Mimura N, Asano A. Further characterization of a conserved actin-binding 27-kDa fragment of actinogelin and alpha-actinins and mapping of their binding sites on the actin molecule by chemical cross-linking. *J Biol Chem*. 1987;262:4717–4723.
15. Otey CA, Pavalko FM, Burridge K. An interaction between alpha-actinin and the beta 1 integrin subunit in vitro. *J Cell Biol*. 1990;111:721–729.
16. Calderwood DA, Huttenlocher A, Kiosses WB, et al. Increased filamin binding to beta-integrin cytoplasmic domains inhibits cell migration. *Nat Cell Biol*. 2001;3: 1060–1068.
17. Gardel ML, Nakamura F, Hartwig JH, Crocker JC, Stossel TP, Weitz DA. Prestressed F-actin networks cross-linked by hinged filamins replicate mechanical properties of cells. *Proc Natl Acad Sci USA*. 2006;103:1762–1767.
18. Loo DT, Kanner SB, Aruffo A. Filamin binds to the cytoplasmic domain of the beta1-integrin. Identification of amino acids responsible for this interaction. *J Biol Chem*. 1998;273:23304–23312.
19. Pfaff M, Liu S, Erle DJ, Ginsberg MH. Integrin beta cytoplasmic domains differentially bind to cytoskeletal proteins. *J Biol Chem*. 1998;273:6104–6109.
20. del Rio A, Perez-Jimenez R, Liu R, Roca-Cusachs P, Fernandez JM, Sheetz MP. Stretching single talin rod molecules activates vinculin binding. *Science*. 2009;323: 638–641.
21. Bachmann C, Fischer L, Walter U, Reinhard M. The EVH2 domain of the vasodilator-stimulated phosphoprotein mediates tetramerization, F-actin binding, and actin bundle formation. *J Biol Chem*. 1999;274:23549–23557.

22. Bubeck P, Pistor S, Wehland J, Jockusch BM. Ligand recruitment by vinculin domains in transfected cells. *J Cell Sci.* 1997;110(Pt 12):1361–1371.
23. Reinhard M, Zumbrunn J, Jaquemar D, Kuhn M, Walter U, Trueb B. An alpha-actinin binding site of zyxin is essential for subcellular zyxin localization and alpha-actinin recruitment. *J Biol Chem.* 1999;274:13410–13418.
24. Huang H, Li L, Wu C, et al. Defining the specificity space of the human SRC homology 2 domain. *Mol Cell Proteomics.* 2008;7:768–784.
25. Pawson T, Gish GD, Nash P. SH2 domains, interaction modules and cellular wiring. *Trends Cell Biol.* 2001;11:504–511.
26. Alexandropoulos K, Cheng G, Baltimore D. Proline-rich sequences that bind to Src homology 3 domains with individual specificities. *Proc Natl Acad Sci USA.* 1995;92:3110–3114.
27. Wang DS, Miller R, Shaw R, Shaw G. The pleckstrin homology domain of human beta I sigma II spectrin is targeted to the plasma membrane in vivo. *Biochem Biophys Res Commun.* 1996;225:420–426.
28. Wang DS, Shaw G. The association of the C-terminal region of beta I sigma II spectrin to brain membranes is mediated by a PH domain, does not require membrane proteins, and coincides with a inositol-1,4,5 triphosphate binding site. *Biochem Biophys Res Commun.* 1995;217:608–615.
29. Pearson MA, Reczek D, Bretscher A, Karplus PA. Structure of the ERM protein moesin reveals the FERM domain fold masked by an extended actin binding tail domain. *Cell.* 2000;101:259–270.
30. Kloeker S, Major MB, Calderwood DA, Ginsberg MH, Jones DA, Beckerle MC. The Kindler syndrome protein is regulated by transforming growth factor-beta and involved in integrin-mediated adhesion. *J Biol Chem.* 2004;279:6824–6833.
31. Montanez E, Ussar S, Schifferer M, et al. Kindlin-2 controls bidirectional signaling of integrins. *Genes Dev.* 2008;22:1325–1330.
32. Moser M, Nieswandt B, Ussar S, Pozgajova M, Fassler R. Kindlin-3 is essential for integrin activation and platelet aggregation. *Nat Med.* 2008;14:325–330.
33. Shi X, Ma YQ, Tu Y, et al. The MIG-2/integrin interaction strengthens cell-matrix adhesion and modulates cell motility. *J Biol Chem.* 2007;282:20455–20466.
34. Castresana J, Saraste M. Does Vav bind to F-actin through a CH domain? *FEBS Lett.* 1995;374:149–151.
35. Kadrmas JL, Beckerle MC. The LIM domain: from the cytoskeleton to the nucleus. *Nat Rev Mol Cell Biol.* 2004;5:920–931.
36. Wang Y, Gilmore TD. Zyxin and paxillin proteins: focal adhesion plaque LIM domain proteins go nuclear. *Biochim Biophys Acta.* 2003;1593:115–120.
37. Chrzanowska-Wodnicka M, Burridge K. Rho-stimulated contractility drives the formation of stress fibers and focal adhesions. *J Cell Biol.* 1996;133:1403–1415.
38. Nobes CD, Hall A. Rho, rac, and cdc42 GTPases regulate the assembly of multimolecular focal complexes associated with actin stress fibers, lamellipodia, and filopodia. *Cell.* 1995;81:53–62.
39. Rottner K, Hall A, Small JV. Interplay between Rac and Rho in the control of substrate contact dynamics. *Curr Biol.* 1999;9:640–648.
40. Ellenbroek SI, Collard JG. Rho GTPases: functions and association with cancer. *Clin Exp Metastasis.* 2007;24:657–672.
41. Hall A. Rho GTPases and the control of cell behaviour. *Biochem Soc Trans.* 2005;33:891–895.
42. Du S, Rubin A, Klepper S, et al. Calcium influx and activation of calpain I mediate acute reactive gliosis in injured spinal cord. *Exp Neurol.* 1999;157:96–105.
43. Bhatt A, Kaverina I, Otey C, Huttenlocher A. Regulation of focal complex composition and disassembly by the calcium-dependent protease calpain. *J Cell Sci.* 2002;115:3415–3425.

44. Carragher NO, Levkau B, Ross R, Raines EW. Degraded collagen fragments promote rapid disassembly of smooth muscle focal adhesions that correlates with cleavage of pp 125(FAK), paxillin, and talin. *J Cell Biol.* 1999;147:619–630.
45. Franco SJ, Huttenlocher A. Regulating cell migration: calpains make the cut. *J Cell Sci.* 2005;118:3829–3838.
46. Du X, Saido TC, Tsubuki S, Indig FE, Williams MJ, Ginsberg MH. Calpain cleavage of the cytoplasmic domain of the integrin beta 3 subunit. *J Biol Chem.* 1995;270: 26146–26151.
47. Pfaff M, Du X, Ginsberg MH. Calpain cleavage of integrin beta cytoplasmic domains. *FEBS Lett.* 1999;460:17–22.
48. Franco SJ, Rodgers MA, Perrin BJ, et al. Calpain-mediated proteolysis of talin regulates adhesion dynamics. *Nat Cell Biol.* 2004;6:977–983.
49. Carragher NO, Westhoff MA, Fincham VJ, Schaller MD, Frame MC. A novel role for FAK as a protease-targeting adaptor protein: regulation by p42 ERK and Src. *Curr Biol.* 2003;13:1442–1450.
50. Chan KT, Bennin DA, Huttenlocher A. Regulation of adhesion dynamics by calpain-mediated proteolysis of focal adhesion kinase (FAK). *J Biol Chem.* 2010;285: 11418–11426.
51. Sato K, Kawashima S. Calpain function in the modulation of signal transduction molecules. *Biol Chem.* 2001;382:743–751.
52. Burridge K, Chrzanowska-Wodnicka M. Focal adhesions, contractility, and signaling. *Annu Rev Cell Dev Biol.* 1996;12:463–518.
53. Gupton SL, Waterman-Storer CM. Spatiotemporal feedback between actomyosin and focal-adhesion systems optimizes rapid cell migration. *Cell.* 2006;125:1361–1374.
54. Schwartz MA. Integrins and extracellular matrix in mechanotransduction. *Cold Spring Harb Perspect Biol.* 2010;2:a005066.
55. Zaidel-Bar R, Cohen M, Addadi L, Geiger B. Hierarchical assembly of cell-matrix adhesion complexes. *Biochem Soc Trans.* 2004;32:416–420.
56. Zamir E, Geiger B, Kam Z. Quantitative multicolor compositional imaging resolves molecular domains in cell-matrix adhesions. *PLoS One.* 2008;3:e1901.
57. Burnette DT, Manley S, Sengupta P, et al. A role for actin arcs in the leading-edge advance of migrating cells. *Nat Cell Biol.* 2011;13:371–381.
58. Choi CK, Vicente-Manzanares M, Zareno J, Whitmore LA, Mogilner A, Horwitz AR. Actin and alpha-actinin orchestrate the assembly and maturation of nascent adhesions in a myosin II motor-independent manner. *Nat Cell Biol.* 2008;10:1039–1050.
59. Even-Ram S, Doyle AD, Conti MA, Matsumoto K, Adelstein RS, Yamada KM. Myosin IIA regulates cell motility and actomyosin-microtubule crosstalk. *Nat Cell Biol.* 2007;9:299–309.
60. Pollard TD, Borisy GG. Cellular motility driven by assembly and disassembly of actin filaments. *Cell.* 2003;112:453–465.
61. Small JV, Stradal T, Vignal E, Rottner K. The lamellipodium: where motility begins. *Trends Cell Biol.* 2002;12:112–120.
62. Ridley AJ, Schwartz MA, Burridge K, et al. Cell migration: integrating signals from front to back. *Science.* 2003;302:1704–1709.
63. Helfman DM, Levy ET, Berthier C, et al. Caldesmon inhibits nonmuscle cell contractility and interferes with the formation of focal adhesions. *Mol Biol Cell.* 1999;10:3097–3112.
64. Kaverina I, Krylyshkina O, Small JV. Regulation of substrate adhesion dynamics during cell motility. *Int J Biochem Cell Biol.* 2002;34:746–761.
65. Parsons JT, Horwitz AR, Schwartz MA. Cell adhesion: integrating cytoskeletal dynamics and cellular tension. *Nat Rev Mol Cell Biol.* 2010;11:633–643.

66. Ishizaki T, Naito M, Fujisawa K, et al. p160ROCK, a Rho-associated coiled-coil forming protein kinase, works downstream of Rho and induces focal adhesions. *FEBS Lett* 1997;404:118–124.
67. Schneider IC, Hays CK, Waterman CM. Epidermal growth factor-induced contraction regulates paxillin phosphorylation to temporally separate traction generation from de-adhesion. *Mol Biol Cell.* 2009;20:3155–3167.
68. Shen B, Zhao X, O'Brien KA, et al. A directional switch of integrin signalling and a new anti-thrombotic strategy. *Nature.* 2013;503:131–135.
69. Wrighton KH. Cell adhesion: The 'ins' and 'outs' of integrin signalling. *Nat Rev Mol Cell Biol.* 2013;14:752–753.
70. Leung T, Chen XQ, Manser E, Lim L. The p160 RhoA-binding kinase ROK alpha is a member of a kinase family and is involved in the reorganization of the cytoskeleton. *Mol Cell Biol.* 1996;16:5313–5327.
71. Maekawa M, Ishizaki T, Boku S, et al. Signaling from Rho to the actin cytoskeleton through protein kinases ROCK and LIM-kinase. *Science.* 1999;285:895–898.
72. Engler AJ, Sen S, Sweeney HL, Discher DE. Matrix elasticity directs stem cell lineage specification. *Cell.* 2006;126:677–689.
73. Galbraith CG, Yamada KM, Sheetz MP. The relationship between force and focal complex development. *J Cell Biol.* 2002;159:695–705.
74. Wang HB, Dembo M, Hanks SK, Wang Y. Focal adhesion kinase is involved in mechanosensing during fibroblast migration. *Proc Natl Acad Sci USA.* 2001;98: 11295–11300.
75. Wang N, Butler JP, Ingber DE. Mechanotransduction across the cell surface and through the cytoskeleton. *Science.* 1993;260:1124–1127.
76. Wolfenson H, Bershadsky A, Henis YI, Geiger B. Actomyosin-generated tension controls the molecular kinetics of focal adhesions. *J Cell Sci.* 2011;124:1425–1432.
77. Bershadsky AD, Balaban NQ, Geiger B. Adhesion-dependent cell mechanosensitivity. *Annu Rev Cell Dev Biol.* 2003;19:677–695.
78. Kuo JC, Han X, Hsiao CT, Yates III JR, Waterman CM. Analysis of the myosin-II-responsive focal adhesion proteome reveals a role for beta-Pix in negative regulation of focal adhesion maturation. *Nat Cell Biol.* 2011;13:383–393.
79. Schiller HB, Friedel CC, Boulegue C, Fassler R. Quantitative proteomics of the integrin adhesome show a myosin II-dependent recruitment of LIM domain proteins. *EMBO Rep.* 2011;12:259–266.
80. Gallegos L, Ng MR, Brugge JS. The myosin-II-responsive focal adhesion proteome: a tour de force? *Nat Cell Biol.* 2011;13:344–346.
81. Grashoff C, Hoffman BD, Brenner MD, et al. Measuring mechanical tension across vinculin reveals regulation of focal adhesion dynamics. *Nature.* 2010;466:263–266.
82. Jiang G, Giannone G, Critchley DR, Fukumoto E, Sheetz MP. Two-piconewton slip bond between fibronectin and the cytoskeleton depends on talin. *Nature.* 2003;424:334–337.
83. Sawada Y, Tamada M, Dubin-Thaler BJ, et al. Force sensing by mechanical extension of the Src family kinase substrate p130Cas. *Cell.* 2006;127:1015–1026.
84. Geiger B. A role for p130Cas in mechanotransduction. *Cell.* 2006;127:879–881.
85. Stossel TP, Condeelis J, Cooley L, et al. Filamins as integrators of cell mechanics and signalling. *Nat Rev Mol Cell Biol.* 2001;2:138–145.
86. Ehrlicher AJ, Nakamura F, Hartwig JH, Weitz DA, Stossel TP. Mechanical strain in actin networks regulates FilGAP and integrin binding to filamin A. *Nature.* 2011;478:260–263.
87. Gad A, Lach S, Crimaldi L, Gimona M. Plectin deposition at podosome rings requires myosin contractility. *Cell Motil Cytoskeleton.* 2008;65:614–625.

88. Zamir E, Katz M, Posen Y, et al. Dynamics and segregation of cell-matrix adhesions in cultured fibroblasts. *Nat Cell Biol.* 2000;2:191–196.
89. Zaidel-Bar R, Ballestrem C, Kam Z, Geiger B. Early molecular events in the assembly of matrix adhesions at the leading edge of migrating cells. *J Cell Sci.* 2003;116: 4605–4613.
90. DeMali KA, Barlow CA, Burridge K. Recruitment of the Arp2/3 complex to vinculin: coupling membrane protrusion to matrix adhesion. *J Cell Biol.* 2002;159:881–891.
91. Meves A, Stremmel C, Gottschalk K, Fassler R. The Kindlin protein family: new members to the club of focal adhesion proteins. *Trends Cell Biol.* 2009;19:504–513.
92. Tu Y, Wu S, Shi X, Chen K, Wu C. Migfilin and Mig-2 link focal adhesions to filamin and the actin cytoskeleton and function in cell shape modulation. *Cell.* 2003;113:37–47.
93. Bai CY, Ohsugi M, Abe Y, Yamamoto T. ZRP-1 controls Rho GTPase-mediated actin reorganization by localizing at cell-matrix and cell-cell adhesions. *J Cell Sci.* 2007;120:2828–2837.
94. Griffith E, Coutts AS, Black DM. RNAi knockdown of the focal adhesion protein TES reveals its role in actin stress fibre organisation. *Cell Motil Cytoskeleton.* 2005;60:140–152.
95. Ren Y, Li R, Zheng Y, Busch H. Cloning and characterization of GEF-H1, a microtubule-associated guanine nucleotide exchange factor for Rac and Rho GTPases. *J Biol Chem.* 1998;273:34954–34960.
96. Otey CA, Carpen O. Alpha-actinin revisited: a fresh look at an old player. *Cell Motil Cytoskeleton.* 2004;58:104–111.
97. Schroeter MM, Beall B, Heid HW, Chalovich JM. In vitro characterization of native mammalian smooth-muscle protein synaptopodin 2. *Biosci Rep.* 2008;28:195–203.
98. Chen Y, Takizawa N, Crowley JL, et al. F-actin and myosin II binding domains in supervillin. *J Biol Chem.* 2003;278:46094–46106.
99. Wulfkuhle JD, Donina IE, Stark NH, et al. Domain analysis of supervillin, an F-actin bundling plasma membrane protein with functional nuclear localization signals. *J Cell Sci.* 1999;112(Pt 13):2125–2136.
100. Hirata H, Tatsumi H, Sokabe M. Zyxin emerges as a key player in the mechanotransduction at cell adhesive structures. *Commun Integr Biol.* 2008;1:192–195.
101. Hirata H, Tatsumi H, Sokabe M. Mechanical forces facilitate actin polymerization at focal adhesions in a zyxin-dependent manner. *J Cell Sci.* 2008;121:2795–2804.
102. Harris BZ, Lim WA. Mechanism and role of PDZ domains in signaling complex assembly. *J Cell Sci.* 2001;114:3219–3231.
103. Schuh M, Ellenberg J. A new model for asymmetric spindle positioning in mouse oocytes. *Curr Biol.* 2008;18:1986–1992.
104. Broussard JA, Webb DJ, Kaverina I. Asymmetric focal adhesion disassembly in motile cells. *Curr Opin Cell Biol.* 2008;20:85–90.
105. Crowley E, Horwitz AF. Tyrosine phosphorylation and cytoskeletal tension regulate the release of fibroblast adhesions. *J Cell Biol.* 1995;131:525–537.
106. Caswell PT, Vadrevu S, Norman JC. Integrins: masters and slaves of endocytic transport. *Nat Rev Mol Cell Biol.* 2009;10:843–853.
107. Ezratty EJ, Bertaux C, Marcantonio EE, Gundersen GG. Clathrin mediates integrin endocytosis for focal adhesion disassembly in migrating cells. *J Cell Biol.* 2009;187: 733–747.
108. Legg K. Cell migration: keeping young and mobile with beta-PIX. *Nat Rev Mol Cell Biol.* 2011;12(278).
109. ten Klooster JP, Jaffer ZM, Chernoff J, Hordijk PL. Targeting and activation of Rac1 are mediated by the exchange factor beta-Pix. *J Cell Biol.* 2006;172:759–769.
110. Innocenti M, Tenca P, Frittoli E, et al. Mechanisms through which Sos-1 coordinates the activation of Ras and Rac. *J Cell Biol.* 2002;156:125–136.

111. Rendon BE, Roger T, Teneng I, et al. Regulation of human lung adenocarcinoma cell migration and invasion by macrophage migration inhibitory factor. *J Biol Chem.* 2007;282:29910–29918.
112. O'Connor KL, Mercurio AM. Protein kinase A regulates Rac and is required for the growth factor-stimulated migration of carcinoma cells. *J Biol Chem.* 2001;276: 47895–47900.
113. Miki H, Yamaguchi H, Suetsugu S, Takenawa T. IRSp53 is an essential intermediate between Rac and WAVE in the regulation of membrane ruffling. *Nature.* 2000;408:732–735.
114. Eden S, Rohatgi R, Podtelejnikov AV, Mann M, Kirschner MW. Mechanism of regulation of WAVE1-induced actin nucleation by Rac1 and Nck. *Nature.* 2002;418: 790–793.
115. Weed SA, Karginov AV, Schafer DA, et al. Cortactin localization to sites of actin assembly in lamellipodia requires interactions with F-actin and the Arp2/3 complex. *J Cell Biol.* 2000;151:29–40.
116. Oser M, Condeelis J. The cofilin activity cycle in lamellipodia and invadopodia. *J Cell Biochem.* 2009;108:1252–1262.
117. Bertling E, Hotulainen P, Mattila PK, Matilainen T, Salminen M, Lappalainen P. Cyclase-associated protein 1 (CAP1) promotes cofilin-induced actin dynamics in mammalian nonmuscle cells. *Mol Biol Cell.* 2004;15:2324–2334.
118. Hersel U, Dahmen C, Kessler H. RGD modified polymers: biomaterials for stimulated cell adhesion and beyond. *Biomaterials.* 2003;24:4385–4415.
119. Dalby MJ, Riehle MO, Johnstone H, Affrossman S, Curtis AS. In vitro reaction of endothelial cells to polymer demixed nanotopography. *Biomaterials.* 2002;23: 2945–2954.
120. Chen CS, Tan J, Tien J. Mechanotransduction at cell-matrix and cell-cell contacts. *Annu Rev Biomed Eng.* 2004;6:275–302.
121. Cukierman E, Pankov R, Stevens DR, Yamada KM. Taking cell-matrix adhesions to the third dimension. *Science.* 2001;294:1708–1712.
122. Rumpler M, Woesz A, Dunlop JW, van Dongen JT, Fratzl P. The effect of geometry on three-dimensional tissue growth. *J R Soc Interface.* 2008;5:1173–1180.
123. Thery M, Pepin A, Dressaire E, Chen Y, Bornens M. Cell distribution of stress fibres in response to the geometry of the adhesive environment. *Cell Motil Cytoskeleton.* 2006;63:341–355.
124. McBeath R, Pirone DM, Nelson CM, Bhadriraju K, Chen CS. Cell shape, cytoskeletal tension, and RhoA regulate stem cell lineage commitment. *Dev Cell.* 2004;6:483–495.
125. Delon I, Brown NH. Integrins and the actin cytoskeleton. *Curr Opin Cell Biol.* 2007;19:43–50.
126. Geiger B, Bershadsky A, Pankov R, Yamada KM. Transmembrane crosstalk between the extracellular matrix–cytoskeleton crosstalk. *Nat Rev Mol Cell Biol.* 2001;2:793–805.
127. Legate KR, Fassler R. Mechanisms that regulate adaptor binding to beta-integrin cytoplasmic tails. *J Cell Sci.* 2009;122:187–198.
128. Humphries JD, Byron A, Bass MD, et al. Proteomic analysis of integrin-associated complexes identifies RCC2 as a dual regulator of Rac1 and Arf6. *Sci Signal.* 2009;2: ra51.
129. Webb DJ, Parsons JT, Horwitz AF. Adhesion assembly, disassembly and turnover in migrating cells—over and over and over again. *Nat Cell Biol.* 2002;4:E97–E100.
130. Levental KR, Yu H, Kass L, et al. Matrix crosslinking forces tumor progression by enhancing integrin signaling. *Cell.* 2009;139:891–906.
131. Yang C, Zeisberg M, Mosterman B, et al. Liver fibrosis: insights into migration of hepatic stellate cells in response to extracellular matrix and growth factors. *Gastroenterology.* 2003;124:147–159.

CHAPTER FOUR

Mechanosensation: A Basic Cellular Process

Wolfgang H. Goldmann
Center for Medical Physics and Technology, Biophysics Group, Friedrich-Alexander-University of Erlangen-Nuremberg, Erlangen, Germany

Contents

Abstract

It has been shown that focal adhesion proteins are crucial for the ability of cells to transmit external forces and to generate cytoskeletal tension. Force transmission over considerable distances and stress focusing at the focal adhesion sites make them prime candidates for mechanosensors. Temporal and spatial changes in the cytoskeletal protein configuration due to mechanical stimulation have been detected and characterized by a wide range of biophysical techniques, including magnetic twisting, magnetic tweezer, traction microscopy, atomic force microscopy, nanoscale particle tracking, and many more. The combination of these techniques will help to understand force transmission and structural remodeling in cells under loading conditions. Force transmission

Progress in Molecular Biology and Translational Science, Volume 126
ISSN 1877-1173
http://dx.doi.org/10.1016/B978-0-12-394624-9.00004-X

and force sensing represent basic biological processes that are crucial for a variety of higher fundamental cell functions including cell division, motility, and differentiation that have implications in medicine and biology.

1. INTRODUCTION

1.1. Historical development

Julius Wolff hypothesized more than 100 years ago that bone adapts under mechanical load by remodeling itself. He proposed that the specific effect on bone structure depends on the duration, magnitude, and rate of loading.[1] As we know today, the remodeling of bone in response to loading is achieved via a cascade of different steps including mechano- and biochemical coupling, signal transmission, and cell response, which are part of the mechanotransduction pathway. Specifically, upon sensing a load, osteocytes regulate bone remodeling either via molecule signaling or direct contact. Osteo-progenitor cells, which may differentiate into osteoblasts or osteoclasts, are regarded as mechanosensors.

Cecil D. Murray laid down in the 1920s a formula that relates the blood vessel radius to the required minimum energy by the organism that is, larger vessels lower the energy required in pumping blood because the pressure drop in the vessels reduces with increasing diameter according to the Hagen–Poiseuille equation.[2] In a seminal study, West *et al.*[3] showed that the allometric scaling relation ultimately leads to the minimization of energy consumption in blood pumps. In a certain sense, frictional forces in the blood vessels therefore dictate the body size of a living being.

Today, Murray's formula has gained increasing use as a biomimetic design tool in the field of mechanosensing/-transduction. It is, for example, applied in the design of minimum mass vascular networks carrying a liquid healing agent to areas of damage in a self-healing material.

1.2. Mechanosensation/-transduction

A typical example of mechanosensing/-transduction can be found in sensory cells of the inner ear.[4–6] Here, so-called "hair" cells transduce the mechanical vibration of the inner ear fluid into an electrical signal that propagates to the brain. Specifically, at the apical surface of hair cells, stereo cilia form bundles, which are able to slide relative to one another when the bundle is pushed one way or the other. An adaption motor that moves along the internal actin filaments, which are tethered to the ion channel, modulates the tension. When filaments slide relative to each other, a force is generated to the point where filaments are connected to the side of the stereo cilium. This force changes the conformation of a transmembrane protein that acts as an ion channel, causing it to open thus allowing the transient entry of calcium ions. The flux of cations initiates the electrical signal that eventually reaches the brain where it is perceived as sound. Details of force transmission to the ion channel of hair cell excitation are still unknown. This is a classic example of the many ways a cell can physically "feel" its surroundings.

Other mechanisms are only now being explored, including (i) conformational modification of intracellular proteins associated with transmission of external forces to the cell interior, leading to the modulation in reaction rates through a change in binding affinity; (ii) changes in the viscosity of the cell membrane, altering the rate of diffusion of transmembrane proteins and consequently their reaction rates; and (iii) direct transmission of forces to the nucleus, affecting expression of specific genes. These mechanisms are less well understood than mechanosensitive channels, and it is likely that other mechanisms exist that have not yet been identified.[7–11] Although the detailed mechanisms remain undeciphered, the consequences of force applied to cells are well documented.[12–19]

1.3. Effects of extracellular matrix stiffness

There is growing evidence that the link between the mechanical properties of the extracellular environment and cellular decision-making mechanotransduction processes are important. Our current understanding of adhesion-mediated environmental sensing is still fragmentary and several design principles have emerged from experiments. For example, surface chemistry, namely, the presence of diverse matrix proteins, has a strong effect on the selection of specific integrin receptors and, consequently, on the initial assembly of the integrin complexes. The mechanical properties of the extracellular environment play a much more important role in cellular behavior than originally thought. It has been shown that cells (i) more strongly upregulate the cytoskeleton and cell–matrix adhesion on stiffer substrates, and (ii) locomote in favor of stiffer or strained substrates, and that (iii) extracellular mechanical properties and cellular decision-making are connected to the internal force developing at cell–matrix contacts due to cellular actomyosin contractility.[20,21]

In recent years, there has been an increased effort to study the effect of externally applied forces on cells. Several different experimental techniques have been applied for this purpose, which include magnetic twisting cytometry, magnetic/laser optical tweezers, atomic force microscopy (AFM), cell poking, rheology, and micropipettes.[22–27] These studies showed that there is a strong correlation between aggregation of cell–matrix contacts, build-up of forces, and triggering of certain signaling cascades determining physiologically important processes, including cell division, migration, and apoptosis (programmed cell death). In particular, there is a close relation between the proper functioning of cell–matrix contacts and certain diseases, such as cancer.[28,29]

Cells may sense anisotropic mechanical properties of the matrix and orient themselves accordingly. Future experiments should, therefore, focus on the relation between structural versus mechanical cues for cell organization in hydrogels, while modeling is needed to account for the mechanical (in particular, viscoelastic) properties of hydrogels.[30] Some evidence now indicates that integrin-based cell–matrix contacts act as local mechanosensors that change mechanical information about the environment directly into cellular decision-making answers. It has been suggested that upregulation of cell growth due to matrix contacts in a stiffer environment might originate from the fact that it is triggered by a threshold force. A similar viewpoint is that growth of cell–matrix contacts is faster on stiffer substrates. To test this hypothesis, correlation studies of the growth of cell–matrix contacts and cellular organization are needed, for example, studying areas close to substrate boundaries, where cells can amplify the mechanical input provided by boundary-induced strain through active mechanosensing. Quantitative data about the growth behavior of cell–matrix contacts will allow for further refining of models possibly also including modeling of cellular features, such as morphology and force pattern.[31–34]

1.4. Stress generated by external compression/contractility

The signal transduction pathways that are activated in response to mechanical force include many components and elements that are shared by other signaling pathways. For instance, mechanotransduction in cardiomyocytes is particularly complex, in that individual muscle cells both respond to externally applied mechanical forces and generate internal loads that are transmitted to adjacent cells and to the surrounding extracellular matrix (ECM).[35,36]

Cells in a prestressed environment have been reported to orient perpendicularly to the axis of compression/extension.[37] This is regarded as an effect of strain avoidance parallel to this axis. Vice versa, a cell can exert dipole-like forces on its surroundings, which can be mechanically transmitted to far distances. A neighbor cell will, therefore, upregulate its contractile apparatus aiming at an alignment along the same direction. This scenario constitutes a positive mechanical feedback loop for cell alignment. At low cell densities, a common pattern for the organization of elastically interacting cells will therefore be the formation of strings of cells, similar to the case of electric dipoles.[9,38–40]

1.5. Stress generated by cell contractility

Forces exerted by mechanically active cells on the environment are mainly due to actomyosin contractility. The cells involve nonequilibrium processes that are tightly regulated by biochemical events inside the cell. Actomyosin contractility is the basis of cell mechanical activity. However, thus far it has not been possible to reconstitute actomyosin contractility in vesicular systems. This is different for the formation of adhesion plaques, which have been reconstituted with lipid vesicles carrying "sticker" and "repeller" molecules that adhere to ligand-coated substrates. Several theoretical studies have been devoted to the possible mechanisms driving plaque formation, including elastically and entropically induced interactions. Motivated by these experiments with elastic substrates, we investigated whether a similar description can be employed for cells. We asked what kind of information a cell can extract from its environment using its contractile machinery (Fig. 4.1).[18,19]

1.6. Biological relevance of external and internal stress

Various forms of force application, whether transmitted via cell membrane adhesion proteins (e.g., integrins and cadherins) or by the effects of fluid shear stress, transmitted either directly to the cell membrane or via the surface receptors elicit a biological response. Known responses to force can be observed in a matter of seconds, such as in channel activation, but can continue for hours after the initiating event, for example, as changes in gene expression, protein synthesis, or morphology. Many signaling pathways that mediate these cellular responses have been identified and have been extensively reviewed.[5,15,41,42]

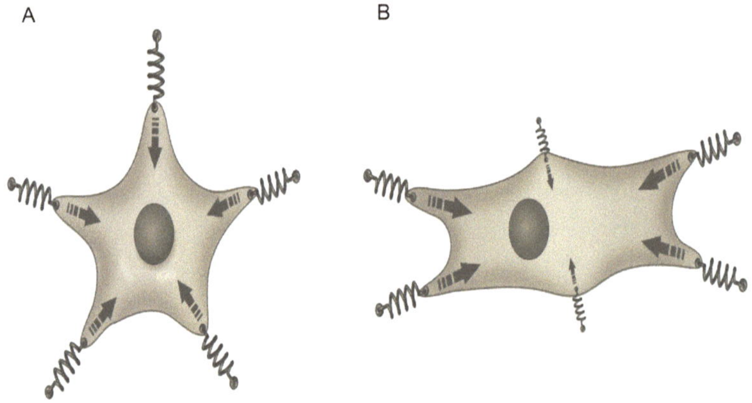

Figure 4.1 An adherent cell actively pulls on its environment through cell–matrix contacts. The cell orients itself in the direction of maximal stiffness of the environment by active mechanosensing. The local elastic environment is represented by linear springs. (A) In an isotropic environment, all spring constants are the same, growth at different contacts is similar, and the cell does not orient. (B) If spring constants are larger in one direction, corresponding contacts outgrow the others and the cell orients in the direction of maximal stiffness of the environment.

The range of stresses (force per area), to which different tissues are naturally exposed, is huge. Cytoskeletal structures are not only responsible for passively providing material strength, but they are also intimately involved in the sensing of external forces and transmitting those forces. How cells respond to mechanical stress depends not only on specific molecular sensors and signaling pathways but also on their internal mechanical properties or rheological parameters. These material properties determine how the cell deforms when subjected to force.[11,43–46] It is assumed that different structures and mechanisms are responsible for mechanical sensing. For example, cartilages typically experience stresses on the order of 20 MPa, and individual chondrocytes alter their expression of glycosaminoglycans and other constituents as they deform in response to such large forces.[47] On the other hand, endothelial cells undergo a wide range of morphological and transcriptional changes in response to shear stresses less than 1 Pa, and neutrophils activate in response to similar or even smaller shear stresses.[48] Not only the magnitude but also the geometry and time course of mechanical perturbations are critical to trigger specific cellular effects. Some tissues, for instance, tendons or skeletal muscle, experience or generate mainly uniaxial forces and deformations, while others, such as the cells lining blood vessels, normally experience shear stresses due to fluid flow. These cells often respond to changes in stress or to oscillatory stress patterns rather than to a specific magnitude of stress.[49,50]

Many cells, including endothelial blood vessel and epithelial cells in the lung, experience *large-area-dilation* forces, and in these settings, both the magnitude and the temporal characteristics of the force are critical to cell response. For example, in vascular endothelial cells, mechanosensing is believed to control the production of protective ECM,[40,51] whereas in bone, mechanosensing is at the basis of bone repair and adaptive restructuring processes.[52] Osteocytes have been studied *in vitro* after extraction from the bone matrix in parallel plate flow chambers. The sites for mechanosensing might be those where strain is high if some large distortion of the sensing element is required to create a signal, in other words, if the sensor is "soft." On the other hand, the sites for sensing might also be those where stress is focused and where little strain occurs if the sensing element requires a small distortion, or is "hard," and functions by having a relatively high force threshold. To understand the mechanobiology of the cell requires a multiscale biophysics view. Externally applied stresses or traction forces are transmitted through focal adhesion (FA) receptors and are distributed throughout the cell, leading to conformational changes, phosphorylation events, and enzymatic activities. In addition, individual mechanosensing proteins may change their binding affinities.

There are also many examples of mechanotransduction that lead to disease forms: (i) arteriosclerosis, i.e., the hardening and narrowing of the arteries, mainly causing shear flow changes, in which endothelial cells sense the level of stress and regulate their behavior concomitantly;[53] (ii) arthritis, i.e., an inflammation of the joints, in which pressure increases are sensed by resident cells;[54] (iii) asthma, i.e., a common inflammatory disease of the airways, in which epithelial cells react to trans-epithelial pressure;[55] and (iv) polycystic kidney disease, i.e., a cystic genetic disorder with massive enlargement of the kidney.[56] All these processes (i–iv) are mediated by an array of signaling cascades that are started by shear stress.

In cell–matrix and cell–cell adhesions, cluster lifetime is usually much longer than the time scale for changes in loading. At the experimental stage, it would, therefore, be helpful to examine FAs as potential mechanosensors. One intriguing possibility is that force at FAs could lead to mechanical opening of domains in certain FA proteins. This might result in certain signaling events leading to the recruitment of additional bonds. If this information could be validated in an independent experiment, it would give support to the notion that FAs (or parts of them) are regulated to be close to critical thresholds, a property which is known from certain signaling pathways.

2. FOCAL ADHESIONS

FAs are large, multiprotein complexes that provide a mechanical link between the cytoskeletal contractile machinery and the ECM. FAs exhibit mechanosensitive properties. They self-assemble and elongate upon application of pulling forces and dissociate when these forces are decreased.[57] A thermodynamic model for the mechanosensitivity of FAs has been proposed. Molecular aggregates, subjected to pulling forces, tend to grow in the direction of force application by incorporating additional molecules.[9] It was demonstrated that this principle is consistent with the phenomenology of FA dynamics and that FA protein aggregates exhibit distinct modes of assembly under force and different regimes of FA assembly, including growth, steady state, and disassembly.[58]

A mechanosensitive behavior of FAs is an important component of cells' ability to spread and move along substrates. The basic observation underlying FA mechanosensitivity is that alterations in the mechanical force applied to these adhesion sites, either by the contractile machinery of the cell or after an external perturbation, have a dramatic effect on FA properties. The effect of external forces on cell–matrix contacts is believed that FAs act as mechanosensors, converting force into biochemistry.[59] Today, it is common practice to monitor FAs in real time and in live cells by using fluorescence constructs for one of the many proteins localized in FAs.[60] In many situations, mature adhesions are characterized by additional active processes, which further increase the mechanical load of the FAs, namely, the build-up of so-called stress fibers.[61] The force-generating activity of the molecular motors requires ATP and is activated by signals from the FAs, in particular, by the small GTPases from the *Rho* family. Stress fibers end in FAs, which are elongated in the direction of the stress fibers. In Fig. 4.2, we show an image of a keratinocyte (A) and a schematic representation of the system of FAs and stress fibers (B), which is characteristic for mature cell–matrix adhesion at substrates.[61,62] FAs are usually connected to stress fibers and are elongated in this direction. Recent observations claim a linear correlation between internal force and size of FAs.[63] Since size in turn correlates with signaling, this indicates that FAs could act as mechanosensors, which convert force into signaling. In other studies, it was also shown that FAs act as mechanosensors with regard to

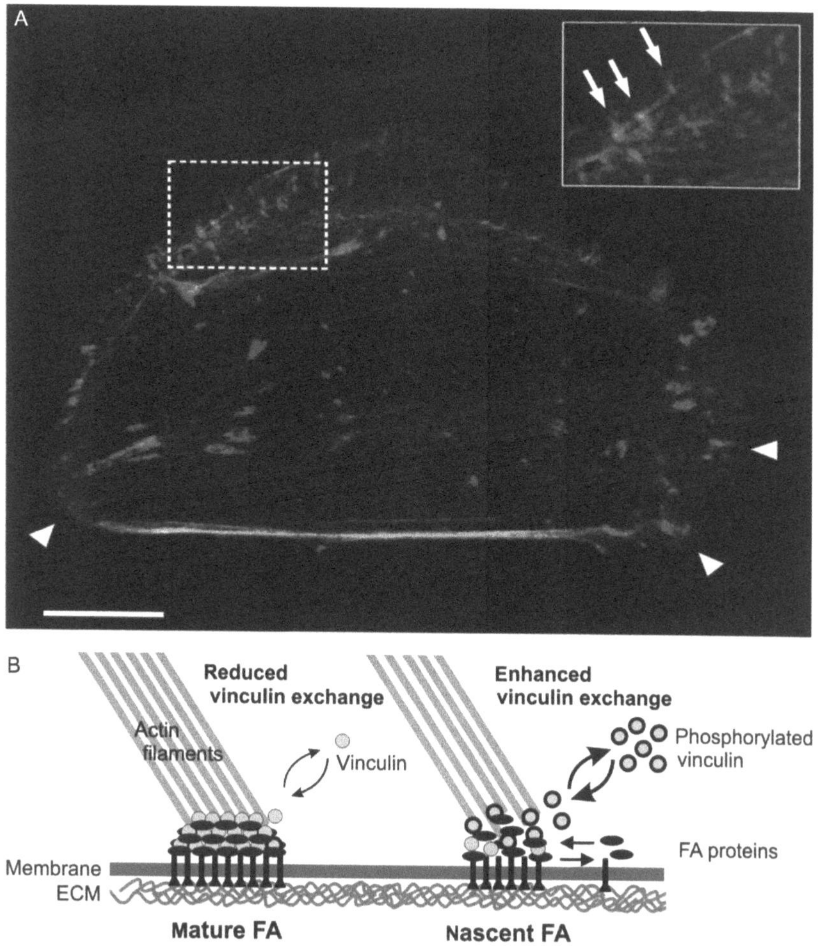

Figure 4.2 (A) Fluorescent staining of a migrating keratinocyte (actin = green; vinculin = red). Nascent adhesions develop in the lamella behind the leading edge and exhibit an elongated shape (inset, arrows). The lamella contains a loose actin meshwork consisting of small bundles that are mostly oriented parallel to the leading edge and perpendicular to focal adhesions (inset). Mature focal adhesions are mainly located at the rear of the cell and are typically bigger than nascent adhesions (arrowheads). Scale bar = 10 μm. (B) Model for vinculin exchange-dependent adhesion maturation. In nascent FAs (right), mainly phosphorylated vinculin is present and bound with high exchange dynamics within the complex. In mature FAs (left), there are stable structures that can resist strong tractions.[61] *With permission from Wiley Press.* (See the color plate.)

external force, that is, after quantitative analysis of elastic extracellular substrates.[51] Further quantitative evaluation of adhesion experiments should be done to better understand the detailed mechanism of the mechanosensor at FAs.

2.1. Mechanotransduction/-signaling

The process of mechanotransduction refers to cellular mechanisms by which load-bearing cells sense physical forces, transduce the forces into biochemical signals, and generate appropriate responses leading to alterations in cellular structure and function. The signal transduction pathways that are activated in response to mechanical forces include many unique components as well as elements shared by other signaling pathways. Mechanotransduction in both atrial and ventricular cardiomyocytes, for instance, not only regulates the *beat-to-beat* cardiac performance but also profoundly affects the proliferation, differentiation, growth, and survival of the cellular components that comprise the human myocardium. Intrinsically generated and externally applied mechanical forces are transmitted bi-directionally to internally situated sarcomeres of the rod-shaped cardiomyocytes. Attachment is clearly one way in which costameres and FAs contribute to mechanotransduction. There is also evidence that mechanical forces (generated by passive stretch and active tension development of cardiomyocytes) are "sensed" by costameres and FA complexes and transduced into biochemical signals, leading to sarcomeric assembly and altered gene expression. Understanding the cellular and molecular basis for mechanotransduction is, therefore, central to our overall understanding of cardiac structure and function in the normal and diseased heart.[64]

2.2. Focal adhesion proteins

2.2.1 Vinculin

Vinculin is one of the major proteins of the submembrane plaque of FAs. It can be tagged with green-fluorescent protein (GFP) at its amino terminal. GFP-vinculin localizes and marks FAs with very high optical quality. One intriguing possibility is that force at FAs leads to mechanical opening-up of vinculin domains. Recent computer analyses have shown that subtle conformational changes may lead to protein activation.[65–67] Vinculin has binding sites on talin, normally buried within a five-helix bundle in the talin rod domain that can rotate out of the helix core when force is applied. Once degraded, this domain becomes accessible for binding. In the case of talin, it is likely that unfolding is not required, since the structure of the bound complex is consistent with a simple helix exchange from talin to vinculin. An alternative mechanism has also been proposed by other authors,[68] involving a breakdown of the rod domain in order to expose the vinculin-binding site. Vinculin residing in a "closed" conformation can be mechanically triggered to reveal cryptic binding sites. Similarly, small conformational changes may also change the binding affinity or enzyme activity. For example, when protein binding occurs through hydrophobic site interactions, a conformational change could modify this function and potentially disrupt it totally.

Force transmission from the ECM to the cell interior occurs through a chain of proteins, located in the FA sites, that comprise an integrin–ECM protein bond (with fibronectin, vitronectin, and others), integrin-associated proteins on the intracellular side (paxillin, talin, vinculin, and others), and proteins linking the FA complex to the cytoskeleton.[63] In human foreskin fibroblasts, for instance, it has been shown that the amount of tension generated by a FA correlates directly with the FA size and with the amount of fluorescence of the FA adaptor protein GFP-vinculin.[63]

2.2.2 Zyxin

In other studies, the fluorescence intensity of GFP-zyxin, another FA adaptor protein, has been compared with the traction forces exerted by FAs in migrating fish fibroblasts. In contrast to the analysis with vinculin, the fluorescence intensity of GFP-zyxin in FAs demonstrated an inverse correlation with the generated traction stress in the respective FAs.[69,70] Using a 2D GFP-β3-integrin marker, the two examples of the changing intensities of the FA markers, GFP-vinculin and GFP-zyxin, showed dramatically that FAs are complex structures that require multiple functional parameters to describe their behavior, such as fluorescence intensity, traction forces, and FA mobility (also termed "*sliding*").

2.2.3 Talin

Another candidate for force sensing is talin, one of the FA major components, which undergoes a regulated conformational change upon its interaction with vinculin.[59,71] Talin is essential for early FA reinforcement under force that leads to the recruitment of vinculin, which stabilizes FAs. Both talin and vinculin can exist in closed and open conformations, a fact which might point in the direction of a *mechanosensor* function at FAs.[72] The incubation of cells in serum-free medium greatly reduced the size and intensity of GFP-paxillin spots at cell edges, whereas the "pipette" shift induced formation of typical focal contacts elongated in the direction of pulling.[73] Force-induced formation of focal contacts was accompanied by the recruitment of talin, vinculin, paxillin, and actin and the elongation in interference reflection microscopic images. These results indicate that focal contacts induced by external force have the same structural characteristics as those produced by cells during spreading and locomotion.[73] The force-driven opening of membrane ionic channels may contribute to their mechanosensing. Talin and vinculin might also act as nucleators[74,75] for the actin cytoskeleton, thus locally modulating the effects of the small GTPases *Rac* and *Rho*.

2.2.4 Paxillin, Pyk2

When extracellular tension is reduced, FA sites lose the ability to recruit paxillin and detach from the relaxed substrate, suggesting that continuous generation of intracellular tension (and hence high-density FAs) is required to maintain mechanical signaling. Force-induced changes at adhesion sites visualized in live cells by expression of GFP-paxillin were similar to those visualized by GFP-vinculin fluorescence. Adaptor proteins that are stacked on top of each other within the actin backbone of FAs, such as vinculin and paxillin, represented markers of the FA volume.

Although Pyk2 is predominantly localized to the cytoplasm, a minor component of the enzyme colocalized with paxillin in FAs of cultured neonatal rat ventricular myocytes. Pyk2, like focal adhesion kinase (FAK),[76–78] acts as an important scaffolding protein and transduces signals from G-protein-coupled receptors to downstream MAPK signaling pathways depending on which signaling kinase and adaptor protein binds to the phosphorylated enzyme.[79] Pyk2 has also been shown to link a variety of stressful stimuli, including Ca^{2+} overload, UV irradiation, and tumor necrosis factor treatment to MAPK activation in several cell types. Hirotani *et al.*[80] demonstrated that Pyk2 is an essential signaling component in endothelin- and phenylephrine-induced cardiomyocyte hypertrophy, perhaps acting via the Ca^{2+}- and/or PKC-dependent activation of Rac1.

2.2.5 p130Cas

Comparing stretched to nonstretched cells, FA proteins bind preferably to stretched cytoskeletal networks. Binding of vinculin, p130Cas, and PKB/Akt to actin were all found to be enhanced under stretching. Phosphorylation of p130Cas is achieved by nonreceptor tyrosine kinase (*SRC*) substrate.[81–84] It is interesting to note that the actin cytoskeleton also features crosstalk to the microtubule system when stretched.[85]

2.2.6 Focal adhesion kinase

It has recently been demonstrated that the mechanically stretched, Triton-resistant cytoskeleton of fibroblasts engages signaling molecules such as paxillin and FAK, to stretched FAs.[86] These signaling molecules are recruited to FAs and not to the actin cytoskeleton extended between them. This is of particular importance because it suggests that the mechanical distortion of FAs itself is at the origin of mechanical signaling. However, it has to be shown that the mechanical distortion of integrin receptors or the specific adaptor proteins such as FAK or paxillin, which extend between integrins and the actin cytoskeleton, is involved in mechanical sensing.

Although physical concepts, such as force and elasticity, are essential to understanding active mechanosensing at FAs, the biochemical aspects of these systems are equally important and far from understood. FAK is a protein tyrosine kinase, which has been shown to be a key component of mechanosensing at FAs. It is activated by integrin ligation and one of its main downstream targets is the small GTPase *Rac*, which leads to reorganization of the actin cytoskeleton into an isotropic network structure. At the same time, FAK-activation downregulates another small GTPase, *Rho*, mainly through activation of p190*Rho*GAP.[87] *Rho* promotes the reorganization of the actin cytoskeleton into stress fibers and it often has an antagonistic role to *Rac*.[88] Both small GTPases belong to the *Rho* family and are also activated by pathways involved in cell survival (epidermal growth factor) and lysophosphatidic acid in the cases of *Rac* and *Rho*, respectively). These schemes focus on important downstream targets of integrin signaling to the actin cytoskeleton.

2.3. Force transduction at focal adhesions

Forces have been measured in resting fibroblasts, where intracellular tension gives rise to stresses in FAs of the cells adherent to flexible two-dimensional substrates.[63,89] The contractile forces are associated with intracellular molecular motors of the myosin family. Heidemann *et al.*[90] failed to observe this phenomenon in living fibroblasts, when they applied various mechanical disturbances to the cell surface through integrin receptors. They found that such disturbances produced only local deformations. However, the authors did not confirm the formation of FAs at the points of application of external loading, which is essential for load transfer between cell surface and the interior cytoskeleton.[44] Thus, their results remain controversial. The majority of data, however, indicate in cells, when a force is applied through integrin receptors at the cell surface, FAs are formed at the site of force application. Intermediate filaments appear to be important contributors to cell contractility and prestress.[91] They serve as molecular "guy wires" that facilitate transfer of mechanical loads between the cell surface and the nucleus and stabilize microtubules. These observations provide evidence in support of the cellular tensegrity model.[92,93]

Other researchers have proposed conformational changes in intracellular proteins along the force-transmission pathway, connecting the ECM with the cytoskeleton through FAs as the main mechanotransduction mechanism.[94–96] In particular, the hypothesis that links mechanotransduction phenomena to mechanically induced alterations in the molecular conformation of proteins has been gaining increasing support. Stresses transmitted through adhesion receptors and distributed throughout the cell could cause conformational changes in individual force-transmitting proteins, any of which would be a candidate for force transduction into biochemical signals. The process by which changes in protein conformation give rise to protein clustering at FAs or initiate intracellular signaling, however, remains largely unknown.[97]

Consistent is the notion that FAs act as mechanical sensors of stress. Since FAs are membrane-attached anchoring points for the actin cytoskeleton, recent observations are consistent with a role for FAs in this process. For instance, in FAs of eukaryotic cells, transmembrane receptors of the integrin family and a large set of adaptor proteins form a physical link between the extracellular substrate and the actin cytoskeleton. During cell migration, nascent FAs within filopodia and lamellipodia make the initial exploratory contacts with the cellular environment, whereas maturing FAs pull the cell forward against the resistance of "sliding" FAs at the cell rear (Fig. 4.2).[61] Experimental approaches are available for analyzing the dynamics and interior structure of FAs. Analyzing FA dynamics using green-fluorescent protein-linked integrin led researchers to propose that the actomyosin-controlled density and turnover of integrins in FAs is used to sense the elasticity and spacing of extracellular ligands, regulating cell migration by mechanically transduced signaling.[98]

During recent years our knowledge about FAs and their role in cell spreading, migration, and survival has increased vastly. The ever-increasing number of proteins found to participate in FAs makes them one of the most complex protein aggregates formed in a cell. FAs fulfill mechanical and sensing functions that involve reversible anchorage of the actin cytoskeleton to the ECM during migration, monitoring intracellular or extracellular tension. Understanding the molecular mechanisms that account for these distinct functions of FAs is a major challenge.[62]

Eukaryotic cells have differently sized and shaped cell–substrate adhesion sites. In fibroblasts, the FAs are commonly referred to as focal complexes, focal contacts, and fibrillar adhesions. Many attempts have been made to classify FAs using descriptive features such as shape, size, cellular location, GTPase dependency, and protein composition. Unfortunately, some of these characteristics vary depending on the environment of the cells. The use of functional criteria to classify FAs according to their physiological role, for example, sensing the environment or providing mechanical support, give new definitions to distinguish focal complexes from focal contacts.

The recent use of chimeras comprising GFP attached to various FA proteins has made important contributions to our understanding of FAs. Owing to the stoichiometric fusion of GFP to FA proteins, such GFP chimeras can be used not only as markers for cellular attachment sites but also to provide dynamic and quantitative information about the composition of FAs. One of the emerging ideas from these studies is that FAs are mechanical transducing devices with a mechanical sensor function. The topology of FAs and their sensing ability for elasticity and spacing of extracellular ligands has been reviewed by Walcott *et al.*[99] Their model is based on the quantitative analysis of GFP-tagged FA proteins associated with the two-dimensional plane of the plasma membrane, providing dynamic insight into the interior structure of FAs.

In FAs, the actin cytoskeleton is linked through various adaptor proteins to heterodimeric receptors of the integrin family. In contrast to the analysis of GFP-vinculin, for instance, the fluorescence intensity of GFP-zyxin in FAs demonstrated an inverse correlation with the generated traction stress in the respective FAs. We are, therefore, led to believe that the different FA markers used in studies have distinct functions and are recruited by signals that might not originate from mechanical forces applied to FAs. Due to the complexity and multiple functions of FAs, it is very difficult to assign a specific cause to changes in the fluorescence intensities of any particular GFP marker. Hence, we opt for 2D GFP markers for quantitative analysis of tension-dependent changes in FA structure. These markers should serve a mechanical function and should be a part of the physical link between the ECM and the actin cytoskeleton.

It was shown that when a 2D GFP-β3-integrin marker is used to study FAs, the respective fluorescence intensity correlates directly with the packing density of this particular integrin in each FA. The analysis of GFP-β3-integrin in five different FAs of migrating melanoblasts revealed many important features: (i) FAs can be classified into low-density and high-density forms; (ii) FA density can change dramatically with time; (iii) high- and low-density contacts are located in different cellular compartments; and (iv) only high-density FAs show mobility (sliding). The value of this complex information can be further extrapolated taking into account that low-density FAs form in response to the activity of the GTPases *Rac1* and *Cdc42* and high-density FAs form in a manner dependent on the GTPase *RhoA* and actomyosin contraction. This implies that, at least for β3-integrins, myosin-dependent actin cytoskeleton contraction is at the origin of the formation of high-density FAs. Owing to this mechanical link, either density changes in the actin backbone of FAs or changes in the spacing of extracellular ligands (e.g., induced by extracellular tension) will mechanically distort the link between integrins and actin-bound adaptor proteins.[100,101]

How do cells measure or sense the physical constraints of their environment? It is possible that mechanical sensing occurs inside FAs, considering that the sensing organelles of cells, namely, the filopodia and lamellipodia have low-density FAs that form in a *Rac1*- or *Cdc42*-dependent manner. It has been observed that, on soft substrates, FAs retract, whereas they are reinforced and maintained on a rigid surface, anchoring the cell for forward motion. In addition, brushing against a moving lamellipodium with a microneedle induces the maturation of lamellipodial focal complexes (low-density FAs) into focal contacts (high-density FAs). This maturation of FAs in response to extracellularly applied tension depends on *RhoA* activation and its downstream target Diaphanous (mDia).[102] Because mDia acts as an actin polymerization factor, the observed increase in size and density of FAs could be linked to increased amounts of polymerized actin.

Different fates of low- and high-density FAs with respect to the elasticity of the substrates have been demonstrated. The absence of mechanical signaling on soft substrates is due to the lack of physical distortion during the contraction of FAs. Different densities of FAs can also be extrapolated from the spacing of extracellular ligands. When cells rapidly spread and attach, they were unable to form focal contacts (high-density FAs) and stress fibers on this substrate. Attempts to examine the proteins of adhesion sites, which are believed to be responsible for surface sensing, have thus far focused primarily on the organization of FAs and related structures. The complexity of the ECM and the uncertainty that surrounds the state of exposure and reactivity of the adhesion mediating domains make it difficult to define the sensing mechanisms. Cellular interactions with such surfaces indicate the need to develop synthetic adhesive surfaces with well-defined structures. The physiological significance of space sensing and the mechanisms whereby the cells measure the particular interligand distance remains unclear.[62]

2.4. Protein crosstalk

Crosstalk between the actin cytoskeleton and the mechanoresponsive matrix sensing machinery clearly plays a crucial role in all types of integrin-mediated adhesions.[57] Thus, existing experimental data on mechanosensing in FAs are considerably more detailed than those on the sensory function of any other type of adhesion. It is worth noting that other types of integrin adhesions, such as podosomes and invadopodia, are also mechanosensitive. Thus, mechanical crosstalk between integrins and the actin cytoskeleton is a key feature of environmental sensing. The major features of the actin–integrin feedback network, as it is presently understood, have been discussed.[62,103] How these diverse molecular mechanosensing devices are indeed integrated into a single mechanosensing module remains a major challenge. Thermodynamic considerations suggest that the application of stretching force to an aggregate of protein subunits should promote the growth of the aggregate in the direction of force, irrespective of any conformational changes in the subunits.[104] FA mechanosensors might also be regarded as a network of tightly interconnected molecular mechanosensing units that operate in a coordinated fashion in response to mechanical forces. These forces might be applied externally and are usually transmitted by the actin cytoskeleton, thereby rendering the formation and maturation of FAs actin-dependent.

2.5. Cell signaling pathways

Mechanotransduction is viewed as a force-induced process initiating biochemical responses (e.g., changing binding affinity, altering phosphorylation state, and/or conformation change) and initiating signal pathways leading to gene expression, protein synthesis, and cellular phenotype change. Activation is started by mechanical stress via second messengers and gene expression. Other forms of mechanotransduction can be (i) stretch-activated ion channel activation, (ii) membrane mechanotransduction (via G-proteins and G-protein-coupled receptors), and various other proteins that connect to FAs/adherence junctions and the cytoskeleton. Shear stress on the membrane can influence the conformation of transmembrane proteins (stress in tension or bending) that leads to, for example, activation of MAPK, *Rac*, *Rho*, et cetera, or (iii) constrains autocrine signaling, for instance, stresses applied to a layer of airway epithelial cells grown on a porous membrane results in changes in gene expression, signaling, and ERK phosphorylation in lungs.[35]

2.6. Translation of information gathered at focal adhesions

Recent developments have shown how cellular forces are measured at the level of single FAs using a novel soft substrate technique. A correlation exists between force and size of FAs.[105] The mutual regulation of force and aggregation cannot proceed without limits, and recent work suggests that the upper bounds are set by the action of microtubules inserting into mature FAs and delivering some kind of stop signal.[100] In particular, cells can learn about the mechanical properties of their environment by monitoring the build-up of force at FAs while pulling on it (active mechanosensing). It has been shown that cells react in a typical way to the elastic properties of their environment, a phenomenon which has been termed durotaxis. Most cell types upregulate their cytoskeleton and their cell–matrix contacts on stiffer substrates, and locomote in favor of stiff or strained regions.[106] In principle, there are many different physical mechanisms that might be at work as mechanosensor at FAs. In fact, another recent study has shown that the aggregation response at FAs under force persists even for permeabilized cells without any plasma membrane.[107]

We have explored how cells behave in a soft environment.[19] It has been suggested that force-induced unbinding of fibronectin on the extracellular side and force-induced unbinding of certain cytoplasmic plaque proteins in FAs (e.g., vinculin) might be involved in mechanosensory processes for tissue cells.[51] At this stage, additional experimental evidence would be very helpful in modeling the mechanosensor at FAs.[103] However, one intriguing possibility is that force at FAs may lead to mechanical opening-up of domains in certain FA proteins such as talin, vinculin, and p130Cas.[59,82,84] In particular, it has been shown that application of external force leads to growth of FAs and, therefore, to strong signaling activity. Protein aggregation has been observed in mature FAs under internally generated force.[61,108]

The dynamics of FAs is also the subject of much current research.[109] Anchorage-dependent cells constantly assemble and disassemble FAs, thereby probing the mechanical properties of their environment. Initial FAs are local processes based on integrin clustering.

2.7. Focal adherence junctions

Cell–matrix adhesions are provided by large FAs, which can contain up to 100 integrin-mediated bonds, while an important part of cell–cell adhesions (focal adherence junctions (FAJs)) are provided by similarly large clusters of cadherin-mediated bonds.[110] Molecular bonds in a cluster can be arranged and loaded in different ways, including in parallel and serial ways. When cells are experimentally probed, the situation is further complicated by relaxation processes in the viscoelastic parts of the cell that act as force transducers.[111] So far the interaction of a force dipole with the boundary has been considered. In an elastic sphere containing many cells, one could separate the contributions to the effective stiffness into a contribution from the boundary-induced field (i.e., a cell-surface interaction)[9,112] and a contribution from the elastic fields of other cells embedded in the sphere (i.e., a cell–cell interaction term). Further work has to be carried out to understand the complex manner of this interaction.

2.8. Measuring mechanotransduction/-sensation

2.8.1 Flow chambers and cone and plate rheometers

In vascular endothelial cells, mechanosensing is believed to control the production of the protective ECM,[113] whereas in bone, mechanosensing is at the basis of bone repair and adaptive restructuring processes.[52] Osteocytes have been studied *in vitro* after extraction from the bone matrix in parallel plate flow chambers. Flow chambers are commonly used to study the adhesion of leukocytes to endothelium-like substrates. For diluted ligands, one usually observes first order dissociation kinetics, which traditionally has been interpreted as the signature of single-molecule events. Although recent results now point to a more complicated situation involving multiple bonds, flow chambers with diluted ligands can indeed be used to study single-molecule unbinding.

2.8.2 Magnetic and optical traps

These methods are called nonphysiological. Exactly how the ECM–integrin–cytoskeletal complex senses mechanical stimuli remains somewhat of a mystery. Seminal observations[24,27,114] using a magnetic tweezer and twisting device to transfer force directly from integrins to the local cytoskeleton suggest that mechanical deformation of one or more FA protein is the proximal step in an intracellular signaling cascade that leads to global cytoskeletal rearrangements and mechanotransduction at multiple, distant sites within the cell.

2.8.3 Atomic force microscopy and biomembrane force probe

The mechanical opening-up of biomolecular bonds has become a subject of extensive research during the last decade, both experimentally and theoretically. The main experimental techniques in this field are AFM and the biomembrane force probe (BFP). In AFM experiments, bonds are attached to sharp tips mounted on soft cantilevers, which are moved on a piezo stage. In BFP experiments, bonds are attached, for example, to red blood cells, which are controlled by micropipette aspiration.[115] The main theoretical approaches in this field are Kramer's theory (which describes thermally assisted escape over a transition state barrier) and steered molecular dynamics (atomic level simulations with force fields and an externally applied force). The field of single-molecule force spectroscopy was opened up by seminal AFM experiments conducted by Hermann Gaub's group[23,116] as well as AFM measurements by Matthias Rief,[117,118] who reported for the first time the mechanically induced unbinding of single biotin–streptavidin bonds, with a binding strength of 140 pN. In an experimental context, binding strength usually means the most frequent rupture force in the spectrum of rupture forces measured in different experiments. The mechanotransduction (i.e., force-transmission pathway via FAs, cell–cell contacts, or cytoskeleton-associated proteins occurs at the single-molecule scale.

2.8.4 Cell stretcher

Experiments can be performed using uniaxial or biaxial strain, as well as oscillatory or static stretch in 2D and 3D possible with cells embedded in gels. However, only a few days of stretching are possible and long-term remodeling or even disease progression cannot be observed. On top of this, the response to strain is complex. Comparing, for instance, stretched to non-stretched cells, proteins bind preferably to stretched cytoskeletal networks: Binding of paxillin, vinculin, FA kinase p130Cas, and PKB/Akt are all enhanced during stretch.[84,119]

2.8.5 Hydrostatic pressure

Physical forces encountered by living cells include membrane stretch, gain and loss of adhesion as well as compression due to an increase in pressure. It is conceivable that different mechanosensors are required to sense transverse versus longitudinal stretch, perhaps accounting for differential signaling and cellular phenotype resulting from pressure versus volume overload.[120]

2.8.6 Stretch-activated ion channels

Ion channels play a central role for mechanotransduction in the sensory systems, but the situation at FAs is very different, since speed of response is not an issue at FAs under force.[121] Moreover, studies have shown that the durotactic response is suppressed when stretch-activated ion channels are blocked with gadolinium. These studies provide an appealing mechanism for signal transduction for mechanically active cells in soft media. However, they are also unspecific and cells might not be able to distinguish between different sources. On the other hand, additional information channels, such as soluble ligands, will certainly supplement elastic signals.[122]

3. CONCLUSIONS

In this review, I have addressed the issue of environmental sensing by cells. I propose that a comprehensive understanding of adhesion-mediated signaling requires the precise characterization of both the sensed surface and the sensory machinery of the cell. In recent years, remarkable progress has been made in both areas. Surface nano-engineering has opened up new possibilities for the systematic modulation of individual surface features, such as surface chemistry, ligand spacing, geometry, and surface rigidity. In parallel, novel techniques of gene modulation enable the selective removal, overexpression, and mutation of individual genes. These effects on the cellular response of the sensory machinery can then be assessed. Although our current understanding of adhesion-mediated environmental sensing is still incomplete, several design principles have emerged from experiments. It seems, for example, that surface chemistry, namely, the presence of diverse matrix proteins, has a strong effect on the selection of specific integrin receptors and consequently on the initial assembly of the integrin nanocomplexes. Indeed, differential activation of integrin's (e.g., $\alpha5\beta1$ integrin compared with $\alpha v\beta3$ integrin) can result in major differences in both the initiation and the progression of the adhesion process.[123] Furthermore, a growing body of evidence implicates mechanical force as central to the regulation of nearly every stage of FA assembly, from the actin polymerization-dependent assembly of the first visible, nascent adhesions, to the myosin-dependent growth and maturation of FAs.[124,125] A deeper understanding of the ongoing interplay between molecular surface design and genetic modulation of the adhesion machinery is likely to reveal the nature of the mechanisms that underlie the sensitivity of living cells to both the chemical and physical characteristics of the surfaces to which they adhere.

FAs are large, multiprotein complexes that provide a mechanical link between the cytoskeletal contractile machinery and the ECM.[46,126,127] FAs exhibit mechanosensitive properties. They self-assemble and elongate upon application of pulling forces and dissociate when these forces are decreased. A thermodynamic model for the mechanosensitivity of FAs, according to which a molecular aggregate, subjected to pulling forces, tends to grow in the direction of force application (by incorporating additional molecules), has been proposed. This principle is consistent with the phenomenology of FA dynamics by considering a one-dimensional protein aggregate subjected to pulling forces and anchored to the substrate.[128] Depending on the force level, force distribution along the aggregate is predicted to exhibit distinct modes of assembly that are largely consistent with the experimentally observed FA behavior.

In all, the mechanosensitive behavior of FAs is an important component of the cell's ability to spread and move along substrates.[57,62,101] The basic observation underlying FA mechanosensitivity is that alterations in the mechanical force applied to these adhesion sites, either by the contractile machinery of the cell or after external perturbation, have a dramatic effect on FA properties and cellular behavior and should be further elucidated.[129]

ACKNOWLEDGMENTS

The author thanks Drs. Claus Metzner and Victoria Jackiw for proofreading the chapter. This work was supported by grants from Deutscher Akademischer Austauschdienst and Deutsche Forschungsgemeinschaft.

REFERENCES

1. Wolff J. *Das Gesetz der Transformation der Knochen*. Berlin: Hirschwald; 1892.
2. Murray CD. The physiological principle of minimum work applied to the angle of branching of arteries. *J Gen Physiol*. 1926;9:835–841.
3. West GB, Brown JH, Enquist BJ. A general model for the origin of allometric scaling laws in biology. *Science*. 1997;276:122–126.
4. Hudspeth AJ. How the ear's works work. *Nature*. 1989;341:397–404.
5. Hamill OP, Martinac B. Molecular basis of mechanotransduction in living cells. *Physiol Rev*. 2001;81:685–740.
6. Martin P, Hudspeth AJ, Julicher F. Comparison of a hair bundle's spontaneous oscillations with its response to mechanical stimulation reveals the underlying active process. *Proc Natl Acad Sci USA*. 2001;98:14380–14385.
7. Ingber DE. Cellular basis of mechanotransduction. *Biol Bull*. 1998;194:323–325, discussion 325–327.
8. Bao G, Suresh S. Cell and molecular mechanics of biological materials. *Nat Mater*. 2003;2:715–725.
9. Chen CS, Tan J, Tien J. Mechanotransduction at cell-matrix and cell-cell contacts. *Annu Rev Biomed Eng*. 2004;6:275–302.
10. Huang H, Kamm RD, Lee RT. Cell mechanics and mechanotransduction: pathways, probes, and physiology. *Am J Physiol Cell Physiol*. 2004;287:C1–C11.
11. Janmey PA, Weitz DA. Dealing with mechanics: mechanisms of force transduction in cells. *Trends Biochem Sci*. 2004;29:364–370.

12. Davies PF. Flow-mediated endothelial mechanotransduction. *Physiol Rev*. 1995;75: 519–560.
13. Goldmann WH, Schindl M, Cardozo TJ, Ezzell RM. Motility of vinculin-deficient F9 embryonic carcinoma cells analyzed by video, laser confocal, and reflection interference contrast microscopy. *Exp Cell Res*. 1995;221:311–319.
14. Goldmann WH, Galneder R, Ludwig M, et al. Differences in elasticity of vinculin-deficient F9 cells measured by magnetometry and atomic force microscopy. *Exp Cell Res*. 1998;239:235–242.
15. Gimbrone Jr MA, Topper JN, Nagel T, Anderson KR, Garcia-Cardena G. Endothelial dysfunction, hemodynamic forces, and atherogenesis. *Ann N Y Acad Sci*. 2000;902:230–239, discussion 239–240.
16. Lehoux S, Tedgui A. Cellular mechanics and gene expression in blood vessels. *J Biomech*. 2003;36:631–643.
17. McCormick SM, Frye SR, Eskin SG, et al. Microarray analysis of shear stressed endothelial cells. *Biorheology*. 2003;40:5–11.
18. Mierke CT, Kollmannsberger P, Paranhos-Zitterbart D, Smith J, Fabry B, Goldmann WH. Mechano-coupling and regulation of contractility by the vinculin tail domain. *Biophys J*. 2008;94:661–670.
19. Mierke CT, Kollmannsberger P, Zitterbart DP, et al. Vinculin facilitates cell invasion into three-dimensional collagen matrices. *J Biol Chem*. 2010;285:13121–13130.
20. Dembo M, Wang YL. Stresses at the cell-to-substrate interface during locomotion of fibroblasts. *Biophys J*. 1999;76:2307–2316.
21. Engler AJ, Sen S, Sweeney HL, Discher DE. Matrix elasticity directs stem cell lineage specification. *Cell*. 2006;126:677–689.
22. Evans E, Yeung A. Apparent viscosity and cortical tension of blood granulocytes determined by micropipet aspiration. *Biophys J*. 1989;56:151–160.
23. Radmacher M, Tillmann RW, Fritz M, Gaub HE. From molecules to cells: imaging soft samples with the atomic force microscope. *Science*. 1992;257:1900–1905.
24. Wang N, Butler JP, Ingber DE. Mechanotransduction across the cell surface and through the cytoskeleton. *Science*. 1993;260:1124–1127.
25. Goldmann WH, Ezzell RM. Viscoelasticity in wild-type and vinculin-deficient (5.51) mouse F9 embryonic carcinoma cells examined by atomic force microscopy and rheology. *Exp Cell Res*. 1996;226:234–237.
26. Goldmann WH. Mechanical manipulation of animal cells: cell indentation. *Biotechnol Lett*. 2000;22:431–435.
27. Fabry B, Maksym GN, Butler JP, Glogauer M, Navajas D, Fredberg JJ. Scaling the microrheology of living cells. *Phys Rev Lett*. 2001;87:148102.
28. Friedl P, Gilmour D. Collective cell migration in morphogenesis, regeneration and cancer. *Nat Rev Mol Cell Biol*. 2009;10:445–457.
29. Friedl P, Locker J, Sahai E, Segall JE. Classifying collective cancer cell invasion. *Nat Cell Biol*. 2012;14:777–783.
30. Lang NR, Münster S, Metzner C, et al. Estimating the 3D pore size distribution of biopolymer networks from directionally biased data. *Biophys J*. 2013;105: 1967–1975.
31. Choquet D, Felsenfeld DP, Sheetz MP. Extracellular matrix rigidity causes strengthening of integrin- cytoskeleton linkages. *Cell*. 1997;88:39–48.
32. Mandeville JT, Lawson MA, Maxfield FR. Dynamic imaging of neutrophil migration in three dimensions: mechanical interactions between cells and matrix. *J Leukoc Biol*. 1997;61:188–200.
33. Shaub A. Unravelling the extracellular matrix. *Nat Cell Biol*. 1999;1:E173–E175.
34. Cukierman E, Pankov R, Stevens DR, Yamada KM. Taking cell-matrix adhesions to the third dimension. *Science*. 2001;294:1708–1712.

35. Gunst SJ, Meiss RA, Wu MF, Rowe M. Mechanisms for the mechanical plasticity of tracheal smooth muscle. *Am J Physiol.* 1995;268:C1267–C1276.
36. Gunst SJ, Tang DD. The contractile apparatus and mechanical properties of airway smooth muscle. *Eur Respir J.* 2000;15:600–616.
37. Faust U, Hampe N, Rubner W, et al. Cyclic stress at mHz frequencies aligns fibroblasts in direction of zero strain. *PLoS One.* 2011;6:e28963.
38. Girton TS, Oegema TR, Tranquillo RT. Exploiting glycation to stiffen and strengthen tissue equivalents for tissue engineering. *J Biomed Mater Res.* 1999;46:87–92.
39. Janmey PA, McCulloch CA. Cell mechanics: integrating cell responses to mechanical stimuli. *Annu Rev Biomed Eng.* 2007;9:1–34.
40. Discher D, Dong C, Fredberg JJ, et al. Biomechanics: cell research and applications for the next decade. *Ann Biomed Eng.* 2009;37:847–859.
41. Malek AM, Izumo S. Mechanism of endothelial cell shape change and cytoskeletal remodeling in response to fluid shear stress. *J Cell Sci.* 1996;109:713–726.
42. Davies PF. Multiple signaling pathways in flow-mediated endothelial mechanotransduction: PYK-ing the right location. *Arterioscler Thromb Vasc Biol.* 2002;22: 1755–1757.
43. Janmey PA. The cytoskeleton and cell signaling: component localization and mechanical coupling. *Physiol Rev.* 1998;78:763–781.
44. Ingber DE. Opposing views on tensegrity as a structural framework for understanding cell mechanics. *J Appl Physiol.* 2000;89:1663–1670.
45. Goldmann WH. Mechanical aspects of cell shape regulation and signaling. *Cell Biol Int.* 2002;26:313–317.
46. Goldmann WH. Mechanotransduction and focal adhesions. *Cell Biol Int.* 2012;36:649–652.
47. Grodzinsky AJ, Levenston ME, Jin M, Frank EH. Cartilage tissue remodeling in response to mechanical forces. *Annu Rev Biomed Eng.* 2000;2:691–713.
48. Fukuda S, Schmid-Schonbein GW. Regulation of CD18 expression on neutrophils in response to fluid shear stress. *Proc Natl Acad Sci USA.* 2003;100:13152–13157.
49. Florian JA, Kosky JR, Ainslie K, Pang Z, Dull RO, Tarbell JM. Heparan sulfate proteoglycan is a mechanosensor on endothelial cells. *Circ Res.* 2003;93:e136–e142.
50. Bacabac RG, Smit TH, Mullender MG, Dijcks SJ, Van Loon JJ, Klein-Nulend J. Nitric oxide production by bone cells is fluid shear stress rate dependent. *Biochem Biophys Res Commun.* 2004;315:823–829.
51. Discher DE, Janmey PA, Wang YL. Tissue cells feel and respond to the stiffness of their substrate. *Science.* 2005;310:1139–1143.
52. Burger EH, Klein-Nulend J. Mechanotransduction in bone–role of the lacuno-canalicular network. *FASEB J.* 1999;13(Suppl):S101–S112.
53. Berger SA, Jou LD. Flows in stenotic vessels. *Annu Rev Fluid Mech.* 2000;32:347–382.
54. Firestein GS. Etiology and pathogenesis of rheumatoid arthritis. In: *Rheumatoid Arthritis.* 8th ed. Philadelphia: Elsevier Inc.; 2008:1035–1086.
55. Fabry B, Fredberg JJ. Mechanotransduction, asthma, and airway smooth muscle. *Drug Discov Today.* 2007;4:131–137.
56. Gonzalez-Perrett S, Kim K, Ibarra C, et al. Polycystin-2, the protein mutated in autosomal dominant polycystic kidney disease (ADPKD), is a Ca2+-permeable nonselective cation channel. *Proc Natl Acad Sci USA.* 2001;98:1182–1187.
57. Bershadsky A, Kozlov M, Geiger B. Adhesion-mediated mechanosensitivity: a time to experiment, and a time to theorize. *Curr Opin Cell Biol.* 2006;18:472–481.
58. Chen CS, Alonso JL, Ostuni E, Whitesides GM, Ingber DE. Cell shape provides global control of focal adhesion assembly. *Biochem Biophys Res Commun.* 2003;307:355–361.
59. Grashoff C, Hoffman BD, Brenner MD, et al. Measuring mechanical tension across vinculin reveals regulation of focal adhesion dynamics. *Nat Lett.* 2010;466:263–267.

60. Danuser G, Waterman-Storer CM. Quantitative fluorescent speckle microscopy of cytoskeleton dynamics. *Annu Rev Biophys Biomol Struct.* 2006;35:361–387.
61. Möhl C, Kirchgessner N, Schaefer C, et al. Becoming stable and strong: the interplay between vinculin exchange dynamics and adhesion strength during adhesion site maturation. *Cell Motil Cytoskeleton.* 2009;66:350–364.
62. Geiger B, Spatz JP, Bershadsky AD. Environmental sensing through focal adhesions. *Nat Rev Mol Cell Biol.* 2009;10:21–33.
63. Balaban NQ, Schwarz US, Riveline D, et al. Force and focal adhesion assembly: a close relationship studied using elastic micropatterned substrates. *Nat Cell Biol.* 2001;3: 466–472.
64. Samarel AM. Costameres, focal adhesions, and cardiomyocyte mechanotransduction. *Am J Physiol Heart Circ Physiol.* 2005;289:H2291–H2301.
65. Golji J, Mofrad MR. A molecular dynamics investigation of vinculin activation. *Biophys J.* 2010;99:1073–1081.
66. Golji J, Mofrad MR. Vinculin activation is necessary for complete talin binding. *Biophys J.* 2011;100:332–340.
67. Golji J, Wendorff T, Mofrad MR. Phosphorylation primes vinculin for activation. *Biophys J.* 2012;102:2022–2030.
68. Ziegler WH, Liddington RC, Critchley DR. The structure and regulation of vinculin. *Trends Cell Biol.* 2006;16:453–460.
69. Beningo KA, Dembo M, Kaverina I, Small JV, Wang YL. Nascent focal adhesions are responsible for the generation of strong propulsive forces in migrating fibroblasts. *J Cell Biol.* 2001;153:881–888.
70. Rottner K, Krause M, Gimona M, Small JV, Wehland J. Zyxin is not colocalized with vasodilator-stimulated phosphoprotein (VASP) at lamellipodial tips and exhibits different dynamics to vinculin, paxillin, and VASP in focal adhesions. *Mol Biol Cell.* 2001;12:3103–3113.
71. Thievessen I, Thompson PM, Berlemont S, et al. Vinculin-actin interaction couples actin retrograde flow to focal adhesions, but is dispensable for focal adhesion growth. *J Cell Biol.* 2013;202:163–177.
72. Schwarz US, Erdmann T, Bischofs IB. Focal adhesions as mechanosensors: the two-spring model. *Biosystems.* 2006;83:225–232.
73. Riveline D, Zamir E, Balaban NQ, et al. Focal contacts as mechanosensors: externally applied local mechanical force induces growth of focal contacts by an mDia1-dependent and ROCK-independent mechanism. *J Cell Biol.* 2001;153: 1175–1186.
74. Kaufmann S, Piekenbrock T, Goldmann WH, Bärmann M, Isenberg G. Talin binds to actin and promotes filament nucleation. *FEBS Lett.* 1991;284:187–191.
75. Goldmann WH, Niggli V, Kaufmann S, Isenberg G. Probing actin and liposome interaction of talin and talin-vinculin complexes: a kinetic, thermodynamic and lipid labeling study. *Biochemistry.* 1992;31:7665–7671.
76. Klemm AH, Kienle S, Rheinlaender J, Schäffer TE, Goldmann WH. The influence of Pyk2 on the mechanical properties in fibroblasts. *Biochem Biophys Res Commun.* 2010;393:694–697.
77. Fabry B, Klemm A, Kienle S, Schäffer TE, Goldmann WH. Focal adhesion kinase stabilizes the cytoskeleton. *Biophys J.* 2011;101:2131–2138.
78. Lange J, Auernheimer V, Strissel PL, Goldmann WH. Influence of focal adhesion kinase on the mechanical behavior of cell populations. *Biochem Biophys Res Commun.* 2013;436:246–251.
79. Lim Y, Lim ST, Tomar A, et al. PyK2 and FAK connections to p190Rho guanine nucleotide exchange factor regulate RhoA activity, focal adhesion formation, and cell motility. *J Cell Biol.* 2008;180:187–203.

80. Hirotani S, Higuchi Y, Nishida K, et al. Ca(2+)-sensitive tyrosine kinase Pyk2/CAK beta-dependent signaling is essential for G-protein-coupled receptor agonist-induced hypertrophy. *J Mol Cell Cardiol.* 2004;36:799–807.
81. Boutahar N, Guignandon A, Vico L, Lafage-Proust MH. Mechanical strain on osteoblasts activates autophosphorylation of focal adhesion kinase and proline-rich tyrosine kinase 2 tyrosine sites involved in ERK activation. *J Biol Chem.* 2004;279: 30588–30599.
82. Sawada Y, Tamada M, Dubin-Thaler BJ, et al. Force sensing by mechanical extension of the Src family kinase substrate p130Cas. *Cell.* 2006;127:1015–1026.
83. Janostiak R, Tolde O, Bruhova Z, et al. Tyrosine phosphorylation within the SH3 domain regulates CAS subcellular localization, cell migration, and invasiveness. *Mol Biol Cell.* 2011;22:4256–4267.
84. Janostiak R, Brabek J, Auernheimer V, et al. CAS directly interacts with vinculin to control mechanosensing and focal adhesion dynamics. *Cell Mol Life Sci.* 2014;71:727–744.
85. del Rio A, Perez-Jimenez R, Liu R, Roca-Cusachs P, Fernandez JM, Sheetz MP. Stretching single talin rod molecules activates vinculin binding. *Science.* 2009;323: 638–641.
86. Pasapera AM, Schneider IC, Rericha E, Schlaepfer DD, Waterman CM. Myosin II activity regulates vinculin recruitment to focal adhesions through FAK-mediated paxillin phosphorylation. *J Cell Biol.* 2010;188:877–890.
87. Parsons JT, Horwitz AR, Schwartz MA. Cell adhesion: integrating cytoskeletal dynamics and cellular tension. *Nat Rev Mol Cell Biol.* 2010;11:633–643.
88. Parri M, Chiarugi P. Rac and Rho GTPases in cancer cell motility control. *Cell Commun Signal.* 2010;8:23.
89. Pelham Jr RJ, Wang Y. Cell locomotion and focal adhesions are regulated by substrate flexibility. *Proc Natl Acad Sci USA.* 1997;94:13661–13665.
90. Heidemann SR, Kaech S, Buxbaum RE, Matus A. Direct observations of the mechanical behaviors of the cytoskeleton in living fibroblasts. *J Cell Biol.* 1999;145:109–122.
91. Na S, Chowdhury F, Tay B, et al. Plectin contributes to mechanical properties of living cells. *Am J Physiol Cell Physiol.* 2009;296:C868–C877.
92. Ingber DE. Tensegrity: the architectural basis of cellular mechanotransduction. *Annu Rev Physiol.* 1997;59:575–599.
93. Ingber DE. Tensegrity I. Cell structure and hierarchical systems biology. *J Cell Sci.* 2003;116:1157–1173.
94. Mack PJ, Kaazempur-Mofrad MR, Karcher H, Lee RT, Kamm RD. Force-induced focal adhesion translocation: effects of force amplitude and frequency. *Am J Physiol Cell Physiol.* 2004;287:C954–C962.
95. Mofrad MR, Kamm RD, eds. *Cytoskeletal Mechanics: Models and Measurements.* Cambridge University Press; 2006. 1–244. ISBN 978-0-521-84637-0.
96. Mofrad MR, Kamm RD, eds. *Cellular Mechanotransduction: Diverse Perspectives Form Molecules to Tissues.* Cambridge University Press; 2010. 1–445. ISBN 978-0-521-89523-1.
97. Geiger B, Bershadsky A. Assembly and mechanosensory function of focal contacts. *Curr Opin Cell Biol.* 2001;13:584–592.
98. Schiller HB, Hermann MR, Polleux J, et al. Beta1- and alphav-class integrins cooperate to regulate myosin II during rigidity sensing of fibronectin-based microenvironments. *Nat Cell Biol.* 2013;15:625–636.
99. Walcott S, Sun SX. A mechanical model of actin stress fiber formation and substrate elasticity sensing in adherent cells. *Proc Natl Acad Sci USA.* 2010;107:7757–7762.
100. Wehrle-Haller B, Imhof BA. Actin, microtubules and focal adhesion dynamics during cell migration. *Int J Biochem Cell Biol.* 2003;35:39–50.

101. Wehrle-Haller B. Assembly and disassembly of cell matrix adhesions. *Curr Opin Cell Biol*. 2012;24:569–581.
102. McBeath R, Pirone DM, Nelson CM, Bhadriraju K, Chen CS. Cell shape, cytoskeletal tension, and RhoA regulate stem cell lineage commitment. *Dev Cell*. 2004;6:483–495.
103. Geiger B, Bershadsky A. Exploring the neighborhood: adhesion-coupled cell mechanosensors. *Cell*. 2002;110:139–142.
104. Dill KA, Ghosh K, Schmit JD. Physical limits of cells and proteomes. *Proc Natl Acad Sci USA*. 2011;108:17876–17882.
105. Stricker J, Aratyn-Schaus Y, Oakes PW, Gardel ML. Spatiotemporal constraints on the force-dependent growth of focal adhesions. *Biophys J*. 2011;100:2883–2893.
106. Koch TM, Münster S, Bonakdar N, Buttler JP, Fabry B. 3D traction forces in cancer cell invasion. *PLoS One*. 2012;7:e33476.
107. Tang X, Wen Q, Kuhlenschmidt TB, Kuhlenschmidt MS, Janmey PA, Saif TA. Attenuation of cell mechanosensitivity in colon cancer cells during in vitro metastasis. *PLoS One*. 2012;7:e50443.
108. Vogel V. Mechanotransduction involving multimodular proteins: converting force into biochemical signals. *Annu Rev Biophys Biomol Struct*. 2006;35:459–488.
109. Wolfenson H, Bershadsky A, Henis YI, Geiger B. Actomyosin-generated tension controls the molecular kinetics of focal adhesions. *J Cell Sci*. 2011;124:1425–1432.
110. Gumbiner BM. Regulation of cadherin-mediated adhesion in morphogenesis. *Nat Rev Mol Cell Biol*. 2005;6:622–634.
111. Yonemura S, Wada Y, Watanabe T, Nagafuchi A, Shibata M. Alpha-Catenin as a tension transducer that induces adherens junction development. *Nat Cell Biol*. 2010;12:533–542.
112. Mason BN, Califano JP, Reinhart-King CA. Matrix stiffness: a regulator of cellular behavior and tissue formation. In: *Engineering Biomaterials for Regenerative Medicine*. 2011. ISBN: 978-1-4614-1079-9:19–37.
113. Weinbaum S, Zhang X, Han Y, Vink H, Cowin SC. Mechanotransduction and flow across the endothelial glycocalyx. *Proc Natl Acad Sci USA*. 2003;100:7988–7995.
114. Ezzell RM, Goldmann WH, Wang N, Parasharama N, Ingber DE. Vinculin promotes cell spreading by mechanically coupling integrins to the cytoskeleton. *Exp Cell Res*. 1997;231:14–26.
115. Evans EA. New membrane concept applied to the analysis of fluid shear- and micropipette-deformed red blood cells. *Biophys J*. 1973;13:941–954.
116. Moy VT, Florin EL, Gaub HE. Intermolecular forces and energies between ligands and receptors. *Science*. 1994;266:257–259.
117. Rief M, Clausen-Schaumann H, Gaub HE. Sequence-dependent mechanics of single DNA molecules. *Nat Struct Biol*. 1999;6:346–349.
118. Schlierf M, Rief M. Single-molecule unfolding force distributions reveal a funnel-shaped energy landscape. *Biophys J*. 2006;90:L33–L35.
119. Sai X, Naruse K, Sokabe M. Activation of pp 60(src) is critical for stretch-induced orienting response in fibroblasts. *J Cell Sci*. 1999;112:1365–1373.
120. Stewart MP, Helenius J, Toyoda Y, Ramanathan SP, Muller DJ, Hyman AA. Hydrostatic pressure and the actomyosin cortex drive mitotic cell rounding. *Nature*. 2011;469:226–230.
121. Lansman JB, Hallam TJ, Rink TJ. Single stretch-activated ion channels in vascular endothelial cells as mechanotransducers? *Nature*. 1987;325:811–813.
122. Tuszynski JA, Portet S, Dixon JM, Luxford C, Cantiello HF. Ionic wave propagation along actin filaments. *Biophys J*. 2004;86:1890–1903.
123. Aplin AE, Howe A, Alahari SK, Juliano RL. Signal transduction and signal modulation by cell adhesion receptors: the role of integrins, cadherins, immunoglobulin-cell adhesion molecules, and selectins. *Pharmacol Rev*. 1998;50:197–263.

124. Alonso JL, Essafi M, Xiong JP, Stehle T, Arnaout MA. Does the integrin *alphaA* domain act as a ligand for its *betaA* domain? *Curr Biol.* 2002;12:R320–R322.
125. Gallant ND, Michael KE, Garcia AJ. Cell adhesion strengthening: contributions of adhesive area, integrin binding, and focal adhesion assembly. *Mol Biol Cell.* 2005;16: 4329–4340.
126. Bershadsky AD, Balaban NQ, Geiger B. Adhesion-dependent cell mechanosensitivity. *Annu Rev Cell Dev Biol.* 2003;19:677–695.
127. Goldmann WH. Mechanotransduction in cells. *Cell Biol Int.* 2012;36:567–570.
128. Shemesh T, Geiger B, Bershadsky AD, Kozlov MM. Focal adhesions as mechanosensors: a physical mechanism. *Proc Natl Acad Sci USA.* 2005;102: 12383–12388.
129. Goldmann WH, Auernheimer V, Thievessen I, Fabry B. Vinculin, cell mechanics, and tumor cell invasion. *Cell Biol Int.* 2013;37:397–405.

CHAPTER FIVE

Mechanical Cues Direct Focal Adhesion Dynamics

Kristina Haase*, Zeinab Al-Rekabi*,†, Andrew E. Pelling*,‡,§

*Centre for Interdisciplinary NanoPhysics, Department of Physics, University of Ottawa, Ottawa, Ontario, Canada

†Department of Mechanical Engineering, University of Washington, Seattle, Washington, USA

‡Department of Biology, University of Ottawa, Ottawa, Ontario, Canada

§Institute for Science Society and Policy, University of Ottawa, Ottawa, Ontario, Canada

Contents

Abstract

Focal adhesions play a fundamental role in force sensing, which influences a variety of cellular processes and functions, particularly migration and the cell cycle. They consist of large macromolecular assemblies of proteins that associate with integrins, in order to serve as anchor points between the cell and the extracellular matrix. These dynamic regions act as a hub for sensing and transmission of mechanical cues between cells and their surrounding microenvironments. A number of techniques have been used to study focal adhesions, including optical microscopy, substrate micropatterning techniques, and tools which can directly manipulate cells, such as the atomic force microscope. Mechanical stimulation of cells leads to changes in cell contractility, stress fiber remodeling, and focal adhesion position and size; several of the responses explored in this chapter.

Progress in Molecular Biology and Translational Science, Volume 126
ISSN 1877-1173
http://dx.doi.org/10.1016/B978-0-12-394624-9.00005-1

1. INTRODUCTION

In vivo, cells are exposed to a variety of mechanical forces, which are now well known to direct the fate and function of cells.[1–5] Not only do cells sense and respond to these physical forces, but also they too exert forces on their surrounding microenvironments.[6–9] Force transmission between cells and the underlying extracellular matrix (ECM) occurs at highly localized junctions, known as focal adhesions (Fig. 5.1). These protein-rich regions transmit and sense mechanical cues arising from both the cell's interior and exterior via integrins.[11] Moreover, these anchor points act as dynamic adhesion sites on the underlying matrix, as seen during cell migration,[12] and tension-mediated structural changes.[13]

Due to their complex composition, understanding the functional mechanisms behind focal adhesions remains challenging. With the use of fluorescent proteins and optical microscopy, a large number of studies have used visual means to examine and quantify the size, spacing, and turnover kinetics of focal adhesions at high levels of spatial and temporal resolution.[5,6,14–16] In tandem, the study of cellular-generated traction forces has also led to a significant depth of understanding on how cells employ adhesion and respond to varying microenvironmental conditions.[6,8,9,17]

The shape of adherent cells is directly affected by focal adhesion size and density, which depends on both the biochemical and physical properties of the ECM. As well, extracellular forces, such as mechanical tension or compression transmitted via the ECM, are also "sensed" by focal adhesions.[18] The dynamics of focal adhesion development can be monitored by exposing cells to a variety of mechanical cues *in vitro*. Recently, researchers have employed atomic force microscopy (AFM) to apply precise forces to cells apically.[5,6] Methods such as these, in tandem with optical techniques, have provided further insight into traction force and focal adhesion dynamics, in response to mechanical cues.

This chapter provides an overview of focal adhesions as sensors of mechanical cues. First, we briefly describe the form and function of focal adhesions, while providing a background on adhesion dynamics, and subsequently, how they act as mechanical sensors. Second, we will provide a brief introduction on techniques used to study focal adhesions. Specifically, we will focus on how AFM is used in conjunction with optical microscopy techniques in a combined effort to study focal adhesions. A comprehensive review of two leading examples of these combined techniques will follow. Finally, we discuss future directions of research in this field.

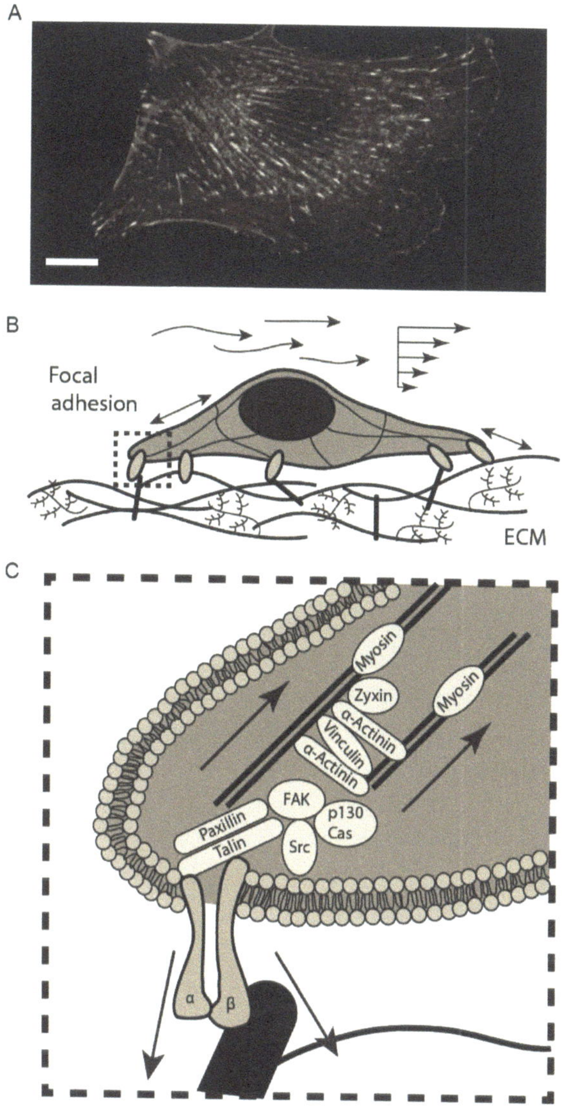

Figure 5.1 Form and function of focal adhesions. (A) Immunofluorescent image of a C2C12 myoblast cell cultured on a glass substrate coated with 0.1% gelatin. The cell was fixed and stained for DNA (blue), actin filaments (red) and a well-known focal adhesion protein, vinculin (green). Scale bar is 10 μm. (B) Schematic of a cell experiencing shear forces from fluid flow, and compression and tension between the cell's inner cytoskeleton and the underlying extracellular matrix. These forces are transmitted between the cell and ECM at focal adhesions. (C) Enlarged depiction of a focal adhesion. Integrin isoforms bind to talin and paxillin, linking them to the actin cytoskeleton. These proteins associate with focal adhesion kinase (FAK) and other adaptor proteins (Src and p130Cas). Actin-associated proteins such as α-actinin not only bind actin filaments together but also recruit other proteins, such as vinculin and zyxin, which are associated with mature focal adhesions. Tension generated in the contractile actomyosin network further activates stress sensitive proteins, such as zyxin. *(C) Adapted from Ref. 10.* (See the color plate.)

2. FORM AND FUNCTION OF FOCAL ADHESIONS

Proliferation,[19,20] migration,[7,21–23] differentiation,[1,24,25] and even apoptosis,[26,27] are dependent upon cellular adhesions to the ECM. These contacts, known as focal adhesions, are comprised of clusters of integrins linked to the actin cytoskeleton by associated proteins and by ligand–receptor connections to the ECM. Over the years, studies have provided valuable insight into the vast number of molecules associated with focal adhesions (Fig. 5.1C).[28,29] However, due to their complexity, how these macromolecular structures sense and respond to their surroundings are only partially understood. The physical interactions of adhesion sites are regulated by signaling and binding events[30–32]; however, mechanical cues generated either internally or from the substrate also influence their behavior and formation. In the following subsections, we describe how the ECM, integrins, and actin all contribute to the behavior and development of focal adhesions. Examples discussed herein use one of two main approaches of study: one method involves examining the molecular machinery, where the other involves changing the chemical and physical properties of the ECM, in an attempt to trigger a certain cellular response.[18]

2.1. Influence of the ECM

The form and function of focal adhesions are largely influenced by the composition of its underlying ECM. Focal adhesions have been observed to assemble/disassemble in response to variations in substrate elasticity,[15,25,27,31,33] geometry,[13,24,34,35] and ligand spacing,[21,36,37] thus, altering the form of the cytoskeleton and overall cell shape. Alterations in the cell's ability to sense these changes may be at the root of transformation into malignant phenotypes.[38] Indeed, transformed cells have been shown to generate weak traction forces,[9] and in some cases no longer require attachment to a substrate for growth.[27]

Systematic adjustment of substrate composition and geometry demonstrates the influence of the ECM on integrin-mediated adhesion. Researchers have used bioinert surfaces, such as polyethylene glycol (PEG), to test adhesion between certain cell–surface receptors and surface ligands.[39,40] Considering that the inclusion of ECM proteins, such as fibronectin, onto biomaterials can be complicated,[39] many studies employ biomimetic short peptides such as RGD (arginine–glycine–aspatate) onto these surfaces, as a means of studying known ECM signaling domains.[41] RGD, in particular, has been used to functionalize glass,[42] PEG,[39] and hyaluronic acid substrates,[43] and, more recently, supported lipid bilayers (formed by adsorption/fusion of small lipid vesicles).[44,45] These techniques used in conjunction with microcontact printing (with micrometer resolution) have shown that cell adhesion and motility are directly related to ligand spacing,[21] size, and arrangement.[46]

Although microcontact printing has been successful at controlling cell shape and viability,[46] protein clustering *in vivo* is on the order of 5–200 nm.[37,47,48] Block copolymer micelle nanolithography provides an improvement on ligand spacing resolution.[49] This method involves functionalization of metallic nanoparticles, usually gold, inserted at tunable spacing (10–200 nm). One study employed hexagonally spaced gold nanodots (~8 nm in size) coated with an RGDfk peptide.[47] This small size restricted binding to only one integrin per bead. Based on their findings, a universal spacing for integrin clustering was proposed to be in the range of 58–73 nm, since spacing >73 nm resulted in reduced cell spreading and adhesion, and <58 nm resulted in erratic cell growth and migratory behavior.[47] Nevertheless, cells have been shown to be sensitive to nanometer spaced integrin ligands.[36]

Besides the arrangement of adhesive ligands, cells also sense the rigidity of their substrates (Fig. 5.2). A prominent example is that of Engler *et al.* who demonstrated that nonmuscle myosin II was necessary for sensing

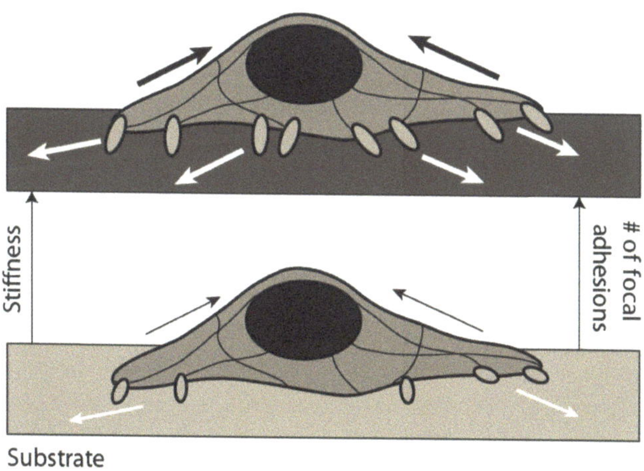

Figure 5.2 Mechanical cues, such as substrate stiffness, affect focal adhesion formation. Increasing substrate stiffness results in an increase in the number and size of focal adhesions. The reasoning behind this is that cells are able to pull with stronger traction forces on a less flexible substrate.

substrate stiffness, which was shown to direct the fate of stem cells.[1] When cultured on soft substrates mimicking brain tissue (0.1–1 kPa), mesenchymal stem cells differentiated into neurons, whereas when cultured on stiffer substrates mimicking muscle (8–17 kPa) or bone (25–40 kPa), they differentiated into myoblasts and osteoblasts, respectively.[1] Committed cells, however, are also influenced by substrate stiffness. For example, fibroblasts exposed to a stiffness gradient on polyacrylamide gels migrated from soft to stiff substrate regions.[25] Cells generated increased traction and cell spread area on their preferential stiff substrates. Moreover, when a strain was applied to the flexible substrate using a microneedle, the cells changed their polarization and migratory response, moving toward substrate tension and away from compression.[25] Application of controlled compression/tension transmitted through flexible substrates *in vitro* has also revealed that cells reorient themselves perpendicular to applied stretch.[50–52] It is clear from these examples that mechanical cues transmitted via the substrate influence adhesion, migration, along with cell polarization, and importantly cell fate.

Although substrate stiffness is important, geometric influences should also be considered. Geometric constraints on 2D substrates, limiting cell spreading and tension, have also been shown to direct the fate of stem cells.[13,24,34,35] More recently, cross-linked gels have been used to provide a 3D scaffold for cell culture. The differentiation of stem cells in these 3D environments appears to be controversial among researchers.[8,53] New evidence suggests that stem cell fate is regulated by different mechanisms that depend on the type of hydrogel employed, but that it may involve the degradation of ECM proteins in an effort to generate tension.[53]

Arrays of elastomeric pillars have also been used to examine cellular forces during adhesion. In a recent study,[33] microposts of varying stiffness and diameters were used to examine forces during the initial cell–ECM contact event. Micron-sized diameter posts resulted in global contractions toward the cell center, while submicron diameter pillars resulted in localized force balancing. The authors found that cells sensed substrate rigidity through periodic contractions. By adjusting their contractility, mouse embryonic fibroblasts maintained relatively stable deformations on the order of ~60 nm every minute. Moreover, the authors demonstrated that paxillin and phosphorylated myosin light chain are key components involved in this contractility, which occurs at a length scale of ~1 μm.[33] This particular example again demonstrates that cells respond to their substrate properties, by regulating cellular tension through focal adhesions.

2.2. Integrins are integral

As previously discussed, the ECM largely influences cellular adhesion; however, it does so by interacting with integrins (see Fig. 5.1B and C). Upon binding to the ECM, integrins form clusters resulting in the recruitment and grouping of scaffolding proteins, activation of tyrosine kinase and phosphatase signaling, and coupling of cell-generated forces.[31] The "integrin adhesome" (including its accessory proteins) are key regulators of physical phenomena such as cell motility, adhesion, matrix deformation, and remodeling.[29] Integrins also act as sensors, by responding to cues from the ECM, along with other signaling molecules such as kinases, phosphatases, and adaptor proteins.

Focal adhesion complexes are comprised of more than 150 proteins,[29] which all play a role in their formation and maintenance. Talins are particularly important, as they have been shown to bind β-integrins to the actin cytoskeleton (see Fig. 5.1C).[32] Interestingly, their recruitment is amplified by cytoskeletal tension. Some researchers hypothesize that the early formation of a nascent focal adhesion is reliant upon the interaction of two talin proteins connecting α and β integrins to actin.[54] Kindlins 2 and 3 are necessary along with talin to activate integrins.[55–58] Following this, many additional proteins must be recruited toward these sites in order to form the complex structures that comprise a fully mature focal adhesion site. Vinculin in particular, has been shown to play a role in the activation of integrin clustering following its attachment with talin.[59] Vinculin also attaches to actin, and so reinforces the focal adhesion site.

2.3. Formation of focal adhesions

Formation of nascent focal adhesions, or focal complexes, begins with force sensing. Protrusions at the cell edge, namely filopodia and lamellipodia, actively explore their surrounding microenvironment (Fig. 5.3). Filopodia are composed of actin bundles and are covered by a veil-like structure, known as lamellipodia, which are formed by Arp 2/3 dependent actin polymerization.[60] Nascent adhesion sites associated with these structures are ~100 nm in diameter, although there have been reports of smaller integrin-associated structures (~30–40 nm).[61,62] Focal adhesion formation occurs under these protruding structures, which is not surprising, since adhesion is strongly associated with motility.

Cells propel forward by actin polymerization in the direction of motion; however, there is an overall backward flow of actin, known as actin

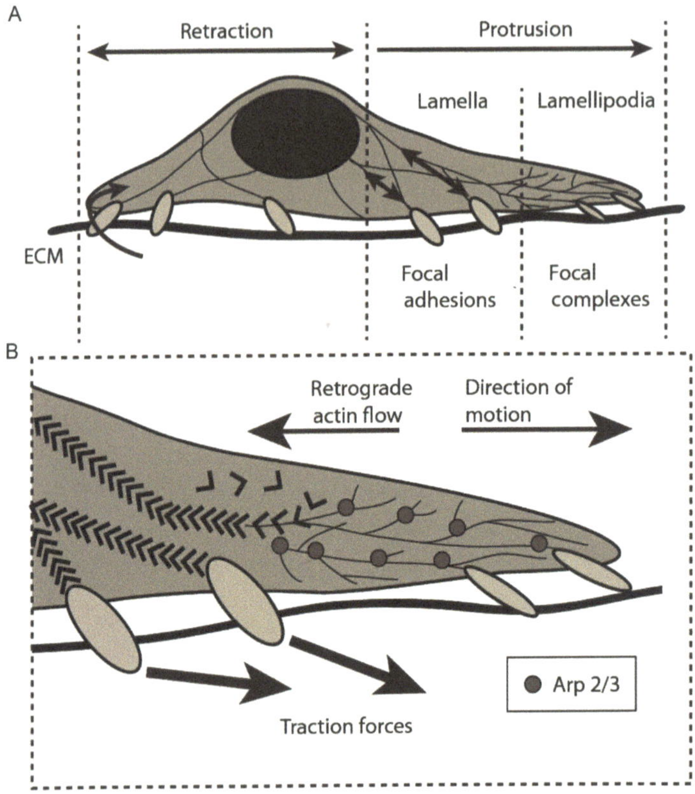

Figure 5.3 Cellular adhesion is a key process involved in motility. (A) Schematic representation of a motile cell demonstrating protrusion in the leading edge and retraction of the cell rear. Small nascent adhesions (also known as focal complexes) are formed in the lamellipodia, whereas mature focal adhesions form in the lamella. Tension generated in stress fibers is necessary for the promotion of focal complexes into mature focal adhesions. (B) Magnified view of the leading edge of a motile cell. Actin polymerization occurs in two forms: actin branching is promoted by Arp 2/3 which occurs in the lamellipodia, and actin polymerization of stress fibers occurs in the lamella. These regions are associated with the retrograde flow of actin, which occurs in the opposite direction to that of motion. Actin flow decreases dramatically upon the formation of a new focal adhesion (and so is slower in the lamella).

retrograde flow.[14] The rate of polymerization in the direction of motion is slower than that of retrograde flow, yet fibroblasts and epithelial cells are still able to migrate at rates on the order of a few millimeters per minute.[14,63] A closer look at the leading edge (Fig. 5.3B) reveals a decreasing gradient of actin, from the lamellipodia tip region to the lamella. The outer lamellipodia is a region of fast retrograde actin flow, whereas the inner lamella is a distinctly slower flow region. Alexandrova *et al.*[14] demonstrated that flow drastically slows, within seconds, following the formation of a new focal adhesion site. Through its integration with the actomyosin network, the lamella is what drives the cell forward.[63] Components of actin branching (Arp 2/3 complex and ADF/cofilin) are no longer present in the lamella; however, components of actin bundles are (topomyosin and myosin II).[63,64] Myosin II contractility is responsible for the retrograde flow of actin in the lamellar region. These contractile forces occur above stationary focal complexes, and so there is a transmission of force to these elements.[65] Retrograde flow provides the forces necessary to develop these nascent sites into fully mature focal adhesions.[18]

As cells spread or migrate, focal complexes transition into mature focal adhesions under lamellar regions, not at the cell periphery. Mature adhesions are usually identified by their size, as they appear elongated and visibly connected to actin fibers (see Fig. 5.1A). Well-known proteins, such as zyxin, can also be used for identification, as they appear only upon maturation.[12] Interestingly, nascent adhesions have been shown to generate stronger traction than mature adhesions.[12] Although this seems counterintuitive, strong forces are necessary only for the initiation of focal adhesion growth. In a recent study, high-resolution time lapse traction force microscopy (TFM) demonstrated that traction forces are asymmetric within a single mature focal adhesion.[15] The distribution of traction indicated a peak near the center, usually directed toward the cell front. TFM performed on a number of mature focal adhesions resulted in the observation of two distinct modes of traction: a stable mode and a tugging mode. Moreover, loss of tension, either by a reduction in substrate stiffness or by inhibition of Rho-dependent kinase (ROCK) resulted in increased numbers of focal adhesions presenting tugging traction.[15]

It is clear from the above examples, that without actin polymerization or myosin II contractility focal adhesions cannot form.[66] These highly localized growth patterns require a range of force magnitudes, which must be sensed by the focal complex. Potential mechanosensors include integrins, mechanosensitive ion channels, and stretch sensitive proteins, such as p130Cas.[67] Other known force-sensitive proteins include the Src family kinase,[68] focal adhesion kinase (FAK) (reviewed in Ref. 69), Src homology 2 (SH2) domain-containing phosphatase,[70] and receptor-like protein tyrosine phosphatases (reviewed in Ref. 71). FAK/Src signaling in particular has been shown to regulate turnover of adhesions at the leading edge.[72–75]

Interestingly, although forces at the leading edge are what drive the formation of nascent adhesions, forces at the rear of the cell also cause their degradation.[76–78] Traction forces exerted by the cell onto its substrate are likely responsible for triggering their disassembly. In addition, the Rho/ROCK pathways have been implicated in cell retraction.[75] Two types of disassembly have been proposed: weakening of the integrin–actin cytoskeleton bonds or of the integrin–ECM bonds. Interestingly, FAK, Src, and SH2, which promote phosphorylation and assembly of focal adhesions, may also play a role in their disassembly.[79] Moreover, differences in molecular compositions, specifically the activation of myosin IIA in the rear of the cell, may influence cell polarity and direct migration.[80] A comprehensive review on nonmuscle myosin II and its role in adhesion can be found in Ref. 81.

It is quite evident that the progression of growth, from nascent to mature adhesion sites, is reliant upon mechanical cues. It is these mechanical cues that initiate integrin clustering, protein recruitment, and conformational changes of proteins, such as talin, allowing for binding of vinculin, and the growth of these sites. Vinculin is also an important regulator of mechanical forces within adhesion sites (reviewed in Ref. 82). Considering the large number of possible mechanosensors, it is difficult to understand the exact mechanisms behind their concerted response to mechanical force. Wozniak *et al.* have provided a thorough overview of the various effects of different focal adhesion molecules on integrin activity, mechanosensing, cell turnover, migration, and proliferation.[83] Focal adhesions are certainly sensitive to forces—whether generated externally (e.g., from the ECM) or internally (actin-generated forces). Moreover, actin polymerization is known to play a major role in the formation and maintenance of these sites.

2.4. Cytoskeletal interplay

Although actin polymerization influences the formation and growth of focal adhesions, these sites also affect the growth of actin (see Fig. 5.4). In fact, actin nucleation has been shown to occur mainly at the ends of stress fibers associated with focal adhesion sites.[84] Stress fibers have been categorized into three major classes: ventral, dorsal, and transverse arcs (a nice overview can be found in Ref. 85). The most commonly observed are ventral stress fibers, which appear in the basal region of the cell, wherein bundles of actin filaments are joined end to end by focal adhesions.[28] Studies have shown that dorsal fibers form by elongation at focal adhesions, resulting in short

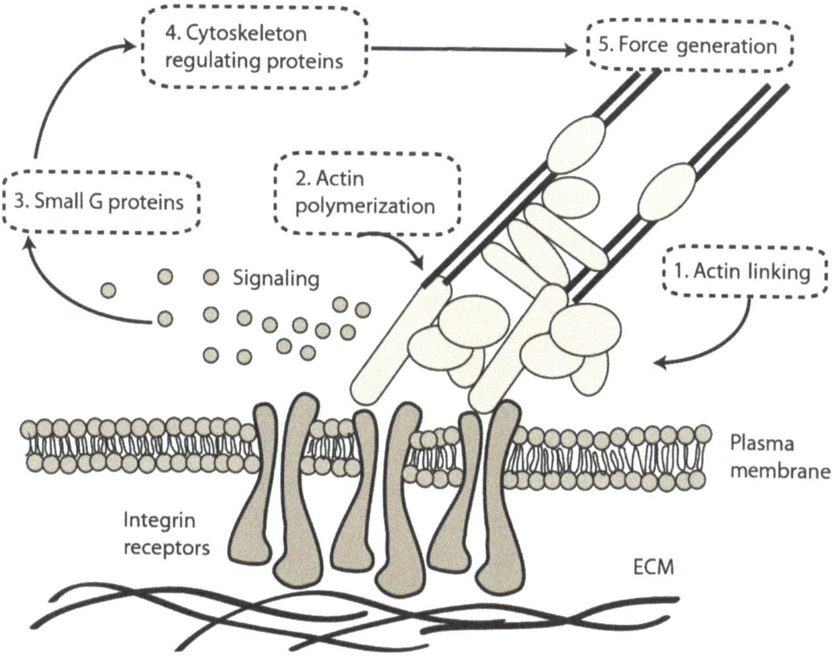

Figure 5.4 Both focal adhesions and actin regulate one another as part of a feedback loop. Integrins are transmembrane receptors, which cluster together and attach to various ECM ligands. These contacts attract actin-linking molecules and promote actin polymerization at the attachment site. However, the rate of polymerization is affected by the unified response of the entire focal adhesion system to interactions with the ECM (stiffness, composition, and ligand density all affect integrin adhesions). Actin polymerization further promotes signaling of GEFs and GTPase by FAK and Src, thus activating small G proteins (e.g., Rho and Rac). Subsequently, G proteins initiate cytoskeleton regulating proteins, and thus force generation of the actomyosin network. Furthermore, tension along stress fibers also activates mechanosensitive molecules, such as vinculin and zyxin. *Adapted from Ref. 18.* (See the color plate.)

filaments containing α-actinin, which then join to myosin clusters.[86] Ventral fibers then form by the connection of two dorsal fiber ends.[86] By culturing cells on specifically shaped micropatterned surfaces, Thery *et al.* have shown that stress fiber growth follows focal adhesion formation along the edges of these adhesive geometries.[87] Stress fibers are also associated with focal adhesion formation at the lamella/lamellipodia boundary.[14] Cells treated with Cytochalasin D, a well-known actin depolymerizer, results in a reduced actin flow rate, leading to inhibition of focal adhesion formation.[14] There is a clear interplay between tension in stress fibers, and focal adhesions, that is likely also influenced by signaling mechanisms.

The major regulators of the cytoskeleton are the small Rho family of GTPases: Rho and Rac. Rho is activated by guanine nucleotide exchange factors (GEFs), which catalyze the exchange of GDP for GTP.[88–91] Rac activation follows adhesion of the GEF known as the DOCK180–ELMO complex, which is activated by paxillin and p130CAS (known FA proteins).[18] Besides GEFs, integrin adhesion also negatively affects RhoA activation through GTPase-activating proteins (GAPs). FAKs are known to regulate the transduction of integrin signals to GEFs and GAPs through binding, phosphorylation, and activation alongside other proteins such as Src (reviewed in Refs. 10,69). It is clear that factors allowing for the transmission of integrin signals to GEFs and GAPs are important; however, many questions still surround their specificity and activation.[18]

Stress fibers are highly cross-linked structures and are associated with myosin II. These structures are under continuous tensile and compressive stresses and provide a constant source of mechanical force to mature focal adhesions. These stresses are transferred via integrins to the ECM. Cells have been shown to produce drastic forces upon their substrates, which has been shown *in vitro* by substrate wrinkling.[92] Moreover, as discussed earlier, traction within a particular focal adhesion can be modulated by ECM rigidity or actomyosin tension.[15] Cells appear to maintain a quasi-homeostatic tension with their substrate that was found to be dependent upon extracellular signal-related kinase and Rho.[38] These examples demonstrate the existence of a feedback mechanism associated with focal adhesions, wherein actomyosin tension is regulated in response to matrix stiffness.

Although we have only discussed actin, the role of microtubules is indispensible in the regulation of focal adhesions. In brief, microtubules have been shown to target and disassemble focal adhesions, a process dependent upon FAK (a known activator of adhesion turnover) and dynamin (a GTPase involved with endocytosis).[93] In this way, microtubules disrupt local adhesion components, likely resulting in local tension relief. Independently, activation of actomyosin contractility (using calyculin A),[94,95] or depolymerization of microtubules (using nocodazole) has been shown to result in increased cell spreading, morphological changes, and the formation of focal adhesions.[96,97] However, a recent study demonstrated that when employed consecutively, these treatments do not have an additive effect. In contrast, there were time dependent changes in cell cortical elasticity and long-term decreases in cell height and focal adhesion size.[97] Clearly, myosin II-dependent contractility and microtubules are intimately linked in regards to their influence on focal adhesion dynamics. Both microtubules and actin appear to be key players in the maintenance of focal adhesion turnover.

3. AFM AS A TOOL TO STIMULATE A CELLULAR RESPONSE

As we have discussed in the preceding section, mechanical forces are necessary for the initiation, maturation, and disassembly of focal adhesions. Integrin-mediated adhesion sites allow for the transmission of internally generated forces to the ECM and vice versa. Both actomyosin generated forces and those arising from the substrate influence one another and focal adhesion formation. Considering that mechanical forces are well known to influence cellular fate and function, recent decades have given rise to numerous studies employing a variety of tools and techniques to mechanically manipulate cells. Tools such as micropipette aspiration,[98] magnetic bead twisting cytometry,[99] optical stretchers,[100] optical traps,[101–103] and AFM,[104] are just a few of the tools used to induce a cellular response. Although each technique has its own advantages and disadvantages; here, we focus on AFM as a unique tool for examining the mechanisms behind cellular adhesion.

The AFM is a versatile tool for studying the cellular response to force. Besides its capacity for high-resolution imaging, AFM can be used to apply precise pico- to nano-Newton forces in order to measure both the elastic and viscoelastic properties of cells.[104–106] Depending on the shape and size of the cantilever probe, the cell can be subjected to either whole-cell or localized forces. Changes in Young's modulus can be measured by fitting force–indentation curves to modified versions of the Hertz contact model.[107] Alternatively, one can employ the cantilever in a raster-scanning mode across the sample to generate force maps. Force maps of cells can be used to visualize variations in cell height, adhesion strength, and to measure variations in elasticity across the cell.[6,105,108] Measurements of the elastic properties of cells can lend information to the physical changes of the cell, as they are exposed to various chemical or mechanical cues. Considering that actomyosin contractility is directly related to focal adhesion dynamics, elasticity measurements are both relevant and important.

In the following subsections, we focus on the discussion of two studies which examine the behavior of cells following direct application of mechanical force using AFM. The first study by Guolla *et al.*[5] looks at how stress fibers deform in response to force, and how their strain leads to actin and focal adhesion remodeling. The second study by Al-Rekabi *et al.*[6] examines changes in traction forces in response to an applied force and altered substrate rigidity. These examples demonstrate the efficacy of using the AFM as a tool to apply precise forces, and to measure cellular elastic properties, all while using visual methods to elucidate the mechanics and function of cells, with particular focus on stress fibers and focal adhesions here.

3.1. Cytoskeletal strain directs focal adhesion formation

In vitro application of mechanical forces applied directly to cells has resulted in direct stress fiber and focal adhesion remodeling, over both short and long durations.[109–111] Considering that tension plays a direct role in the formation and remodeling of focal adhesions,[15,66] measurements of deformation and strain of stress fibers may provide information relevant to the development of adhesion sites. In a recent study,[5] both lateral stress fiber deformations and longitudinal strains were measured following local cell perturbations using an AFM. In particular, nanoscale forces (0–20 nN) were applied to NIH3T3 fibroblasts transiently transfected with actin–EGFP (Fig. 5.5A). Laser scanning confocal microscopy (LSCM) was used to observe the deformation of actin fibers following several minutes of force application. Lateral displacement vectors were measured by tracking regularly spaced (~5 μm) points along the stress fibers over time (Fig. 5.5B).

Applying a force apically to cells resulted in nonisotropic deformations of stress fibers.[5] Although many stress fibers displayed no lateral motion; fibers that deformed did so proportionally to the amount load magnitude (Fig. 5.5C). While unperturbed cells also displayed stress fiber movements over time, an applied force resulted in a significant increase in displacements. For instance, stress fibers near the point of contact (within 5–10 μm) resulted in an average lateral motion bulging outwardly (~100–300 nm). This force-dependent displacement occurred within 60 s of force application and increased with increasing time. On the other hand, stress fibers distant from the point of loading (30–50 μm away) demonstrated a retraction of the cell edge toward the center of the cell, following 1 min of force application, indicative of actin remodeling (see Fig. 5.6).

Importantly, the remodeling and relaxation of the deformed stress fibers were shown to be directly correlated to focal adhesion movement. This was directly visualized by cells simultaneously expressing actin–EGFP and zyxin–mRFP (Fig. 5.7).[5] Zyxin, as was discussed in preceding sections, is a well-known focal adhesion protein that has been described as a tension sensor. Zyxin has been shown to appear with increasing tension, and

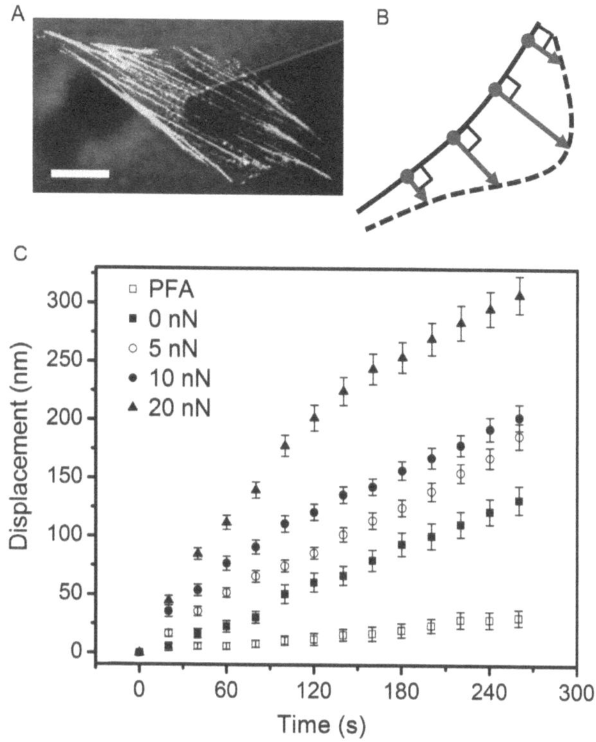

Figure 5.5 Lateral stress fiber deformations in response to an applied force. (A) AFM tip is used to apply a local force, centered over the nucleus, to NIH3T3 fibroblast expressing actin-EGFP. Scale bar is 15 μm. (B) Stress fiber deformations were tracked by mapping perpendicular displacement vectors (red) (gray in print) from an initial point on a filament at time zero (solid line) to a new point on the same filament at the next time step (dashed line). (C) Average displacement of stress fibers over time in response to locally applied forces. Stress fiber displacement is proportional to the applied force and occurs within 20 s. Displacement occurring in cells in the absence of force (0 nN) represents normal stress fiber remodeling dynamics. Displacements for cells treated with paraformaldehyde (PFA) can be attributed to microscope drift or error within the tracking method. Displacement is shown as mean + s.e.m. *Reproduced with permission from Ref. 5.*

conversely disappear with decreasing tension at both mature and nascent focal adhesion sites.[111] Guolla *et al.*[5] used kymographs to demonstrate that zyxin remained visibly linked with stationary stress fibers (in the presence of an applied force). Moreover, only large deformations of stress fibers resulted in the movement of focal adhesions. Consistent fluorescence intensity of zyxin–mRFP within the kymographs suggests that tension remains relatively stable among the stress fibers, as the concentration of this force-sensitive protein is neither increased nor decreased.

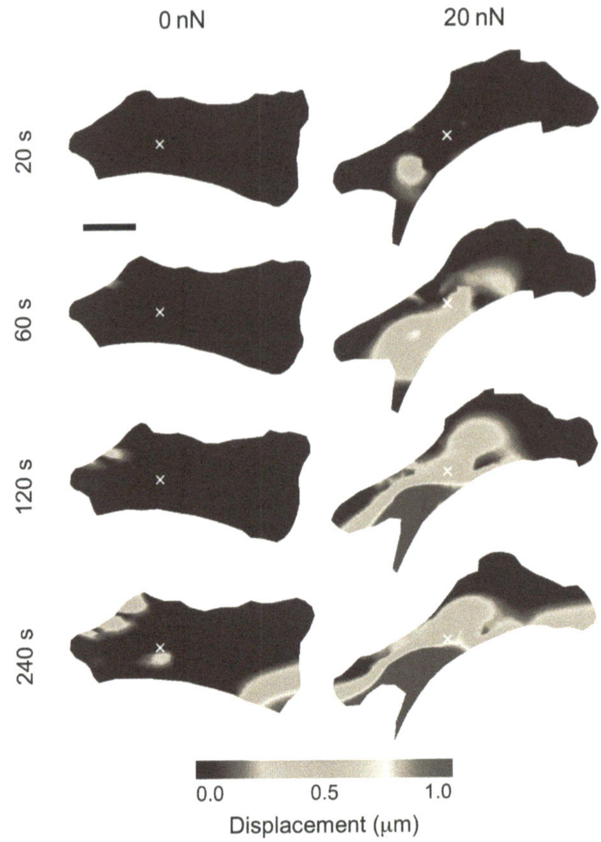

Figure 5.6 Spatial heat maps of stress fiber displacement in fibroblast cells experiencing 0 or 20 nN of applied force. The magnitude of the displacement vectors was plotted as spatial heat maps, demonstrating the increased motion of stress fibers in response to force. Stress fiber displacements occurred immediately (within 20 s) and near the point of force application (the contact point of the AFM cantilever is marked with a white "*x*" and is over the center of the nucleus). At later times, retraction sometimes takes place at distant cell edges (red areas at 120 and 240 s after the application of 20 nN). Scale bar: 15 μm. *Reproduced with permission from Ref. 5.* (See the color plate.)

Although lateral displacement vectors provided general information about stress fiber motion,[5] tension along the length of these fibers is what has been shown previously to contribute to remodeling of focal adhesion sites.[112] In order to measure strain, a regularly spaced (5 μm) grid was photobleached over images of the actin stress fibers. Deformations in the repeated bleached and nonbleached segments were measured over time, and measurable strains along the fibers. Basal measurements (0 nN load) showed a highly

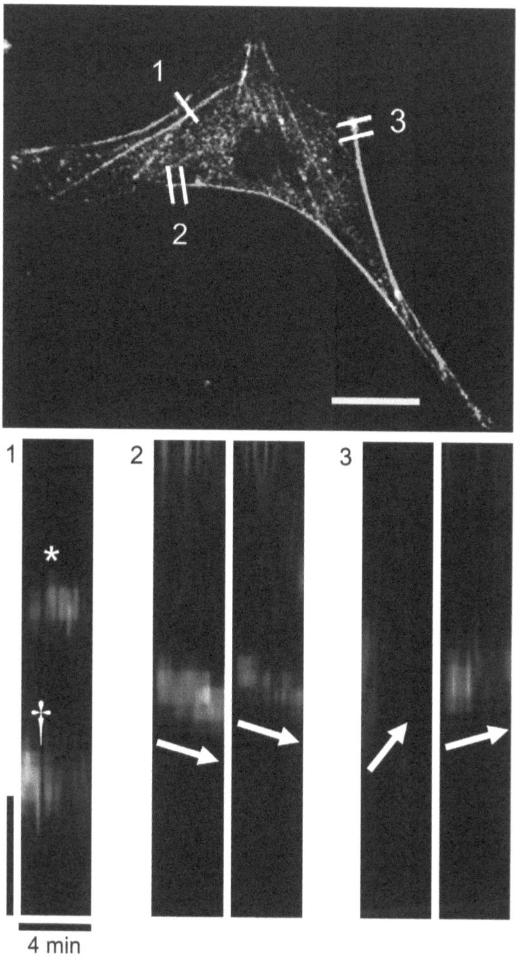

Figure 5.7 Correlated movement of focal adhesions and stress fibers in response to local forces. An LSCM image of an NIH3T3 cell expressing both actin–GFP and zyxin–mRFP (scale bar = 24 μm). Kymographs were generated for three separate locations on the cell (vertical scale bar = 2 μm). It can be seen in (1) that these fibers are sliding along their length (but not displacing laterally) which results in the movement of one focal adhesion into the image at ~120 s (*) and another out of the image at ~60 s (†) after force application. In (2) and (3), a correlated downward lateral displacement of F-actin filaments and focal adhesions are observed (upward and downward arrows). In (3), the F-actin filament begins to retract downward after force application, resulting in the movement of an associated focal adhesion. *Reproduced with permission from Ref. 5.* (See the color plate.)

heterogeneous distribution of tensile and compressive strains along the length of stress fibers (see Fig. 5.8). Moreover, upon force application (20 nN), strain variations increased by nearly 50% and resulted in more contractile strains. This was hypothesized to be the result of force-induced upregulation of myosin II.[5]

In order to examine their role in the transmission of force, the authors employed selective inhibitors of microtubule polymerization (nocodazole) and actin polymerization (Y27632—specifically inhibits ROCK). Treatment with nocodazole resulted in a loss of induced stress fiber remodeling and demonstrated the necessity of microtubules for any force transmission through the cell.[5] Alternately, treatment with Y27632 inhibited the formation of actin stress fibers and caused a reduction in the cell's elastic modulus (1 kPa, compared to 7 kPa for untreated). Despite the loss of intact stress fibers, forces were propagated throughout the cell, but were observed to be isotropic in nature, and noncontractile at the cell edge.[5]

The importance of this study is that it demonstrated heterogeneous displacement and strain among actin stress fibers, which was directly correlated to focal adhesion movements. Local forces applied to the cell were shown to result in immediate deformations within the vicinity of the load, and contractile movements occurring later on in more distal regions. These results clearly demonstrate that mechanical cues initiate long-term actin and focal adhesion remodeling.

3.2. Forces and substrate elasticity influence traction

Many studies, such as the example previously discussed,[5] have shown that mechanical cues influence cellular structure,[17,113,114] as well as their function (reviewed in Ref. 115). Moreover, both the mechanical and biochemical properties of the ECM are known to direct cell fate and function.[23,116,117] For example, cells actively probe their surrounding microenvironment, and can sense changes in matrix elasticity through a complex interplay of actomyosin dynamics and focal adhesion remodeling.[18] Through these processes, the cell applies dynamic traction forces on its underlying substrate.[118,119] We can naturally anticipate then, that both extracellular forces and the material properties of the cellular microenvironment should concurrently affect critical cellular processes, including the formation of focal adhesions and generation of traction forces.

In a recent study, Al-Rekabi and Pelling studied the response of cells to both external force and substrate elasticity using simultaneous AFM and

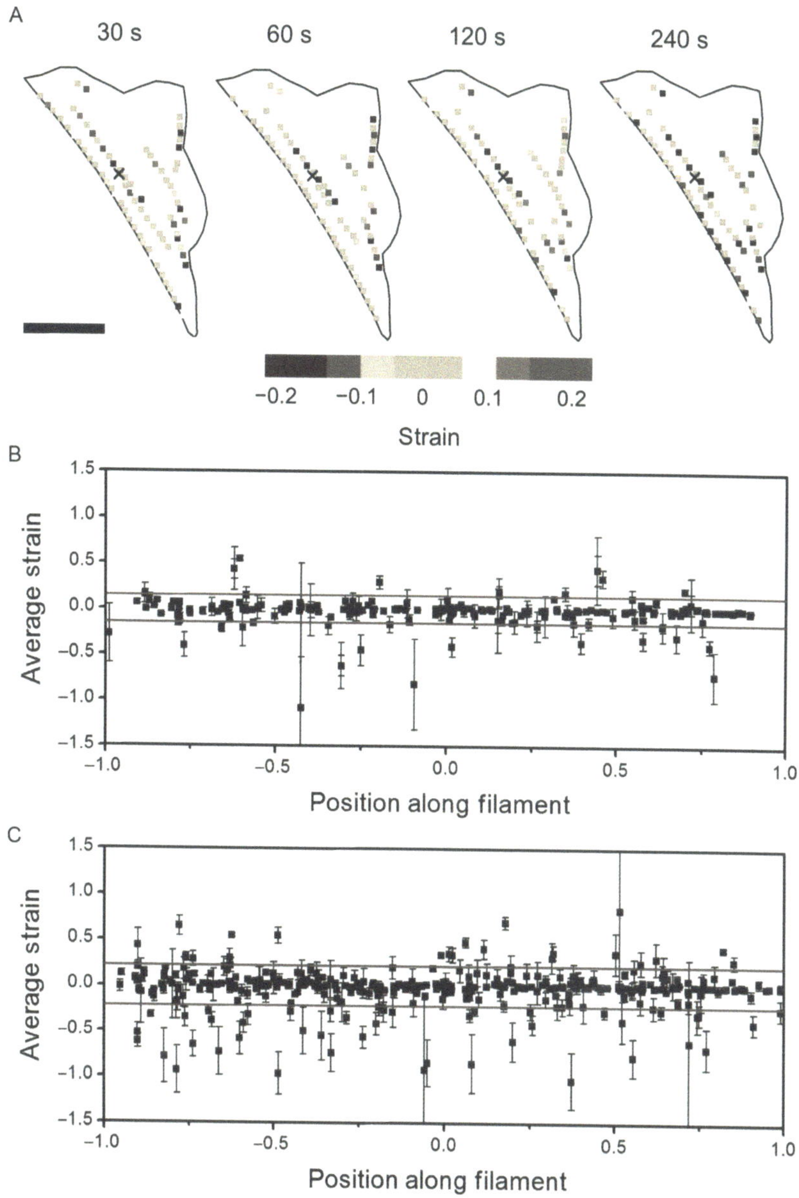

Figure 5.8 Heterogeneous strain occurs along the length of stress fibers. (A) A diagram of stress fiber strain for a cell undergoing mechanical stimulation (20 nN). Strain along each stress fiber is shown as a colored spot according to magnitude (the contact point of the AFM is marked with a black "*x*"). The strain magnitude and sign is heterogeneous along the length of any given filament and fluctuates in time, indicating rapid stretch/contraction dynamics. Scale bar: 20 μm. (B and C) Plots of normalized stress fiber strain as a function of position along the fiber after 240 s. The measured strain tended to fluctuate around zero. However, the standard deviation (red lines) for nonstimulated cells (B) is 0.15 and mechanically stimulated cells (C) is 0.22; a 50% increase. These strain fluctuations might reflect an increase in actomyosin activity in response to applied force. Error bars are s.e.m. *Reproduced with permission from Ref. 5.* (See the color plate.)

TFM on C2C12 mouse myoblasts.[6] Considering that muscles *in vivo* involve highly dynamic tensile and compressive forces and changes to the microenvironment, these muscle precursor cells are ideal for studying these two effects simultaneously. This is particularly true as they have been shown to provide traction forces while withstanding large substrate stretching and actin depolymerization.[120] The elasticity of muscle tissue has been shown to increase nearly eightfold during its transition from resting to active muscle tissue.[121–125] Muscle contractions promote changes in the microenvironment of myoblast cells, which are necessary for the formation of myotubes, a process known as myogenesis.[126–130] Numerous studies have shown that traction forces generated by C2C12 cells vary drastically and are dependent on the present stage of myogenesis.[121] Interestingly, the onset of myogenesis begins on stiffer substrates (~38–97 kPa) in muscle tissue.[121,122,129] Conversely, myotube formation has been shown to occur optimally on softer substrates (~12–39 kPa),[119] which can be linked to resting muscle tissue. With this in mind, the authors hypothesized that myoblast-generated traction forces would increase during the application of an externally applied force, however, only on substrates imitating active muscle tissue.

In order to test the hypothesis, C2C12s were grown on glutaraldehyde cross-linked gelatine (GXG) substrates of varying stiffness, ranging from soft to stiff (~16–89 kPa), mimicking resting and active tissue, respectively.[121–124] The stiffness of the substrates was increased by accumulative gelatine content, from 3% to 7% (w/v), and was verified by measuring force–indentation curves and fitting them to the modified Hertz model, as well as by AFM force-mapping in some cases. Force–indentation measurements of myoblasts revealed a relatively constant stiffness (~7 kPa) despite the cells having been cultured on substrates with increasing stiffness.[6]

Many researchers have used TFM techniques to quantify the level of force exerted by cells onto their substrates.[7–9,17] In order to perform TFM, fluorescent beads are embedded into a flexible, clear substrate, allowing for their movement to be tracked by optical microscopy. For the study by Al-Rekabi *et al.*, 200-nm red fluorescent beads were embedded within the cross-linked gelatine.[6] TFM requires the capture of two simultaneous images: a phase-contrast image of the cell, and a fluorescent image of the beads. The cross-linked gelatine substrates used in this particular study were ~500 μm thick and clear, allowing for both confocal and phase-contrast imaging (Fig. 5.9A). Images were captured every 30 s for 120 s using a CCD camera. Two methods of analysis were carried out:

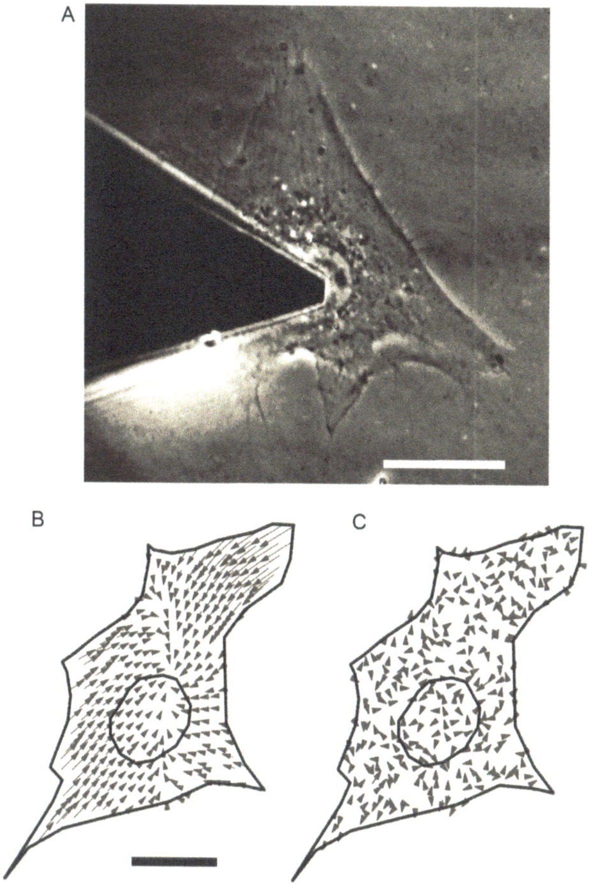

Figure 5.9 AFM and TFM measurements performed on C2C12 cells cultured on a tunable hydrogel. (A) A phase-contrast image of a C2C12 cell on a ~64 kPa glutaraldehyde-cross-linked gelatin substrate with an AFM tip pushing on the nucleus with a constant force (10 nN). (B and C) Two methods were used to measure the traction forces in this experiment: (B) absolute TFM reveals a smooth traction force field with force vectors pointing toward the cell center, indicative of cell contractility and (C) differential TFM illustrates that the change in traction forces between sequential times can be spatially heterogeneous. *Reproduced with permission from Ref. 6.*

absolute TFM and differential TFM. In the absolute method, bead displacements are tracked over time and then compared to the corresponding null image—wherein the cell is removed from the substrate (with EDTA in this case), thus relieving the substrate of elastic strain. In the differential method, TFM is first carried out on the cell prior to applying mechanical stimulation, and then while applying a continuous load. Thus, the differential method allows for measurement of the change in traction forces between sequential images.

The absolute method resulted in traction vectors directed toward the centroid of the cell (Fig. 5.9B), indicating an overall contractile behavior.[6] Differential TFM, on the other hand, resulted in varied traction vectors, revealing the dynamic temporal fluctuations of locally generated traction forces (Fig. 5.9C).

By stimulating the cells with a local force (10 nN applied via AFM), absolute TFM revealed that myoblast cells responded as hypothesized; traction forces increased dramatically, however, only on stiff substrates (Fig. 5.10A and B).[6] Absolute traction forces remained heightened for long durations (120 s) of applied force. This increase was shown to be transient in nature when analyzed with the differential method, as traction forces increased within 30 s, and then diminished to basal levels following 60 s (Fig. 5.10C and D). Interestingly, stimulation of cells with mechanical force appeared to have no significant influence on traction force generation when cells were grown on soft substrates (16–51 kPa GXG substrates), as visualized by traction force heat maps (Fig. 5.11).[6] These results demonstrate the complex interplay between substrate stiffness and external force on cellular traction. This work also highlights the different findings presented by absolute and differential TFM analysis methods.

In order to examine the cytoskeleton's role in traction force generation, the authors employed Y27632 to selectively inhibit ROCK, and blebbistatin to inhibit myosin II, separately.[6] Both treatments totally impaired the contractility of myoblasts, even after 240 s of applied force. The once elongated actin stress fibers of untreated cells became drastically reduced or altogether absent, due to these drug treatments, regardless of substrate. Moreover, the focal adhesions became smaller in size and less well defined.[6] Basal levels of traction, however, were unaltered, which is inconsistent with others.[7,17] It is possible that the response is cell type or dose dependent, or could differ due to the range of substrate stiffness employed, as others tend to implicate much softer (<10 kPa) or stiffer (glass) substrates. Considering that there was no increase in traction on stiff substrates either, this finding may suggest an inability of the cell to transmit the locally applied force throughout its cytoskeleton on soft substrates, which also resulted in a 20% decrease in focal adhesion size. Considering that traction force magnitude is directly related to size,[131,132] the authors speculated that focal adhesions must have a threshold below which they cannot resist contractility.

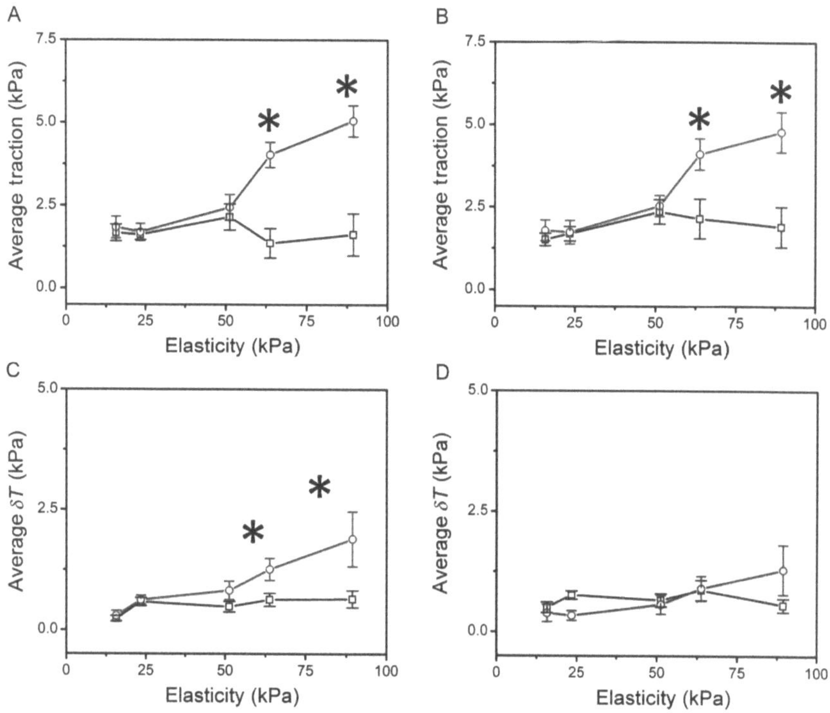

Figure 5.10 Absolute and differential traction force dynamics on substrates mimicking resting and active muscle tissue in response to a local force. (A) Average absolute traction as a function of substrate elasticity and force: control/0 nN (blue squares) (dark gray in print) and 10 nN (red circles) (light gray in print) after 30 s. C2C12s clearly show an increase in traction when subjected to a 10 nN force on substrates with an elasticity of ~64 kPa ($P = 0.003$) and ~89 kPa ($P = 0.004$). (B) Average absolute traction after 120 s (~64 kPa ($P = 0.019$) and ~89 kPa ($P = 0.004$)). (C) Differential traction reveals a statistically significant increase after 30 s of stimulation with a constant 10 nN force (in agreement with (A)) on ~64 kPa ($P = 0.032$) and ~89 kPa ($P = 0.022$) substrates: control (blue squares) (dark gray in print) and 10 nN (red circles) (gray in print). (D) The average change in traction forces dissipates to basal levels for 10 nN after 60 s. *Reproduced with permission from Ref. 6.*

In conclusion, this work demonstrated that mechanical cues (by AFM) and substrate material properties both affect the outcome of cellular traction forces of C2C12s, in a complex manner. In particular, myoblasts responded to mechanical force with a transient increase in traction forces, only apparent on substrates mimicking stiff tissue. Local traction hotspots (Fig. 5.11) appeared throughout the basal cell region, confirming that the localized force from the AFM tip results in remodeling far from the point of contact, as also seen in the previously described work in Section 3.1.[5] As well, both ROCK

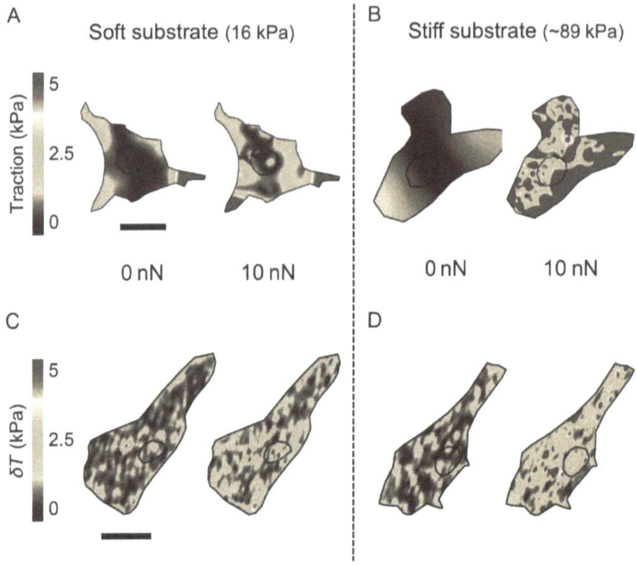

Figure 5.11 Heat maps for representative cells for the absolute and differential traction force measurements. (A) Absolute traction for a cell exposed to 0 and 10 nN of force on a soft (~16 kPa) substrate. (B) Absolute traction for a cell exposed to 0 and 10 nN of force on a stiff (~89 kPa) substrate. A significant increase in traction is observed alongside the appearance of local traction hotspots throughout the contact area. (C) Differential traction for a cell exposed to 0 and 10 nN of force on a soft (~16 kPa) substrate. (D) Differential traction for a cell exposed to 0 and 10 nN of force on a stiff (~89 kPa) substrate. A significant increase in traction fluctuations are observed as well as local hotspots throughout the contact area. These results demonstrate that both absolute and differential traction forces are profoundly affected when a mechanical force is applied to these cells, particularly on stiff substrates. *Reproduced with permission from Ref. 6.* (See the color plate.)

and myosin II were necessary mechanotransduction pathways for the initiation of increased traction force generation on stiff substrates. In the context of *in vivo* myogenesis, the notion stands that myotube formation occurs optimally on soft substrates,[125,133,134] yet follows primary initiation on stiff substrates.[135,136] Muscle contraction during stretching or exercise is known to produce many changes *in vivo* with regards to myogenesis.[136–139] The increase in traction forces on stiff substrates in response to applied force, as seen in this reviewed work, could implicate the combination of these two stimuli for the onset of myogenesis. Of course, the results are limited to the experimental conditions; however, it does highlight the necessity for research into the response of cells to multiple stimuli.

4. FUTURE DIRECTIONS

Although many focal adhesion-associated proteins have been identified over the years, the inner mechanics of this important cell feature remain poorly understood. It is apparent that both biochemical and physical mechanisms are at play and that both the cell and its microenvironment act in unison. The composition and mechanical properties of the ECM largely influence the formation and growth of focal adhesions. Rigidity sensing of focal complexes, along with other internal mechanical cues, play a large role in the formation and maturation of focal adhesions. Mechanical forces are at play during a number of processes including: conformational changes of proteins in order to expose hidden binding sites, actin-generated tension-initiating nascent adhesions, myosin II-dependent contractility promoting the maturation of adhesions, and recruitment of associated proteins. Moreover, examples discussed in this chapter demonstrate the existence of a feedback mechanism associated with focal adhesions, wherein actomyosin tension is regulated in response to matrix stiffness.

Through a detailed discussion of two recent studies, we have seen that the combined use of AFM and optical microscopy techniques, allows for the investigation of the mechanics behind focal adhesion formation and traction force generation. Local deformations on the apical regions of cells were shown to transfer through the cytoskeleton to basal cell regions, influencing stress fibers and focal adhesions. Mechanical forces increased stress fiber deformations, and initiated long-term remodeling of stress fibers, and focal adhesion sites, as was demonstrated visually.

Traction forces, on the other hand, provide a measurable indication of a mechanotransductive response. It was demonstrated herein that different modes of analysis can also provide different insight into cellular-generated traction forces. For instance, it was seen that absolute TFM lends to an overall picture of cellular behavior—one that is generally contractile in nature. On the other hand, differential TFM analysis provides insight into the temporal behavior of local regions, revealing traction force hot spots.

Moreover, it was seen that when a cell is exposed to multiple forms of stimuli, the response can be complex. In the preceding example, it was shown that external mechanical stimulation affects traction alongside substrate elasticity. In particular, myoblasts were only sensitive to the applied forces on stiff substrates, reacting with a transient increase in generated traction.[6] Although studying a single effect, such as substrate elasticity is useful in providing certain insight—it does not provide the whole picture, as clearly multiple mechanical influences simultaneously affect cellular adhesion *in vivo*. Future work will be necessary to expose the inner complexities of focal adhesion formation and remodeling. It will be necessary to involve a multitude of stimuli, in order to understand the complex interplay of the response mechanisms associated with focal adhesions.

ACKNOWLEDGMENTS

Z. A. R. was supported by the uOttawa NSERC CREATE program in Quantitative Biomedicine. A.E.P. gratefully acknowledges generous support from the Canada Research Chairs (CRC) program. This work was supported by a Natural Sciences and Engineering Research Council (NSERC) Discovery Grant, an NSERC Discovery Accelerator Supplement, a CRC Operating Grant and the Canadian Foundation for Innovation Leaders Opportunity Fund.

REFERENCES

1. Engler AJ, Sen S, Sweeney HL, Discher DE. Matrix elasticity directs stem cell lineage specification. *Cell.* 2006;126(4):677–689.
2. Buxboim A, Ivanovska IL, Discher DE. Matrix elasticity, cytoskeletal forces and physics of the nucleus: how deeply do cells "feel" outside and in? *J Cell Sci.* 2010;123(3):297–308.
3. Janmey PA, Miller RT. Mechanisms of mechanical signaling in development and disease. *J Cell Sci.* 2011;124(1):9–18.
4. Janmey PA, Weitz DA. Dealing with mechanics: mechanisms of force transduction in cells. *Trends Biochem Sci.* 2004;29(7):364–370.
5. Guolla L, Bertrand M, Haase K, Pelling AE. Force transduction and strain dynamics in actin stress fibres in response to nanonewton forces. *J Cell Sci.* 2012;125(3):603–613.
6. Al-Rekabi Z, Pelling AE. Cross talk between matrix elasticity and mechanical force regulates myoblast traction dynamics. *Phys Biol.* 2013;10(6):066003.
7. Gardel ML, Sabass B, Ji L, Danuser G, Schwarz US, Waterman CM. Traction stress in focal adhesions correlates biphasically with actin retrograde flow speed. *J Cell Biol.* 2008;183(6):999–1005.
8. Huebsch N, Arany PR, Mao AS, et al. Harnessing traction-mediated manipulation of the cell/matrix interface to control stem-cell fate. *Nat Mater.* 2010;9(6):518–526.
9. Munevar S, Wang YL, Dembo M. Traction force microscopy of migrating normal and H-ras transformed 3T3 fibroblasts. *Biophys J.* 2001;80(4):1744–1757.
10. Mitra SK, Hanson DA, Schlaepfer DD. Focal adhesion kinase: in command and control of cell motility. *Nat Rev Mol Cell Biol.* 2005;6(1):56–68.
11. Geiger B, Bershadsky A, Pankov R, Yamada KM. Transmembrane extracellular matrix-cytoskeleton crosstalk. *Nat Rev Mol Cell Biol.* 2001;2(11):793–805.
12. Beningo KA, Dembo M, Kaverina I, Small JV, Wang YL. Nascent focal adhesions are responsible for the generation of strong propulsive forces in migrating fibroblasts. *J Cell Biol.* 2001;153(4):881–887.
13. McBeath R, Pirone DM, Nelson CM, Bhadriraju K, Chen CS. Cell shape, cytoskeletal tension, and RhoA regulate stem cell lineage commitment. *Dev Cell.* 2004;6(4):483–495.
14. Alexandrova AY, Arnold K, Schaub S, et al. Comparative dynamics of retrograde actin flow and focal adhesions: formation of nascent adhesions triggers transition from fast to slow flow. *PLoS ONE.* 2008;3(9):e3234. http://dx.doi.org/10.1371/journal.pone.0003234.

15. Plotnikov SV, Pasapera AM, Sabass B, Waterman CM. Force fluctuations within focal adhesions mediate ECM-rigidity sensing to guide directed cell migration. *Cell.* 2012;151(7):1513–1527.
16. Wolfenson H, Bershadsky A, Henis YI, Geiger B. Actomyosin-generated tension controls the molecular kinetics of focal adhesions. *J Cell Sci.* 2011;124(9):1425–1432.
17. Beningo KA, Hamao K, Dembo M, Wang YL, Hosoya H. Traction forces of fibroblasts are regulated by the Rho-dependent kinase but not by the myosin light chain kinase. *Arch Biochem Biophys.* 2006;456(2):224–231.
18. Geiger B, Spatz JP, Bershadsky AD. Environmental sensing through focal adhesions. *Nat Rev Mol Cell Biol.* 2009;10(1):21–33.
19. Provenzano PP, Keely PJ. Mechanical signaling through the cytoskeleton regulates cell proliferation by coordinated focal adhesion and Rho GTPase signaling. *J Cell Sci.* 2011;124(8):1195–1205.
20. Ulrich TA, Pardo EMD, Kumar S. The mechanical rigidity of the extracellular matrix regulates the structure, motility, and proliferation of glioma cells. *Cancer Res.* 2009;69(10):4167–4174.
21. Maheshwari G, Brown G, Lauffenburger DA, Wells A, Griffith LG. Cell adhesion and motility depend on nanoscale RGD clustering. *J Cell Sci.* 2000;113(10):1677–1686.
22. Small JV, Geiger B, Kaverina I, Bershadsky A. How do microtubules guide migrating cells? *Nat Rev Mol Cell Biol.* 2002;3(12):957–964.
23. Hadjipanayi E, Mudera V, Brown RA. Guiding cell migration in 3D: a collagen matrix with graded directional stiffness. *Cell Motil Cytoskeleton.* 2009;66(3):121–128.
24. Guvendiren M, Burdick JA. The control of stem cell morphology and differentiation by hydrogel surface wrinkles. *Biomaterials.* 2010;31(25):6511–6518.
25. Lo CM, Wang HB, Dembo M, Wang YL. Cell movement is guided by the rigidity of the substrate. *Biophys J.* 2000;79(1):144–152.
26. Wang HB, Wang YL. Substrate flexibility regulates growth and apoptosis of normal but not h-ras-transformed cells. *Mol Biol Cell.* 1999;10:66a.
27. Wang HB, Dembo M, Wang YL. Substrate flexibility regulates growth and apoptosis of normal but not transformed cells. *Am J Physiol Cell Physiol.* 2000;279(5):C1345–C1350.
28. Burridge K, Fath K, Kelly T, Nuckolls G, Turner C. Focal adhesions—transmembrane junctions between the extracellular-matrix and the cytoskeleton. *Annu Rev Cell Biol.* 1988;4:487–525.
29. Zaidel-Bar R, Itzkovitz S, Ma'ayan A, Iyengar R, Geiger B. Functional atlas of the integrin adhesome. *Nat Cell Biol.* 2007;9(8):858–868.
30. Tilghman RW, Parsons JT. Focal adhesion kinase as a regulator of cell tension in the progression of cancer. *Semin Cancer Biol.* 2008;18(1):45–52.
31. Giannone G, Sheetz MP. Substrate rigidity and force define form through tyrosine phosphatase and kinase pathways. *Trends Cell Biol.* 2006;16(4):213–223.
32. Nayal A, Webb DJ, Horwitz AF. Talin: an emerging focal point of adhesion dynamics. *Curr Opin Cell Biol.* 2004;16(1):94–98.
33. Ghassemi S, Meacci G, Liu SM, et al. Cells test substrate rigidity by local contractions on submicrometer pillars. *Proc Natl Acad Sci USA.* 2012;109(14):5328–5333.
34. Kilian KA, Bugarija B, Lahn BT, Mrksich M. Geometric cues for directing the differentiation of mesenchymal stem cells. *Proc Natl Acad Sci USA.* 2010;107(11):4872–4877.
35. Ruiz SA, Chen CS. Emergence of patterned stem cell differentiation within multicellular structures. *Stem Cells.* 2008;26(11):2921–2927.
36. Le Saux G, Magenau A, Gunaratnam K, et al. Spacing of integrin ligands influences signal transduction in endothelial cells. *Biophys J.* 2011;101(4):764–773.
37. Massia SP, Hubbell JA. An Rgd spacing of 440 nm is sufficient for integrin alpha-V-beta-3-mediated fibroblast spreading and 140 nm for focal contact and stress fiber formation. *J Cell Biol.* 1991;114(5):1089–1100.

38. Paszek MJ, Zahir N, Johnson KR, et al. Tensional homeostasis and the malignant phenotype. *Cancer Cell.* 2005;8(3):241–254.
39. Benoit DSW, Anseth KS. The effect on osteoblast function of colocalized RGD and PHSRN epitopes on PEG surfaces. *Biomaterials.* 2005;26(25):5209–5220.
40. Kim P, Kim DH, Kim B, et al. Fabrication of nanostructures of polyethylene glycol for applications to protein adsorption and cell adhesion. *Nanotechnology.* 2005;16(10): 2420–2426.
41. Hersel U, Dahmen C, Kessler H. RGD modified polymers: biomaterials for stimulated cell adhesion and beyond. *Biomaterials.* 2003;24(24):4385–4415.
42. Tocce EJ, Broderick AH, Murphy KC, et al. Functionalization of reactive polymer multilayers with RGD and an antifouling motif: RGD density provides control over human corneal epithelial cell–substrate interactions. *J Biomed Mater Res A.* 2012;100A(1):84–93.
43. Shu XZ, Ghosh K, Liu YC, et al. Attachment and spreading of fibroblasts on an RGD peptide-modified injectable hyaluronan hydrogel. *J Biomed Mater Res A.* 2004;68A(2): 365–375.
44. Ananthanarayanan B, Little L, Schaffer DV, Healy KE, Tirrell M. Neural stem cell adhesion and proliferation on phospholipid bilayers functionalized with RGD peptides. *Biomaterials.* 2010;31(33):8706–8715.
45. Yu CH, Law JBK, Suryana M, Low HY, Sheetz MP. Early integrin binding to Arg-Gly-Asp peptide activates actin polymerization and contractile movement that stimulates outward translocation. *Proc Natl Acad Sci USA.* 2011;108(51):20585–20590.
46. Chen CS, Mrksich M, Huang S, Whitesides GM, Ingber DE. Geometric control of cell life and death. *Science.* 1997;276(5317):1425–1428.
47. Arnold M, Cavalcanti-Adam EA, Glass R, et al. Activation of integrin function by nanopatterned adhesive interfaces. *Chemphyschem.* 2004;5(3):383–388.
48. Jiang FZ, Horber H, Howard J, Muller DJ. Assembly of collagen into microribbons: effects of pH and electrolytes. *J Struct Biol.* 2004;148(3):268–278.
49. Deng C, Chen XS, Yu HJ, Sun J, Lu TC, Jing XB. A biodegradable triblock copolymer poly(ethylene glycol)-b-poly(L-lactide)-b-poly(L-lysine): synthesis, self-assembly and RGD peptide modification. *Polymer.* 2007;48(1):139–149.
50. Huang YL, Nguyen NT. A polymeric cell stretching device for real-time imaging with optical microscopy. *Biomed Microdevices.* 2013;15(6):1043–1054.
51. Tremblay D, Chagnon-Lessard S, Mirzaei M, Pelling AE, Godin M. A microscale anisotropic biaxial cell stretching device for applications in mechanobiology. *Biotechnol Lett.* 2013;16:16.
52. Wang D, Xie YY, Yuan B, Xu J, Gong PY, Jiang XY. A stretching device for imaging real-time molecular dynamics of live cells adhering to elastic membranes on inverted microscopes during the entire process of the stretch. *Integr Biol (UK).* 2010;2(5–6):288–293.
53. Khetan S, Guvendiren M, Legant WR, Cohen DM, Chen CS, Burdick JA. Degradation-mediated cellular traction directs stem cell fate in covalently crosslinked three-dimensional hydrogels. *Nat Mater.* 2013;12(5):458–465.
54. Giannone GP, Jiang G, Critchley D, Sheetz MP. Talin-1 is a critical component in early force-dependent reinforcement of integrin-cytoskeleton connections. *Mol Biol Cell.* 2002;13:350a.
55. Montanez E, Ussar S, Schifferer M, et al. Kindlin-2 controls bidirectional signaling of integrins. *Gene Dev.* 2008;22(10):1325–1330.
56. Moser M, Nieswandt B, Ussar S, Pozgajova M, Fassler R. Kindlin-3 is essential for integrin activation and platelet aggregation. *Nat Med.* 2008;14(3):325–330.
57. Ma YQ, Qin J, Wu CY, Plow EF. Kindlin-2 (Mig-2): a co-activator of beta(3) integrins. *J Cell Biol.* 2008;181(3):439–446.

58. Ye F, Petrich BG, Anekal P, et al. The mechanism of kindlin-mediated activation of integrin alpha IIb beta 3. *Curr Biol.* 2013;23(22):2288–2295.
59. Humphries JD, Wang P, Streuli C, Geiger B, Humphries MJ, Ballestrem C. Vinculin controls focal adhesion formation by direct interactions with talin and actin. *J Cell Biol.* 2007;179(5):1043–1057.
60. Small JV, Stradal T, Vignal E, Rottner K. The lamellipodium: where motility begins. *Trends Cell Biol.* 2002;12(3):112–120.
61. Shroff H, Galbraith CG, Galbraith JA, Betzig E. Live-cell photoactivated localization microscopy of nanoscale adhesion dynamics. *Nat Methods.* 2008;5(5):417–423.
62. Shroff H, Galbraith CG, Galbraith JA, et al. Dual-color superresolution imaging of genetically expressed probes within individual adhesion complexes (vol 104, pg 20308, 2007). *Proc Natl Acad Sci USA.* 2008;105(39):15220.
63. Ponti A, Machacek M, Gupton SL, Waterman-Storer CM, Danuser G. Two distinct actin networks drive the protrusion of migrating cells. *Science.* 2004;305(5691): 1782–1786.
64. Vallotton P, Danuser G, Bohnet S, Meister JJ, Verkhovsky AB. Tracking retrograde flow in keratocytes: news from the front. *Mol Biol Cell.* 2005;16(3):1223–1231.
65. Guo WH, Wang YL. Retrograde fluxes of focal adhesion proteins in response to cell migration and mechanical signals. *Mol Biol Cell.* 2007;18(11):4519–4527.
66. Giannone G, Dubin-Thaler BJ, Rossier O, et al. Lamellipodial actin mechanically links myosin activity with adhesion-site formation. *Cell.* 2007;128(3):561–575.
67. Geiger B. A role for p130Cas in mechanotransduction. *Cell.* 2006;127(5):879–881.
68. Sawada Y, Tamada M, Dubin-Thaler BJ, et al. Force sensing by mechanical extension of the Src family kinase substrate p130Cas. *Cell.* 2006;127(5):1015–1026.
69. Parsons JT. Focal adhesion kinase: the first ten years. *J Cell Sci.* 2003;116(8): 1409–1416.
70. von Wichert G, Haimovich B, Feng GS, Sheetz MP. Force-dependent integrin-cytoskeleton linkage formation requires downregulation of focal complex dynamics by Shp2. *EMBO J.* 2003;22(19):5023–5035.
71. Stoker AW. Protein tyrosine phosphatases and signalling. *J Endocrinol.* 2005;185(1): 19–33.
72. Fincham VJ, Frame MC. The catalytic activity of Src is dispensable for translocation to focal adhesions but controls the turnover of these structures during cell motility. *EMBO J.* 1998;17(1):81–92.
73. Schober M, Raghavan S, Nikolova M, et al. Focal adhesion kinase modulates tension signaling to control actin and focal adhesion dynamics. *J Cell Biol.* 2007;176(5): 667–680.
74. Hsia DA, Mitra SK, Hauck CR, et al. Differential regulation of cell motility and invasion by FAK. *J Cell Biol.* 2003;160(5):753–767.
75. Webb DJ, Donais K, Whitmore LA, et al. FAK-Src signalling through paxillin, ERK and MLCK regulates adhesion disassembly. *Nat Cell Biol.* 2004;6(2):154.
76. Parsons JT, Horwitz AR, Schwartz MA. Cell adhesion: integrating cytoskeletal dynamics and cellular tension. *Nat Rev Mol Cell Biol.* 2010;11(9):633–643.
77. Broussard JA, Webb DJ, Kaverina I. Asymmetric focal adhesion disassembly in motile cells. *Curr Opin Cell Biol.* 2008;20(1):85–90.
78. Webb DJ, Parsons JT, Horwitz AF. Adhesion assembly, disassembly and turnover in migrating cells—over and over and over again. *Nat Cell Biol.* 2002;4(4):E97–E100.
79. Huttenlocher A, Horwitz AR. Integrins in cell migration. *Cold Spring Harb Perspect Biol.* 2011;3(9). a005074. http://dx.doi.org/10.1101/cshperspect.a005074.
80. Vicente-Manzanares M, Koach MA, Whitmore L, Lamers ML, Horwitz AF. Segregation and activation of myosin IIB creates a rear in migrating cells. *J Cell Biol.* 2008; 183(3):543–554.

81. Vicente-Manzanares M, Ma XF, Adelstein RS, Horwitz AR. Non-muscle myosin II takes centre stage in cell adhesion and migration. *Nat Rev Mol Cell Biol.* 2009;10(11): 778–790.
82. Mierke CT. The role of vinculin in the regulation of the mechanical properties of cells. *Cell Biochem Biophys.* 2009;53(3):115–126.
83. Wozniak MA, Modzelewska K, Kwong L, Keely PJ. Focal adhesion regulation of cell behavior. *BBA Mol Cell Res.* 2004;1692(2–3):103–119.
84. Endlich N, Otey CA, Kriz W, Endlich K. Movement of stress fibers away from focal adhesions identifies focal adhesions as sites of stress fiber assembly in stationary cells. *Cell Motil Cytoskeleton.* 2007;64(12):966–976.
85. Pellegrin S, Mellor H. Actin stress fibres. *J Cell Sci.* 2007;120(20):3491–3499.
86. Hotulainen P, Lappalainen P. Stress fibers are generated by two distinct actin assembly mechanisms in motile cells. *J Cell Biol.* 2006;173(3):383–394.
87. Thery M, Pepin A, Dressaire E, Chen Y, Bornens M. Cell distribution of stress fibres in response to the geometry of the adhesive environment. *Cell Motil Cytoskeleton.* 2006;63(6):341–355.
88. Tapon N, Hall A. Rho, Rac and Cdc42 GTPases regulate the organization of the actin cytoskeleton. *Curr Opin Cell Biol.* 1997;9(1):86–92.
89. Iwaki N, Karatsu K, Miyamoto M. Role of guanine nucleotide exchange factors for Rho family GTPases in the regulation of cell morphology and actin cytoskeleton in fission yeast. *Biochem Biophys Res Commun.* 2003;312(2):414–420.
90. Begum R, Nur-E-Kamal MSA, Zaman MA. The role of Rho GTPases in the regulation of the rearrangement of actin cytoskeleton and cell movement. *Exp Mol Med.* 2004;36(4):358–366.
91. Sit ST, Manser E. Rho GTPases and their role in organizing the actin cytoskeleton. *J Cell Sci.* 2011;124(5):679–683.
92. Lee S, Zeiger A, Maloney JM, Kotecki M, Van Vliet KJ, Herman IM. Pericyte actomyosin-mediated contraction at the cell–material interface can modulate the microvascular niche. *J Phys-Condens Matter.* 2010;22(19):11. http://dx.doi.org/10.1088/0953-8984/22/19/194115.
93. Ezratty EJ, Partridge MA, Gundersen GG. Microtubule-induced focal adhesion disassembly is mediated by dynamin and focal adhesion kinase. *Nat Cell Biol.* 2005;7(6):U515–U581.
94. Kunda P, Pelling AE, Liu T, Baum B. Moesin controls cortical rigidity, cell rounding, and spindle morphogenesis during mitosis. *Curr Biol.* 2008;18(2):91–101.
95. Chartier L, Rankin LL, Allen RE, et al. Calyculin-a Increases the Level of Protein-Phosphorylation and Changes the Shape of 3t3 Fibroblasts. *Cell Motil Cytoskeleton.* 1991;18(1):26–40.
96. Rape A, Guo WH, Wang YL. Microtubule depolymerization induces traction force increase through two distinct pathways. *J Cell Sci.* 2011;124(24):4233–4240.
97. Al-Rekabi Z, Haase K, Pelling AE. Microtubules mediate changes in membrane cortical elasticity during contractile activation. *Exp Cell Res.* 10 March 2014;322(1):21–29.
98. Hochmuth RM. Micropipette aspiration of living cells. *J Biomech.* 2000;33(1):15–22.
99. Wang N, Ingber DE. Probing transmembrane mechanical coupling and cytomechanics using magnetic twisting cytometry. *Biochem Cell Biol.* 1995;73(7–8):327–335.
100. Guck J, Ananthakrishnan R, Mahmood H, Moon TJ, Cunningham CC, Kas J. The optical stretcher: a novel laser tool to micromanipulate cells. *Biophys J.* 2001;81(2): 767–784.
101. Laurent VM, Henon S, Planus E, et al. Assessment of mechanical properties of adherent living cells by bead micromanipulation: comparison of magnetic twisting cytometry vs optical tweezers. *J Biomech Eng T ASME.* 2002;124(4):408–421.

102. Brown TD. Techniques for mechanical stimulation of cells in vitro: a review. *J Biomech.* 2000;33(1):3–14.
103. Ashkin A, Dziedzic JM, Yamane T. Optical trapping and manipulation of single cells using infrared-laser beams. *Nature.* 1987;330(6150):769–771.
104. Haupt BJ, Pelling AE, Horton MA. Integrated confocal and scanning probe microscopy for biomedical research. *Sci World J.* 2006;6:1609–1618.
105. Haase K, Pelling AE. Resiliency of the plasma membrane and actin cortex to large-scale deformation. *Cytoskeleton.* 2013;70(9):494–514.
106. Radmacher M, Fritz M, Kacher CM, Cleveland JP, Hansma PK. Measuring the viscoelastic properties of human platelets with the atomic force microscope. *Biophys J.* 1996;70(1):556–567.
107. Radmacher M. Measuring the elastic properties of biological samples with the AFM. *IEEE Eng Med Biol.* 1997;16(2):47–57.
108. Haga H, Sasaki S, Kawabata K, Ito E, Ushiki T, Sambongi T. Elasticity mapping of living fibroblasts by AFM and immunofluorescence observation of the cytoskeleton. *Ultramicroscopy.* 2000;82(1–4):253–258.
109. Trache A, Lim SM. Integrated microscopy for real-time imaging of mechanotransduction studies in live cells. *J Biomed Opt.* 2009;14(3). http://dx.doi.org/10.1117/1.3155517. 034024-034024-13.
110. Ueki Y, Uda Y, Sakamoto N, Sato M. Measurements of strain on single stress fibers in living endothelial cells induced by fluid shear stress. *Biochem Biophys Res Commun.* 2010;395(3):441–446.
111. Colombelli J, Besser A, Kress H, et al. Mechanosensing in actin stress fibers revealed by a close correlation between force and protein localization. *J Cell Sci.* 2009;122(10):1665–1679.
112. Kumar S, Maxwell IZ, Heisterkamp A, et al. Viscoelastic retraction of single living stress fibers and its impact on cell shape, cytoskeletal organization, and extracellular matrix mechanics. *Biophys J.* 2006;90(10):3762–3773.
113. Chen C, Krishnan R, Zhou EH, et al. Fluidization and resolidification of the human bladder smooth muscle cell in response to transient stretch. *PLoS One.* 2010;5(8):e12035. http://dx.doi.org/10.1371/journal.pone.0012035.
114. Tzima E, del Pozo MA, Shattil SJ, Chien S, Schwartz MA. Activation of integrins in endothelial cells by fluid shear stress mediates Rho-dependent cytoskeletal alignment. *EMBO J.* 2001;20(17):4639–4647.
115. Vogel V, Sheetz M. Local force and geometry sensing regulate cell functions. *Nat Rev Mol Cell Biol.* 2006;7(4):265–275.
116. Discher DE, Janmey P, Wang YL. Tissue cells feel and respond to the stiffness of their substrate. *Science.* 2005;310(5751):1139–1143.
117. Winer JP, Janmey PA, McCormick ME, Funaki M. Bone marrow-derived human mesenchymal stem cells become quiescent on soft substrates but remain responsive to chemical or mechanical stimuli. *Tissue Eng A.* 2009;15(1):147–154.
118. Holle AW, Engler AJ. More than a feeling: discovering, understanding, and influencing mechanosensing pathways. *Curr Opin Biotechnol.* 2011;22(5):648–654.
119. Schwarz US, Gardel ML. United we stand—integrating the actin cytoskeleton and cell-matrix adhesions in cellular mechanotransduction. *J Cell Sci.* 2012;125(13):3051–3060.
120. Modulevsky DJ, Tremblay D, Gullekson C, Bukoresthliev NV, Pelling AE. The physical interaction of myoblasts with the microenvironment during remodeling of the cytoarchitecture. *PLoS One.* 2012;7(9):e45329. http://dx.doi.org/10.1371/journal.pone.0045329.

121. Collinsworth AM, Zhang S, Kraus WE, Truskey GA. Apparent elastic modulus and hysteresis of skeletal muscle cells throughout differentiation. *Am J Physiol Cell Phys.* 2002;283(4):C1219–C1227.
122. Mathur AB, Collinsworth AM, Reichert WM, Kraus WE, Truskey GA. Endothelial, cardiac muscle and skeletal muscle exhibit different viscous and elastic properties as determined by atomic force microscopy. *J Biomech.* 2001;34(12):1545–1553.
123. Nyland LR, Maughan DW. Morphology and transverse stiffness of Drosophila myofibrils measured by atomic force microscopy. *Biophys J.* 2000;78(3):1490–1497.
124. Ogneva IV. Transversal stiffness of fibers and desmin content in leg muscles of rats under gravitational unloading of various durations. *J Appl Physiol.* 2010;109(6): 1702–1709.
125. Engler AJ, Griffin MA, Sen S, Bonnetnann CG, Sweeney HL, Discher DE. Myotubes differentiate optimally on substrates with tissue-like stiffness: pathological implications for soft or stiff microenvironments. *J Cell Biol.* 2004;166(6):877–887.
126. Akimoto T, Uchida T, Miyaki S, Akaogi H, Tateishi T, Fukubayashi T. Effect of mechanical stretch on TGF-beta 1 expression of C2C12 myogenic cells. *Mater Sci Eng C Biol Sci.* 2004;24(3):387–389.
127. Kook SH, Lee HJ, Chung WT, et al. Cyclic mechanical stretch stimulates the proliferation of C2C12 myoblasts and inhibits their differentiation via prolonged activation of p38 MAPK. *Mol Cells.* 2008;25(4):479–486.
128. Otis JS, Burkholder TJ, Pavlath GK. Stretch-induced myoblast proliferation is dependent on the COX2 pathway. *Exp Cell Res.* 2005;310(2):417–425.
129. Zhan M, Jin BW, Chen SE, Reecy JM, Li YP. TACE release of TNF-alpha mediates mechanotransduction-induced activation of p38 MAPK and myogenesis. *J Cell Sci.* 2007;120(4):692–701.
130. Parker MH, Seale P, Rudnicki MA. Looking back to the embryo: defining transcriptional networks in adult myogenesis. *Nat Rev Genet.* 2003;4(7):497–507.
131. Fu J, Wang YK, Yang MT, et al. Mechanical regulation of cell function with geometrically modulated elastomeric substrates. *Nat Methods.* 2010;7(9):733–736.
132. Han SJ, Bielawski KS, Ting LH, Rodriguez ML, Sniadecki NJ. Decoupling substrate stiffness, spread area, and micropost density: a close spatial relationship between traction forces and focal adhesions. *Biophys J.* 2012;103(4):640–648.
133. Nag AC, Foster JD. Myogenesis in adult mammalian skeletal-muscle in vitro. *J Anat.* 1981;132(Jan):1–18.
134. Yang YF, Creeer A, Jemiolo B, Trappe S. Time course of myogenic and metabolic gene expression in response to acute exercise in human skeletal muscle. *J Appl Physiol.* 2005;98(5):1745–1752.
135. Goldspink DF, Easton J, Winterburn SK, Williams PE, Goldspink GE. The role of passive stretch and repetitive electrical-stimulation in preventing skeletal-muscle atrophy while reprogramming gene-expression to improve fatigue resistance. *J Cardiac Surg.* 1991;6(1):218–224.
136. Goldspink DF, Cox VM, Smith SK, et al. Muscle growth in response to mechanical stimuli. *Am J Physiol Endocrinol Metab.* 1995;268(2):E288–E297.
137. Vandenburgh HH, Hatfaludy S, Karlisch P, Shansky J. Skeletal-muscle growth is stimulated by intermittent stretch-relaxation in tissue-culture. *Am J Physiol.* 1989;256(3): C674–C682.
138. Collinsworth AM, Torgan CE, Nagda SN, Rajalingam RJ, Kraus WE, Truskey GA. Orientation and length of mammalian skeletal myocytes in response to a unidirectional stretch. *Cell Tissue Res.* 2000;302(2):243–251.
139. Vandenburgh HH, Swasdison S, Karlisch P. Computer-aided mechanogenesis of skeletal-muscle organs from single cells-in vitro. *FASEB J.* 1991;5(13):2860–2867.

CHAPTER SIX

Molecular Mechanisms Underlying the Force-Dependent Regulation of Actin-to-ECM Linkage at the Focal Adhesions

Hiroaki Hirata*, Masahiro Sokabe*,†, Chwee Teck Lim*,‡

*Mechanobiology Institute, National University of Singapore, Singapore, Singapore
†Mechanobiology Laboratory, Nagoya University Graduate School of Medicine, Nagoya, Japan
‡Department of Biomedical Engineering, National University of Singapore, Singapore, Singapore

Contents

Abstract

The linkage of the actin cytoskeleton to extracellular matrices (ECMs) at focal adhesions provides a physical path for cells to exert traction forces on substrates during cellular processes such as migration and morphogenesis. Mechanical strength of the actin-to-ECM linkage increases in response to forces loaded at this linkage. This is achieved by local accumulations of actin filaments, as well as linker proteins connecting actins to integrins, at force-bearing adhesion sites, which leads to an increase in the number of molecular bonds between the actin cytoskeleton- and ECM-bound integrins. Zyxin-dependent actin polymerization and filamin-mediated actin bundling are seemingly involved in the force-dependent actin accumulation. Each actin–integrin link is primarily mediated by the linker protein talin, which is strengthened by another linker protein vinculin connecting the actin filaments to talin in a force-dependent manner. This

Progress in Molecular Biology and Translational Science, Volume 126
ISSN 1877-1173
http://dx.doi.org/10.1016/B978-0-12-394624-9.00006-3

eliminates slippage between the actin cytoskeleton and talin (clutch mechanism), thus playing a crucial role in creating cell membrane protrusions mediated by actin polymerization. Finally, each integrin–ECM bond is also strengthened when a force is loaded on it, which ensures force transmission at focal adhesions, contributing to stable cell-substrate adhesion in cell migration.

1. INTRODUCTION

Cell adhesions to neighboring cells and underlying/surrounding extracellular matrices (ECMs) are essential for maintaining the integrity of multicellular organisms. These adhesions experience and transmit mechanical forces originating from deformation of skin, muscle contraction, expansion of lung, bladder, digestive tract and vessels, etc. Individual nonmuscle cells generate contractile forces, which are also transmitted to and sustained at adhesion sites. Cell adhesion structures are never passive and static, but they are dynamically remodeled. A growing body of studies has revealed that forces loaded to adhesion sites play a crucial role in the regulation of adhesion structures.[1,2] For example, cadherin-mediated adherens junctions between cells change their length and width in response to forces exerted to them.[3] The more extensively studied case is the force-responses of the cell-to-ECM adhesion structure called focal adhesion.

Almost all somatic cells form adhesion to ECM, and cell-ECM adhesion is crucial for various cellular functions including morphogenesis, migration, proliferation, and differentiation. Integrins are primary receptors for ECM and clustered at focal adhesions. At the cytoplasmic side of focal adhesions, the actin cytoskeleton is linked to the integrin clusters through a variety of linker proteins including talin and vinculin. The actin cytoskeleton is a major force-generating organelle in nonmuscle cells; actin polymerization generates a force and protrudes the membrane forward, and the actin network-containing myosin II generates a contractile force. The linkage between the actin cytoskeleton, integrins, and ECM experiences and bears both actin-originating forces and external forces caused by deformation of ECM. Importantly, the linkage is strengthened when it is mechanically loaded,[4,5] which enables cells to maintain their morphologies against mechanical perturbations and to exert traction forces to ECM in cell spreading and migration.

The force-dependent regulation of the linkage from actin to integrin and ECM should rely on molecules which sense and transduce forces into changes in molecular interaction. Eventually, every component in the linkage should, more or less, feel a force, and therefore can be a candidate of a force-sensor/transducer. Recent studies have indeed revealed that several component proteins change their conformation in response to the force applied, which modulates their interactions with binding partners. In this chapter, we review the latest knowledge about the molecular mechanism underlying the force-dependent regulation of the linkage between the actin cytoskeleton, integrin, and ECM.

2. MOLECULAR ASSEMBLY IN THE ACTIN–INTEGRIN–ECM LINKAGE

At focal adhesions, transmembrane ECM receptors, integrins as well as syndecan-4,[6] are clustered. Integrins function as heterodimers of α- and β-subunits and bind to specific sets of ECM proteins. The cytoplasmic domain of β integrin binds to several linker proteins (talin, tensin, α-actinin, and filamin) that directly bind to actin filaments.[7] Other actin-binding proteins that do not bind directly to integrins also localize at focal adhesions (e.g., vinculin, Ena/VASP proteins, ezrin–radixin–moesin proteins, parvin/actopaxin, nexilin, and cysteine-rich proteins). Many of these actin-binding proteins can be indirectly linked to integrin through their bindings to the actin–integrin linker proteins and/or adaptor proteins such as paxillin, zyxin, integrin-linked kinase, and focal adhesion kinase (FAK). Therefore, there are multiple potential pathways linking between the actin cytoskeleton and integrin (Fig. 6.1).

2.1. Formation of the initial linkage

During cell spreading and migration, actin monomers are rapidly polymerized beneath the protruding plasma membrane of lamellipodia and filopodia, which pushes the membrane forward.[8,9] Local, spotty increases in actin polymerization along the protruding cell edge initiate the formation of small clusters of activated, but unligated, β1 integrin.[10] These integrin clusters at the cell edge contain VASP,[10] which protects barbed ends of actin filaments from capping,[11] Talin is also likely to be contained, because binding of the N-terminal head domain of talin to the cytoplasmic domain of β integrin is critical for integrin activation.[12,13] The active integrin clusters then bind to

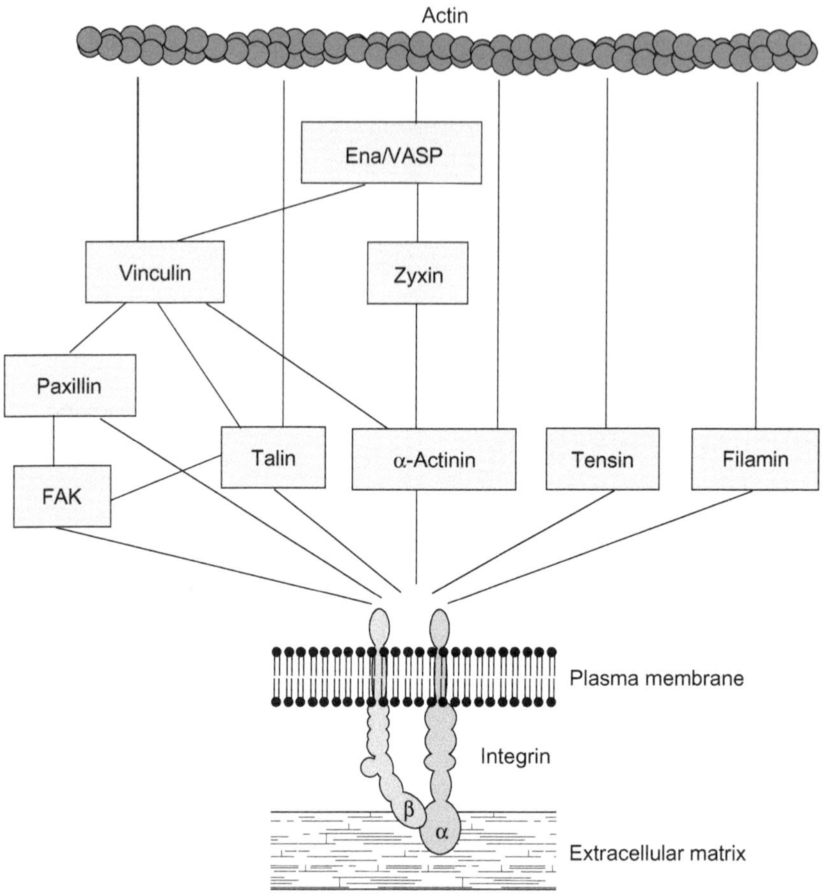

Figure 6.1 Potential molecular connections linking the actin cytoskeleton to integrin at cell–ECM adhesion sites. (See the color plate.)

ECM, forming "nascent adhesions,"[14] the initial points where the actin cytoskeleton is linked to ECM through integrin. The nascent adhesions are formed independently of myosin II activity[14,15] and contain β1 and β3 integrins, paxillin, FAK, and talin.[14,16,17] Among these components, talin1 serves as an initial bond between actin filaments and the ECM-bound integrin cluster.[18,19]

2.2. Force-dependent maturation of the linkage

The cell-protruding region is composed of two distinct domains: lamellipodium (including filopodium extending from the lamellipodium[20]) and lamella.[21] The lamellipodium spans ~2–4 μm from the protruding edge, which is followed by the lamella.[21] Nascent adhesions are formed in the lamellipodium.[14] As the cell edge advances, the lamellipodium moves forward over the stationary adhesions. Nascent adhesions are stable in the lamellipodium; however, once the adhesions reach to the interface between the lamellipodium and the lamella, most of them disassemble.[14] On the other hand, a population of the nascent adhesions is elongated and matured into "focal complexes" and then "focal adhesions."

Elongation and maturation of the adhesions are initiated by the formation of short actin bundles.[14,22] Actin filaments are polymerized at the adhesion sites and elongated toward the cell body in a manner dependent on the actin elongation factor, formin mDia1.[23] The filaments are then crosslinked by α-actinin and myosin II to form short bundles, providing a structural template for elongation of the adhesion complex. Each adhesion complex elongates along this actin bundle,[14,22] where α-actinin mediates association of integrin with the actin bundle.[24] The bundle formation by crosslinking of actin filaments does not require the ATPase activity of myosin II.[14] However, since the formed bundles are connected to the actin network that contains active myosin II in the lamella, the myosin II-based contractile forces generated in the network are transmitted to the actin bundles, adhesion complexes and ECM.[22] In concert with the actin bundle formation, the force loaded to actin bundle–adhesion complex–ECM facilitates elongation and maturation of the adhesion complexes and stabilizes them.[22,25,26] Indeed, full maturation into focal adhesions requires myosin II activity,[27–29] and force exertion from contractile actin stress fibers to connected focal adhesions is needed to maintain molecular compositions of focal adhesions.[30,31]

The maturation of adhesion complexes accompanies a remarkable increase in accumulation of several focal adhesion proteins, including vinculin, FAK, α-actinin, VASP, zyxin, and tensin.[16] Concomitantly with the recruitment of these adaptor proteins, the linkage of the actin cytoskeleton to ECM-bound integrin clusters is strengthened. Strengthening of the linkage is also force dependent. Talin1-mediated initial linkage between the actin cytoskeleton and ECM-bound integrin is relatively weak and easily slips by a force of ~2 pN.[19] However, when a mechanical force is loaded to the linkage by manipulating ECM-coated beads bound to cell surface integrin using the magnetic or optical trapping, the linkage is strengthened.[4,5] The strengthened linkage sustains forces over 20 pN, eliminating slippage between the actin cytoskeleton and integrin clusters.[5] Talin1 plays an essential role in this force-responsive event,[18] and recruitments of vinculin and paxillin around the beads are associated with this.[18,26]

3. FORCE-SENSING/TRANSDUCING MOLECULES IN THE REGULATION OF THE ACTIN–INTEGRIN–ECM LINKAGE

It is thought that strengthening of the linkage between the actin cytoskeleton, integrin, and ECM is mainly based on an increase in the number of bonds in the linkage. Furthermore, strengthening of each bond may also be involved. In this section, we discuss the molecular mechanisms of how forces affect and regulate the number and the strength of bonds in the actin–integrin–ECM linkage (Fig. 6.2).

3.1. Talin and vinculin

3.1.1 Force-dependent vinculin binding with talin

Talin is an elongated (ca. 60 nm) molecule that comprises an N-terminal globular head domain and a flexible rod domain.[32] The head domain contains the binding sites for β integrin, FAK, and actin, and the rod domain also has one β integrin-binding site and two actin-binding sites.[32] Through the bindings with integrin and/or FAK, talin localizes to adhesion sites.[17] When cells adhere to the fibronectin substrate, talin forms a complex with fibronectin-bound α5β1 integrin, a receptor for fibronectin, at adhesion sites independently of myosin II activity.[29,33]

Talin has up to 11 binding sites for vinculin in the rod domain.[34] Deficiency in talin expression causes delocalization of vinculin from focal adhesions.[18,35] On the other hand, the vinculin fragment that consists of only the talin-binding domain (domain D1) of vinculin localizes to focal adhesions.[36,37] Thus, vinculin localization at focal adhesions depends on the binding with talin. However, in contrast to talin, vinculin localizes to adhesion sites in a force-dependent manner; a decrease in the actomyosin-based force loaded to focal adhesions results in delocalization of vinculin from focal adhesions,[29,30] while external forces applied to cell-to-ECM adhesion sites, in turn, induce recruitment of vinculin without affecting the distribution of talin.[25,26,38] The residence time of vinculin at focal adhesions increases linearly with increasing the magnitude of actomyosin-based forces exerted to focal adhesions,[39,40] demonstrating a close relationship between force exertion and vinculin association at focal adhesions.

Some of the vinculin-binding sites (VBSs) of talin are buried in the bundles of amphipathic helices[41,42] and are biochemically inactive in the native form.[41,43] On the other hand, molecular dynamics simulations have predicted that mechanical forces induce exposure of the VBSs.[44,45] This

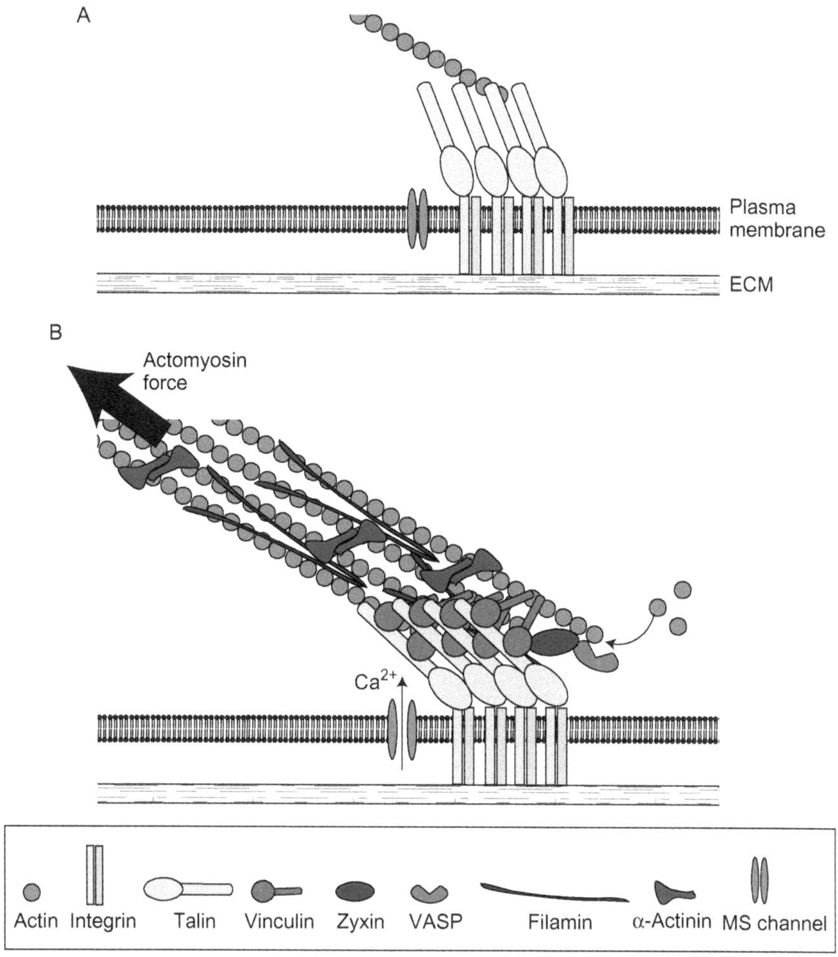

Figure 6.2 Model of force-induced strengthening of the actin–integrin–ECM linkage at adhesion sites. (A) Talin mediates force-independent linkage between the actin cytoskeleton and integrin clusters. (B) Actomyosin forces expose cryptic vinculin-binding sites (VBSs) in talin, which enables the vinculin binding with talin and thereby reinforces talin–actin links. Forces also induce accumulation of actin filaments at adhesion sites through zyxin–VASP-dependent actin polymerization and filamin- and α-actinin-dependent actin bundling, where Ca^{2+} influxes probably through mechanosensitive ion channels (MS channels) are involved. Accumulated actin filaments provide more binding sites for linker proteins connected to integrin. See the text for details. (See the color plate.)

was confirmed by the *in vitro* experiment in which the force applied to a single talin rod domain increased the number of vinculin head domains bound to the rod *in vitro*.[46] Recent *in vivo* studies have indeed revealed that talin molecules at a focal adhesion, which are oriented to connect between the integrin cluster and the actin cytoskeleton that are vertically separated by ~40 nm,[47] are stretched depending on the actomyosin activity in living cells,[48] and that the binding of vinculin with talin is increased at focal adhesions by actomyosin or externally applied forces.[38] Thus, the binding of vinculin with talin at adhesion sites is truly regulated by a force.

A vinculin molecule is composed of an N-terminal headpiece including the talin-binding domain D1 and a C-terminal tail domain, which are linked by a short proline-rich sequence. Full-length vinculin adopts a globular conformation through the intramolecular head–tail interaction.[49] In contrast to full-length vinculin, the deleted form of vinculin that consists of only the domain D1 localizes at focal adhesions regardless of myosin II activity.[36,37] The head–tail interaction-defective mutant form of vinculin, which adopts an extended conformation, also localizes at focal adhesions in an actomyosin-independent manner.[37] These results imply that the globular conformation of full-length vinculin endows vinculin with the force-dependent talin binding; the domain D1 in globular-shaped vinculin, but not in the extended-shaped one, may not gain access to VBSs in talin due to steric constraints unless these sites are fully exposed by a force. On the other hand, a population of vinculin at focal adhesions adopts the extended conformation.[50] Since each vinculin molecule at focal adhesions is loaded with a force of ~2.5 pN between its head and tail domains,[51] it may also be possible that the force switches the vinculin conformation from globular to extended one, which stabilizes the vinculin-binding with talin at focal adhesions.

3.1.2 The talin–vinculin binding in strengthening of the actin–integrin linkage

Contribution of vinculin to mechanical strength of the linkage between the actin cytoskeleton and ECM-bound integrin has been studied; vinculin-deficient cells exhibit a weaker linkage than wild-type cells.[52,53] Since vinculin has an actin-binding site in its C-terminal tail domain, as well as the N-terminal talin-binding domain D1,[49] it can bind to both talin and actin. Therefore, when a talin molecule is stretched and exposes multiple VBSs, the bond between talin and the actin filament will be reinforced by multiple "talin–vinculin–actin" links (Fig. 6.3). In living cells, each vinculin molecule at adhesion sites sustains a force of ~2.5 pN,[51] that is, the talin–actin link reinforced by multiple vinculin molecules could sustain a much larger force than a single talin–actin bond can (ca. 2 pN[19]).

Focal adhesions provide anchor sites of the actin–myosin II network including stress fibers to ECM. When the binding between vinculin and

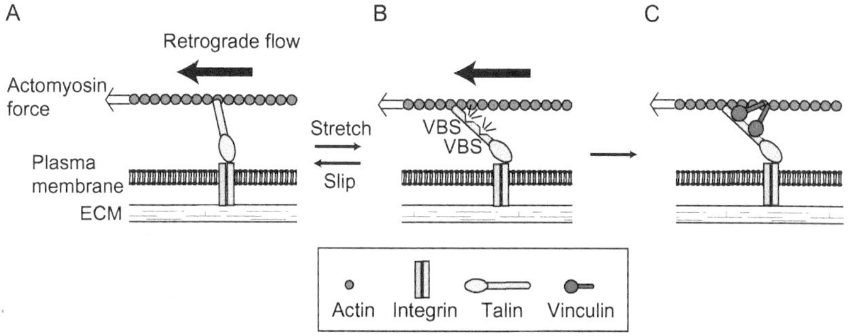

Figure 6.3 Model of force-dependent regulation of vinculin recruitment at adhesion sites and anchoring the actin cytoskeleton. (A and B) The actomyosin-based force drives retrograde flow of actin filaments. The force stretches talin molecules associated with integrin and causes exposure of cryptic VBSs. The actin–talin link slips as the force exceeds ca. 2 pN (B). The state (B) returns to the state (A) when the free actin-binding domain of talin binds again to the actin filament. (C) Vinculin binds to the exposed VBSs via the domain D1 and to the actin filament via its tail domain. This vinculin-reinforced integrin–talin–vinculin–actin linkage may anchor the actin cytoskeleton to adhesion sites. (See the color plate.)

talin is interfered with the overexpressed domain D1 of vinculin, the actin network in the lamella is not anchored but flows backward over stationary clusters of talin at focal adhesions, indicating that the talin–vinculin binding is crucial for anchoring the actin cytoskeleton to focal adhesions (Fig. 6.3).[38] The binding of vinculin with actin is also required for the anchoring.[54] The vinculin-dependent actin anchorage ensures transmission of actomyosin-based forces to ECM,[40,54] but neither the N-terminal nor the C-terminal fragment of vinculin supports the force transmission and cell spreading.[40,55] While force-dependent phosphorylation of paxillin may also provide binding sites for vinculin at focal adhesions,[29] above results show that the physical link of "talin–vinculin–actin" serves as a major path in the actin–integrin–ECM linkage at focal adhesions.

3.2. Zyxin, filamin, and actin assembly

Force application to clustered integrin induces accumulation of bundled actin filaments around the integrin cluster.[56–58] The local actin accumulation is associated with recruitments of actin–integrin linker proteins including talin, vinculin, α-actinin and filamin, and the accumulated actin filaments are connected to the actin cytoskeleton network.[4,18,26,57] Therefore, the force-induced accumulation of actin filaments leads to an increase in the number of links between the actin cytoskeleton and integrin, and thereby contributes to strengthening of the actin–integrin linkage.[4,58]

The local accumulation of actin filaments could, primarily, be based on *de novo* actin polymerization on site and/or the redistribution of actin filaments. As discussed below, it is likely that both mechanisms are involved in the force-induced actin accumulation at integrin clusters.

3.2.1 Zyxin-dependent actin polymerization

Focal adhesions are active sites for actin polymerization.[59,60] In contrast to actin polymerization at the leading edge of a cell, actin polymerization at focal adhesions is strongly dependent on myosin II-based or externally applied forces.[33] Involvement of several actin nucleation or elongation factors in the actin polymerization at focal adhesions has been revealed. Among them, roles of diaphanous-related formins have been extensively studied. When the expression of mDia1 is depleted, elongation of actin bundles from focal adhesions is largely attenuated,[23] suggesting that mDia1 facilitates actin elongation/polymerization at focal adhesions. The ability of mDia1 to facilitate actin polymerization at integrin-based adhesion complexes was more clearly demonstrated *in vitro*; actin polymerization at the isolated adhesion complexes was diminished when the inhibitory domain of mDia1 was present.[61] The contribution of mDia2 to actin polymerization at focal adhesions has also been reported.[62] However, since mDia1 and mDia2 do not exhibit apparent localizations at focal adhesion, the mechanisms of how these proteins locally facilitate actin polymerization at focal adhesions remain unclear. On the other hand, the actin nucleation factor, Arp2/3 complex, localizes at focal adhesions in some extent;[63] but, its contribution to actin polymerization at focal adhesions may not be dominant.[61]

Ena/VASP family proteins, which antagonize actin-capping proteins and promote actin polymerization,[11,64,65] show clear localization at focal adhesions, implying a role of Ena/VASP proteins in local actin polymerization at focal adhesions. Ena/VASP localizes at focal adhesions through the binding with the focal adhesion protein zyxin.[66–68] Importantly, when zyxin is delocalized from focal adhesions by overexpressing the focal adhesion-targeting domain of zyxin, Ena/VASP is delocalized, and actin polymerization at focal adhesions is suppressed.[33] Since zyxin localization at focal adhesions is highly dependent on forces exerted on focal adhesions,[31,33,69] zyxin plays a key role, through the recruitment of Ena/VASP to focal adhesions, in force-induced actin polymerization at focal adhesions.[33,70]

3.2.2 Filamin and bundling of actin filaments

The actin-crosslinking protein, filamin, is also required for the force-induced accumulation of actin filaments around integrin clusters.[57] Filamin accumulates at integrin clusters where forces are loaded[57] and is involved in the force-induced strengthening of the actin–integrin–ECM linkage,[18] while its contribution to the strengthening is not as much as that of talin. Since a high concentration of filamin causes the rearrangement of actin filaments into actin bundles *in vitro*,[71] accumulated filamin would locally remodel a loose network of actin filaments into dense actin bundles that are seen around the force-loaded integrin cluster.[23,57] Another actin-crosslinker, α-actinin, may also contribute to bundling of actin filaments. Concentration of actin filaments around the integrin cluster would give rise to an increase in the number of sites available for linker proteins connected to integrin.

The molecular mechanism(s) that enables filamin to accumulate at force-loaded integrin clusters and to strengthen the linkage between actin and integrin is largely unclear. However, it has been reported that an increase in the intracellular Ca^{2+} concentration and PKC-dependent phosphorylation of filamin are both involved in the force-induced filamin accumulation.[57] Since forces loaded to the integrin-based adhesion sites activate mechanosensitive ion channels and thereby induce influxes of Ca^{2+},[72–74] mechanosensitive channels are likely to be involved in the force-induced activation of the Ca^{2+}–PKC–filamin cascade. Another intriguing possibility is the force-induced binding of filamin with integrin. When a shear force is applied *in vitro* to a network of actin filaments crosslinked with filamin A, the binding of β integrin to filamin A is increased,[75] probably through force-induced exposure of the cryptic β integrin-binding site in filamin A.[76–78] The force-induced binding between filamin and integrin may facilitate the accumulation of filamin at integrin clusters and strengthen the actin-integrin linkage, although further studies are needed to clarify whether the integrin–filamin binding is truly regulated by a force *in vivo*.

3.3. Integrin–fibronectin binding

Extracellular domains of integrin heterodimers adopt either a bent conformation, which corresponds to the low affinity state of integrin against its ECM ligand, or an extended conformation, which corresponds to the high affinity state.[79] The conformation change of integrin from a bend to an extended form takes place at focal adhesions.[80] While the talin binding to integrin triggers the extension of integrin,[81] molecular dynamics simulations have suggested that a force loaded to integrin through the ligand also induces the conformation change into the extended form.[82] Each integrin-ligand bond at focal adhesions in living cells sustains a force of ~40 pN.[83] Single molecule atomic force microscopy experiments clearly demonstrated that the lifetime of the binding between α5β1 integrin and the fragment of its ligand, fibronectin, was prolonged by applying a force in the physiological magnitude range.[84,85] The force-dependent reinforcement of individual α5β1 integrin–fibronectin bonds was observed also in living cells.[86] Although accumulation of α5β1 integrin at focal adhesions does not require myosin II-based forces,[33] the "catch bond" behavior of the α5β1 integrin–fibronectin binding would lead to the force-induced increase in mechanical strength of the integrin–ECM linkage at focal adhesions.

Fibronectin, a ubiquitous ECM protein, is also force sensitive. Soluble fibronectin dimers secreted from cells are assembled into insoluble fibrils. The fibrillogenesis of fibronectin, which depends on the fibronectin–fibronectin interaction, requires the binding to integrin on the cell surface, the intact actin cytoskeleton, and the Rho-dependent myosin II activation.[87,88] A growing body of studies has revealed that forces exerted from cells through integrin play a critical role in the fibronectin fibrillogenesis; cell-generated forces unfold fibronectin molecules, expose cryptic binding sites for fibronectin, and thereby facilitate fibrillogenesis of fibronectin.[88–92] Elongated, very mature focal adhesions called fibrillar adhesions are formed along fibronectin fibrils.[93] However, since fibronectin fibrils do not always colocalize with integrin,[94] it is uncertain whether force-induced fibrillogenesis of fibronectin is involved in the strengthening of the integrin–ECM linkage.

4. DYNAMIC ASPECT OF THE ACTIN–INTEGRIN–ECM LINKAGE: MOLECULAR CLUTCH

Individual focal adhesions have a typical lifespan of tens of minutes.[22,95] By contrast, individual molecules in the actin–integrin linkage, which include talin, vinculin, α-actinin, paxillin, and zyxin, reside at focal adhesions only for a few seconds to a minute.[29,31,36,39,96] The lifetime of the integrin–fibronectin bond is also in a several seconds range.[84] Thus, the linkage of the actin cytoskeleton to integrin and ECM at focal adhesions is mediated by the accumulation of transient, dynamic interactions of constituent proteins. It is conceivable that forces on focal adhesions modulate association/dissociation kinetics of molecular bindings at focal adhesions[31,39] and thereby regulate the mechanical strength of the linkage between the actin cytoskeleton, integrin, and ECM.

The cytoplasmic portion of each focal adhesion is composed of multiple vertical layers with distinct proteins; the most membrane-proximal layer contains cytoplasmic domains of integrin, and actin filaments reside in the uppermost layer.[47,97] Rapid association/dissociation cycles of molecular bonds at focal adhesions make interactions between these protein layers rather dynamic. In particular, centripetally directed forces generated in the actin-myosin II network including stress fibers cause the relative movement of actin filaments against ECM at focal adhesions. This is associated with relative movements which occur at every interface of protein layers (Fig. 6.4).[98–101] Thus, the actin–integrin–ECM linkage at focal adhesions

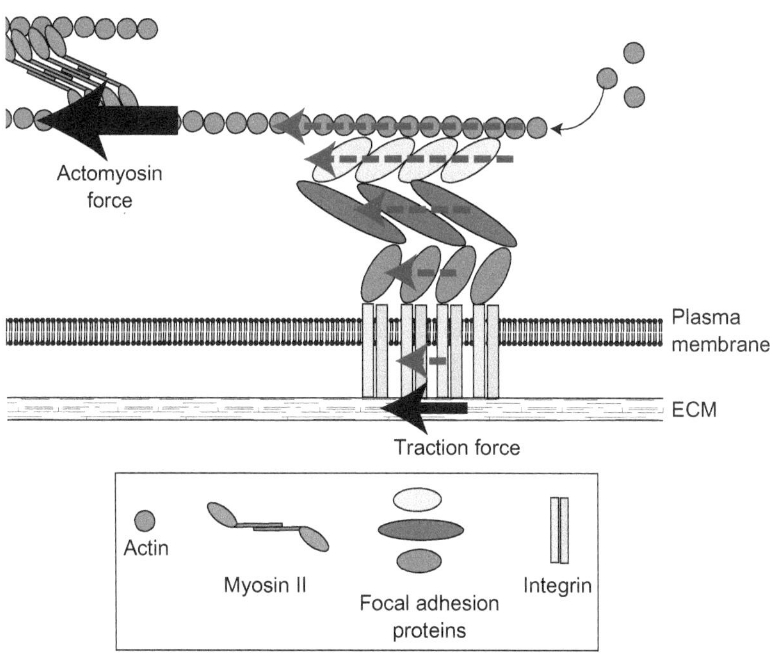

Figure 6.4 Schematic diagram of the multilayer molecular clutch. A focal adhesion is composed of multiple layers with distinct proteins. Since interactions between layers are mediated by association/dissociation cycles of molecular bonds, myosin II-generated actomyosin forces cause "slippage" at layer interfaces including the integrin–ECM interface, and each layer moves toward the force direction with different velocities: antin filaments with the highest velocity and integrin with the lowest one (red-dotted arrows). The slippage leads to dissipation in the force transmitted from actin filaments to ECM. The extent of the slippage, or the efficiency of the force transmission, can be tuned by modulations in the number and association/dissociation kinetics of molecular bonds. (See the color plate.)

is in a dynamic steady state between the moving actin cytoskeleton and stationary ECM.

The dynamic actin–integrin–ECM linkage at focal adhesions has been modeled as a "molecular clutch" (Fig. 6.4).[102] When the clutch is disengaged, the actin cytoskeleton moves without connection to ECM, where forces generated in the actin–myosin II network are not transmitted to ECM. When the clutch is partially engaged, the moving actin cytoskeleton is linked to ECM through slipping connections between protein layers, where the actomyosin forces are partially transmitted to ECM. On the other hand, when the clutch is fully engaged, the actin movement is largely retarded, and forces are efficiently transmitted from the actin cytoskeleton to ECM. A notable feature of the molecular clutch is that its engagement state can be modulated by forces through the regulation of the number and the strength of molecular bonds in the actin–integrin–ECM linkage, as discussed in Section 3. When considered the catch bond behavior of molecular bonds (e.g., talin–vinculin–actin bonds and integrin–fibronectin bond), the mathematical model has predicted that the magnitude of forces transmitted to ECM depends biphasically on the velocity of centripetal movement of the actin cytoskeleton; that is, there is the optimal actin velocity that maximizes the force transmission.[103] Since similar biphasic dependency has been observed also in living cells,[104] molecular bonds with the catch bond nature are likely to play a key role in engaging the molecular clutch at focal adhesions. This notion is consistent with the recent findings that vinculin, in particular the vinculin binding with talin, is crucial for the clutch engagement.[38,40,54,105]

5. CONCLUDING REMARKS

In this chapter, we have reviewed the molecular mechanism underlying the force-dependent regulation of the actin–integrin–ECM linkage at cell adhesion sites. In particular, we focused on the roles and behaviors of structural components in the linkage. However, biochemical reactions, which include tyrosine phosphorylation and dephosphorylation, at focal adhesions also have a significant impact on the actin–integrin–ECM linkage. Focal adhesions are the site where tyrosine-phosphorylated proteins are enriched, and the tyrosine kinase Src[106] and the tyrosine phosphatases SHP-2[107] and RPTP-α[108] are, for example, involved in the force-dependent regulation of the linkage between the actin cytoskeleton and integrin. While phosphorylation of tyrosine residues can affect molecular interactions typically through the binding with SH2 domains, we know very little at present about how tyrosine phosphorylation/dephosphorylation signals affect and cooperate with behaviors of structural components such as talin and vinculin to achieve fine regulation of the actin–integrin–ECM linkage. The dynamic actin–integrin–ECM linkage at adhesion sites is closely related to the mechanism for cells to probe rigidity of extracellular substrates, and protein tyrosine phosphorylation is apparently involved in this mechanosensing process.[102,109,110] Therefore, understanding the interplay between biochemical and mechanical regulations of the actin–integrin–ECM linkage will bring significant insights into molecular and biophysical mechanisms underlying the fact that cell migration, proliferation, and differentiation are influenced by substrate rigidity in both physiological and pathological processes.[111]

ACKNOWLEDGMENTS

We thank Dr. Hitoshi Tatsumi for critical discussion. M. S. was supported by JSPS Grant-in-Aid for Scientific Research (A) (21247021, 24247028) and a grant from Japan Space Forum. C. T. L. is supported by research funding from the Mechanobiology Institute.

REFERENCES

1. Geiger B, Spatz JP, Bershadsky AD. Environmental sensing through focal adhesions. *Nat Rev Mol Cell Biol.* 2009;10:21–33.
2. Gomez GA, McLachlan RW, Yap AS. Productive tension: force-sensing and homeostasis of cell-cell junctions. *Trends Cell Biol.* 2011;21:499–505.
3. Liu Z, Tan JL, Cohen DM, et al. Mechanical tugging force regulates the size of cell-cell junctions. *Proc Natl Acad Sci USA.* 2010;107:9944–9949.
4. Wang N, Butler JP, Ingber DE. Mechanotransduction across the cell surface and through the cytoskeleton. *Science.* 1993;260:1124–1127.
5. Choquet D, Felsenfeld DP, Sheetz MP. Extracellular matrix rigidity causes strengthening of integrin-cytoskeleton linkages. *Cell.* 1997;88:39–48.
6. Woods A, Couchman JR. Syndecan-4 and focal adhesion function. *Curr Opin Cell Biol.* 2001;13:578–583.
7. Geiger B, Bershadsky A, Pankov R, Yamada KM. Transmembrane extracellular matrix-cytoskeleton crosstalk. *Nat Rev Mol Cell Biol.* 2001;2:793–805.
8. Pollard TD, Borisy GG. Cellular motility driven by assembly and disassembly of actin filaments. *Cell.* 2003;112:453–465.
9. Bornschlögl T. How filopodia pull: what we know about the mechanics and dynamics of filopodia. *Cytoskeleton.* 2013;70:590–603.
10. Galbraith CG, Yamada KM, Galbraith JA. Polymerizing actin fibers position integrins primed to probe for adhesion sites. *Science.* 2007;315:992–995.
11. Bear JE, Svitkina TM, Krause M, et al. Antagonism between Ena/VASP proteins and actin filament capping regulates fibroblast motility. *Cell.* 2002;109:509–521.
12. Calderwood DA, Zent R, Grant R, Rees JG, Hynes RO, Ginsberg MH. The talin head domain binds to integrin β subunit cytoplasmic tails and regulates integrin activation. *J Biol Chem.* 1999;274:28071–28074.

13. Tadokoro S, Shattil SJ, Eto K, et al. Talin binding to integrin β tails: a final common step in integrin activation. *Science*. 2003;302:103–106.
14. Choi CK, Vicente-Manzanares M, Zareno J, Whitmore LA, Mogilner A, Horwitz AR. Actin and α-actinin orchestrate the assembly and maturation of nascent adhesions in a myosin II motor-independent manner. *Nat Cell Biol*. 2008;10:1039–1050.
15. Vicente-Manzanares M, Zareno J, Whitmore L, Choi CK, Horwitz AF. Regulation of protrusion, adhesion dynamics, and polarity by myosins IIA and IIB in migrating cells. *J Cell Biol*. 2007;176:573–580.
16. Zaidel-Bar R, Ballestrem C, Kam Z, Geiger B. Early molecular events in the assembly of matrix adhesions at the leading edge of migrating cells. *J Cell Sci*. 2003;116:4605–4613.
17. Lawson C, Lim ST, Uryu S, Chen XL, Calderwood DA, Schlaepfer DD. FAK promotes recruitment of talin to nascent adhesions to control cell motility. *J Cell Biol*. 2012;196:223–232.
18. Giannone G, Jiang G, Sutton DH, Critchley DR, Sheetz MP. Talin1 is critical for force-dependent reinforcement of initial integrin-cytoskeleton bonds but not tyrosine kinase activation. *J Cell Biol*. 2003;163:409–419.
19. Jiang G, Giannone G, Critchley DR, Fukumoto E, Sheetz MP. Two-piconewton slip bond between fibronectin and the cytoskeleton depends on talin. *Nature*. 2003;424:334–337.
20. Svitkina TM, Bulanova EA, Chaga OY, et al. Mechanism of filopodia initiation by reorganization of a dendritic network. *J Cell Biol*. 2003;160:409–421.
21. Ponti A, Gupton MSL, Waterman-Storer CM, Danuser G. Two distinct actin networks drive the protrusion of migrating cells. *Science*. 2004;305:1782–1786.
22. Oakes PW, Beckham Y, Stricker J, Gardel ML. Tension is required but not sufficient for focal adhesion maturation without a stress fiber template. *J Cell Biol*. 2012;196:363–374.
23. Hotulainen P, Lappalainen P. Stress fibers are generated by two distinct actin assembly mechanisms in motile cells. *J Cell Biol*. 2006;173:383–394.
24. Roca-Cusachs P, del Rio A, Puklin-Faucher E, Gauthier NC, Biais N, Sheetz MP. Integrin-dependent force transmission to the extracellular matrix by α-actinin triggers adhesion maturation. *Proc Natl Acad Sci USA*. 2013;110:E1361–E1370.
25. Riveline D, Zamir E, Balaban NQ, et al. Focal contacts as mechanosensors: externally applied local mechanical force induces growth of focal contacts by an mDia1-dependent and ROCK-independent mechanism. *J Cell Biol*. 2001;153:1175–1185.
26. Galbraith CG, Yamada KM, Sheetz MP. The relationship between force and focal complex development. *J Cell Biol*. 2002;159:695–705.
27. Chrzanowska-Wodnicka M, Burridge K. Rho-stimulated contractility drives the formation of stress fibers and focal adhesions. *J Cell Biol*. 1996;133:1403–1415.
28. Rottner K, Hall A, Small JV. Interplay between Rac and Rho in the control of substrate contact dynamics. *Curr Biol*. 1999;9:640–648.
29. Pasapera AM, Schneider IC, Rericha E, Schlaepfer DD, Waterman CM. Myosin II activity regulates vinculin recruitment to focal adhesions through FAK-mediated paxillin phosphorylation. *J Cell Biol*. 2010;188:877–890.
30. Balaban NQ, Schwarz US, Riveline D, et al. Force and focal adhesion assembly: a close relationship studied using elastic micropatterned substrates. *Nat Cell Biol*. 2001;3:466–472.
31. Lele TP, Pendse J, Kumar S, Salanga M, Karavitis J, Ingber DE. Mechanical forces alter zyxin unbinding kinetics within focal adhesions in living cells. *J Cell Physiol*. 2006;207:187–194.
32. Critchley DR. Biochemical and structural properties of the integrin-associated cytoskeletal protein talin. *Annu Rev Biophys*. 2009;38:235–254.

33. Hirata H, Tatsumi H, Sokabe M. Mechanical forces facilitate actin polymerization at focal adhesions in a zyxin-dependent manner. *J Cell Sci*. 2008;121:2795–2804.
34. Gingras AR, Ziegler WH, Frank R, et al. Mapping and consensus sequence identification for multiple vinculin binding sites within the talin rod. *J Biol Chem*. 2005;280:37217–37224.
35. Zhang X, Jiang G, Cai Y, Monkley SJ, Critchley DR, Sheetz MP. Talin depletion reveals independence of initial cell spreading from integrin activation and traction. *Nat Cell Biol*. 2008;10:1062–1068.
36. Humphries JD, Wang P, Streuli C, Geiger B, Humphries MJ, Ballestrem C. Vinculin controls focal adhesion formation by direct interactions with talin and actin. *J Cell Biol*. 2007;179:1043–1057.
37. Carisey A, Tsang R, Greiner AM, et al. Vinculin regulates the recruitment and release of core focal adhesion proteins in a force-dependent manner. *Curr Biol*. 2013;23:271–281.
38. Hirata H, Tatsumi H, Lim CT, Sokabe M. Force-dependent vinculin binding to talin in live cells: a crucial step in anchoring the actin cytoskeleton to focal adhesions. *Am J Physiol Cell Physiol*. 2014;306:C607–C620.
39. Wolfenson H, Bershadsky A, Henis YI, Geiger B. Actomyosin-generated tension controls the molecular kinetics of focal adhesions. *J Cell Sci*. 2011;124:1425–1432.
40. Dumbauld DW, Lee TT, Singh A, et al. How vinculin regulates force transmission. *Proc Natl Acad Sci USA*. 2013;110:9788–9793.
41. Papagrigoriou E, Gingras AR, Barsukov IL, et al. Activation of a vinculin-binding site in the talin rod involves rearrangement of a five-helix bundle. *EMBO J*. 2004;23:2942–2951.
42. Fillingham I, Gingras AR, Papagrigoriou E, et al. A vinculin binding domain from the talin rod unfolds to form a complex with the vinculin head. *Structure*. 2005;13:65–74.
43. Patel B, Ginhras AR, Bobkov AA, et al. The activity of the vinculin binding sites in talin is influenced by the stability of the helical bundles that make up the talin rod. *J Biol Chem*. 2006;281:7458–7467.
44. Lee SE, Kamm RD, Mofrad MRK. Force-induced activation of talin and its possible role in focal adhesion mechanotransduction. *J Biomech*. 2007;40:2096–2106.
45. Hytönen VP, Vogel V. How force might activate talin's vinculin binding sites: SMD reveals a structural mechanism. *PLoS Comput Biol*. 2008;4:e24.
46. del Rio A, Perez-Jimenez R, Liu R, Roca-Cusachs P, Fernandez JM, Sheetz MP. Stretching single talin rod molecules activates vinculin binding. *Science*. 2009;323:638–641.
47. Kanchanawong P, Shtengel G, Pasapera AM, et al. Nanoscale architecture of integrin-based cell adhesions. *Nature*. 2010;468:580–584.
48. Margadant F, Chew LL, Hu X, et al. Mechanotransduction in vivo by repeated talin stretch-relaxation events depends upon vinculin. *PLoS Biol*. 2011;9:e1001223.
49. Ziegler WH, Liddington RC, Critchley DR. The structure and regulation of vinculin. *Trends Cell Biol*. 2006;16:453–460.
50. Chen H, Cohen DM, Choudhury DM, Kioka N, Craig SW. Spatial distribution and functional significance of activated vinculin in living cells. *J Cell Biol*. 2005;169:459–470.
51. Grashoff C, Hoffman BD, Brenner MD, et al. Measuring mechanical tension across vinculin reveals regulation of focal adhesion dynamics. *Nature*. 2010;466:263–266.
52. Ezzell RM, Goldmann WH, Wang N, Parasharama M, Ingber DE. Vinculin promotes cell spreading by mechanically coupling integrins to the cytoskeleton. *Exp Cell Res*. 1997;231:14–26.
53. Alenghat FJ, Fabry B, Tsai KY, Goldmann WH, Ingber DE. Analysis of cell mechanics in single vinculin-deficient cells using a magnetic tweezer. *Biochem Biophys Res Commun* 2000;277:93–99.

54. Thievessen I, Thompson PM, Berlemont S, et al. Vinculin-actin interaction couples actin retrograde flow to focal adhesions, but is dispensable for focal adhesion growth. *J Cell Biol*. 2013;202:163–177.
55. Xu W, Coll JL, Adamson ED. Rescue of the mutant phenotype by reexpression of full-length vinculin in null F9 cells; effects on cell locomotion by domain deleted vinculin. *J Cell Sci*. 1998;111:1535–1544.
56. Glogauer M, Arora P, Yao G, Sokholov I, Ferrier J, McCulloch CAG. Calcium ions and tyrosine phosphorylation interact coordinately with actin to regulate cytoprotective responses to stretching. *J Cell Sci*. 1997;110:11–21.
57. Glogauer M, Arora P, Chou D, Janmey PA, Downey GP, McCulloch CAG. The role of actin-binding protein 280 in integrin-dependent mechanoprotection. *J Biol Chem*. 1998;273:1689–1698.
58. Icard-Arcizet D, Cardoso O, Richert A, Hénon S. Cell stiffening in response to external stress is correlated to actin recruitment. *Biophys J*. 2008;94:2906–2913.
59. Glacy SD. Subcellular distribution of rhodamine-actin microinjected into living fibroblastic cells. *J Cell Biol*. 1983;97:1207–1213.
60. Turnacioglu KK, Sanger JW, Sanger JM. Sites of monomeric actin incorporation in living PTK2 and REF-52 cells. *Cell Motil Cytoskeleton*. 1998;40:59–70.
61. Butler B, Gao C, Mersich AT, Blystone SD. Purified integrin adhesion complexes exhibit actin-polymerization activity. *Curr Biol*. 2006;16:242–251.
62. Gupton SL, Eisenmann K, Alberts AS, Waterman-Storer CM. mDia2 regulates actin and focal adhesion dynamics and organization in the lamella for efficient epithelial cell migration. *J Cell Sci*. 2007;120:3475–3487.
63. DeMali KA, Barlow CA, Burridge K. Recruitment of the Arp2/3 complex to vinculin: coupling membrane protrusion to matrix adhesion. *J Cell Biol*. 2002;159: 881–891.
64. Hansen SD, Mullins RD. VASP is a processive actin polymerase that requires monomeric actin for barbed end association. *J Cell Biol*. 2010;191:571–584.
65. Breitsprecher D, Kiesewetter AK, Linkner J, et al. Molecular mechanism of Ena/VASP-mediated actin-filament elongation. *EMBO J*. 2011;30:456–467.
66. Drees B, Friederich E, Fradelizi J, Louvard D, Beckerle MC, Golsteyn RM. Characterization of the interaction between zyxin and member of the Ena/Vasodilator-stimulated phosphoprotein family of proteins. *J Biol Chem*. 2000;275:22503–22511.
67. Nix DA, Fradelizi J, Bockholt S, et al. Targeting of zyxin to sites of actin membrane interaction and to the nucleus. *J Biol Chem*. 2001;276:34759–34767.
68. Hoffman LM, Jensen CC, Kloeker S, Wang CLA, Yoshigi M, Beckerle MC. Genetic ablation of zyxin causes Mena/VASP mislocalization, increased motility, and deficits in actin remodeling. *J Cell Biol*. 2006;172:771–782.
69. Uemura A, Nguyen TN, Steele AN, Yamada S. The LIM domain of zyxin is sufficient for force-induced accumulation of zyxin during cell migration. *Biophys J*. 2011;101:1069–1075.
70. Hirata H, Tatsumi H, Sokabe M. Zyxin emerges as a key player in the mechanotransduction at cell adhesive structures. *Commun Integr Biol*. 2008;1:192–195.
71. Schmoller KM, Lieleg O, Bausch AR. Structural and viscoelastic properties of actin/filamin networks: cross-linked versus bundled networks. *Biophys J*. 2009;97:83–89.
72. Lee J, Ishihara A, Oxford G, Johnson B, Jacobson K. Regulation of cell movement is mediated by stretch-activated calcium channels. *Nature*. 1999;400:382–386.
73. Hayakawa K, Tatsumi H, Sokabe M. Actin stress fibers transmit and focus force to activate mechanosensitive channels. *J Cell Sci*. 2008;121:496–503.
74. Wei C, Wang X, Chen M, Ouyang K, Song LS, Chen H. Calcium flickers steer cell migration. *Nature*. 2009;457:901–906.

75. Ehrlicher AJ, Nakamura F, Hartwig JH, Weitz DA, Stossel TP. Mechanical strain in actin networks regulates FilGAP and integrin binding to filamin A. *Nature*. 2011;478:260–263.
76. Lad Y, Kiema T, Jiang P, et al. Structure of three tandem filamin domains reveals auto-inhibition of ligand binding. *EMBO J*. 2007;26:3993–4004.
77. Pentikäinen U, Ylänne J. The regulation mechanism for the auto-inhibition of binding of human filamin A to integrin. *J Mol Biol*. 2009;393:644–657.
78. Rognoni L, Stigler J, Pelz B, Ylänne J, Rief M. Dynamic force sensing of filamin revealed in single-molecule experiments. *Proc Natl Acad Sci USA*. 2012;109:19679–19684.
79. Shattil SJ, Kim C, Ginsberg MH. The final steps of integrin activation: the end game. *Nat Rev Mol Cell Biol*. 2010;11:288–300.
80. Askari JA, Tynan CJ, Webb SED, Martin-Fernandez ML, Ballestrem C, Humphries MJ. Focal adhesions are sites of integrin extension. *J Cell Biol*. 2010;188: 891–903.
81. Ye F, Hu G, Taylor D, et al. Recreation of the terminal events in physiological integrin activation. *J Cell Biol*. 2010;188:157–173.
82. Puklin-Faucher E, Gao M, Schulten K, Vogel V. How the headpiece hinge angle is opened: new insights into the dynamics of integrin activation. *J Cell Biol*. 2006;175:349–360.
83. Wang X, Ha T. Defining single molecular forces required to activate integrin and notch signaling. *Science*. 2013;340:991–994.
84. Kong F, García AJ, Mould AP, Humphries MJ, Zhu C. Demonstration of catch bonds between an integrin and its ligand. *J Cell Biol*. 2009;185:1275–1284.
85. Kong F, Li Z, Parks WM, et al. Cyclic mechanical reinforcement of integrin-ligand interactions. *Mol Cell*. 2013;49:1060–1068.
86. Friedland JC, Lee MH, Boettiger D. Mechanically activated integrin switch controls $\alpha_5\beta_1$ function. *Science*. 2009;323:642–644.
87. Wu C, Keivens VM, O'Toole TE, McDonald JA, Ginsberg MH. Integrin activation and cytoskeletal interaction are essential for the assembly of a fibronectin matrix. *Cell*. 1995;83:715–724.
88. Zhong C, Chrzanowska-Wodnicka W, Brown J, Shaub A, Belkin AM, Burridge K. Rho-mediated contractility exposes a cryptic site in fibronectin and induces fibronectin matrix assembly. *J Cell Biol*. 1998;141:539–551.
89. Baneyx G, Baugh L, Vogel V. Fibronectin extension and unfolding within cell matrix fibrils controlled by cytoskeletal tension. *Proc Natl Acad Sci USA*. 2002;99:5139–5143.
90. Gao M, Craig D, Lequin O, Campbell ID, Vogel V, Schulten K. Structure and functional significance of mechanically unfolded fibronectin type III_1 intermediates. *Proc Natl Acad Sci U S A*. 2003;100:14784–14789.
91. Smith ML, Gourdon D, Little WC, et al. Force-induced unfolding of fibronectin in the extracellular matrix in living cells. *PLoS Biol*. 2007;5:e268.
92. Lemmon CA, Chen CS, Romer LH. Cell traction forces direct fibronectin matrix assembly. *Biophys J*. 2009;96:729–738.
93. Zamir E, Katz M, Posen Y, et al. Dynamics and segregation of cell-matrix adhesions in cultured fibroblasts. *Nat Cell Biol*. 2000;2:191–196.
94. Hirata H, Lim CT, Miyata H. 3D coupling of fibronectin fibril arrangement with topology of ventral plasma membrane. *Cell Commun Adhes*. 2012;19:17–23.
95. Gupton SL, Waterman-Storer CM. Spatiotemporal feedback between actomyosin and focal-adhesion systems optimizes rapid cell migration. *Cell*. 2006;125:1361–1374.
96. Herrera Abreu MT, Penton PC, Kwok V, et al. Tyrosine phosphatase PTPα regulates focal adhesion remodeling through Rac1 activation. *Am J Physiol Cell Physiol*. 2008;294:C931–C944.

97. Chen WT, Singer SJ. Immunoelectron microscopic studies of the sites of cell-substratum and cell-cell contacts in cultured fibroblasts. *J Cell Biol.* 1982;95:205–222.
98. Brown CM, Hebert B, Kolin DL, et al. Probing the integrin-actin linkage using high-resolution protein velocity mapping. *J Cell Sci.* 2006;119:5204–5214.
99. Guo WH, Wang YL. Retrograde fluxes of focal adhesion proteins in response to cell migration and mechanical signals. *Mol Biol Cell.* 2007;18:4519–4527.
100. Hu K, Ji L, Applegate KT, Danuser G, Waterman-Storer CM. Differential transmission of actin motion within focal adhesions. *Science.* 2007;315:111–115.
101. Aratyn-Schaus Y, Gardel ML. Transient frictional slip between integrin and the ECM in focal adhesions under myosin II tension. *Curr Biol.* 2010;20:1145–1153.
102. Giannone G, Mège RM, Thoumine O. Multi-level molecular clutches in motile cell processes. *Trends Cell Biol.* 2009;19:475–486.
103. Li Y, Bhimalapuram P, Dinner AR. Model for how retrograde actin flow regulates adhesion traction stresses. *J Phys Condens Matter.* 2010;22:194113.
104. Gardel ML, Sabass B, Ji L, Danuser G, Schwarz US, Waterman CM. Traction stress in focal adhesions correlates biphasically with actin retrograde flow speed. *J Cell Biol.* 2008;183:999–1005.
105. Ji L, Lim J, Danuser G. Fluctuations of intracellular forces during cell protrusion. *Nat Cell Biol.* 2008;10:1393–1400.
106. Felsenfeld DP, Schwartzberg PL, Venegas A, Tse R, Sheetz MP. Selective regulation of integrin-cytoskeleton interactions by the tyrosine kinase Src. *Nat Cell Biol.* 1999;1:200–206.
107. von Wichert G, Haimovich B, Feng GS, Sheetz MP. Force-dependent integrin-cytoskeleton linkage formation requires downregulation of focal complex dynamics by Shp2. *EMBO J.* 2003;22:5023–5035.
108. von Wichert G, Jiang G, Kostic A, De Vos K, Sap J, Sheetz MP. RPTP-α acts as a transducer of mechanical force on α_v/β_3-integrin-cytoskeleton linkages. *J Cell Biol.* 2003;161:143–153.
109. Giannone G, Sheetz MP. Substrate rigidity and force define form through tyrosine phosphatase and kinase pathways. *Trends Cell Biol.* 2006;16:213–223.
110. Kobayashi T, Sokabe M. Sensing substrate rigidity by mechanosensitive ion channels with stress fibers and focal adhesions. *Curr Opin Cell Biol.* 2010;22:669–676.
111. Discher DE, Janmey P, Wang YL. Tissue cells feel and respond to the stiffness of their substrate. *Science.* 2005;310:1139–1143.

PART THREE

Nuclear Mechanisms of Sensing

CHAPTER SEVEN

The Cellular Mastermind(?)—Mechanotransduction and the Nucleus

Ashley Kaminski[1], Gregory R. Fedorchak[1], Jan Lammerding
Department of Biomedical Engineering & Weill Institute for Cell and Molecular Biology, Cornell University, Ithaca, New York, USA
[1]These authors contributed equally

Contents

Progress in Molecular Biology and Translational Science, Volume 126
ISSN 1877-1173
http://dx.doi.org/10.1016/B978-0-12-394624-9.00007-5

Abstract

Cells respond to mechanical stimulation by activation of specific signaling pathways and genes that allow the cell to adapt to its dynamic physical environment. How cells sense the various mechanical inputs and translate them into biochemical signals remains an area of active investigation. Recent reports suggest that the cell nucleus may be directly implicated in this cellular mechanotransduction process. Taken together, these findings paint a picture of the nucleus as a central hub in cellular mechanotransduction—both structurally and biochemically—with important implications in physiology and disease.

ABBREVIATIONS

2-D 2-dimensional
3-D 3-dimensional
BAF barrier to autointegration factor
CB Cajal body
CS-MS cysteine-shotgun mass spectrometry
EDMD Emery-Dreifuss Muscular Dystrophy
ER endoplasmic reticulum
ERK extracellular-signal-regulated kinase
FISH fluorescence *in situ* hybridization
FRET fluorescence resonance energy transfer
GFP green fluorescent protein
HGPS Hutchinson–Gilford progeria syndrome
KASH Klarsicht/ANC-1/Syne homology
LAD lamina-associated domain
LAP lamina-associated protein
LBR lamin B receptor
LEM-domain LAP2, Emerin, and MAN1-domain
LINC linker of nucleoskeleton and cytoskeleton
MAPK mitogen-activated protein kinase
MKL1 megakaryocytic acute leukemia-1
MRTF-A myocardin-related transcription factor-A
PML-NB promyelocytic leukemia nuclear body
Rb retinoblastoma protein
SMN survival of motor neuron
SREBP1 sterol regulatory element-binding protein-1
SRF serum response factor
SUN Sad-1/UNC-84

SYNE synaptic nuclear envelope proteins
TAN lines transmembrane actin-associated nuclear lines
TGF transforming growth factor

1. INTRODUCTION

The nucleus, often considered the "brain" of the cell, contains the cell's vast genome. The nucleus is primarily responsible for regulating gene expression, which in turn controls the instructions for protein synthesis. This transcriptional and translational mechanism enables cells to respond to mechanical, and other, stimuli. Initiating the appropriate mechanoresponse necessitates the nucleus correctly interpreting the type, duration, and magnitude of the signals it receives. These signals can be biochemical, stemming from signaling cascades that started at the cell membrane and cytoskeleton, or they can be mechanical, reaching the nucleus through direct force transmission. Highlighting the importance of the nucleus in mechanotransduction is the fact that defects in the function of the nucleus can result in misregulated expression of mechanosensitive genes, even if the rest of the cell is functioning properly. Thanks to groundbreaking research conducted over the past three decades, much is known about how integrins and focal adhesions respond to substrate stiffness, strain, and other mechanical stimuli, as discussed in detail in the previous chapters. For the nucleus, such insights are only beginning to emerge. It is now evident that the nucleus plays a vital role in cellular mechanotransduction, even though many questions still surround its precise function(s). Specifically, is the nucleus merely a puppet to the surrounding cytoplasm, relaying biochemical signals into changes in gene expression? Or does the "brain of the cell" have a mind of its own when it comes to sensing mechanical inputs? These questions make up the two main hypotheses regarding the role of the nucleus, and its associated proteins and structures, in cellular mechanoresponse.[1] The first scenario is that the nucleus has the primary role of transcriptional regulation in response to biochemical signals coming from the rest of the cell, with little direct involvement in the actual sensing of mechanical forces. In the second scenario, the nucleus is the final element in a cellular Rube Goldberg machine that consists of a chain of structural, force propagating elements which induce nuclear deformations that directly translate into gene expression instructions. Distinguishing between these two scenarios is a major question in mechanobiology, and the answer likely lies somewhere in between these extremes.

Before investigating the role of the nucleus in mechanotransduction, we need to define the components and processes of the cellular mechanoresponse. A transducer, by definition, is a device that converts one form of energy into another. There are numerous transducers that we interact with in everyday life; for instance, a loud speaker transduces electrical energy into sound waves and an electric motor transduces electricity into mechanical energy, i.e., motion. In the context of the cell, we will refer to the specific structure (protein or other molecule) that transduces mechanical signals into biochemical signals as the "mechanotransducer" or, to avoid any confusion, the "mechanosensor." While technically mechanotransduction only describes the immediate process of converting a mechanical signal into a biochemical signal, the term "mechanotransduction" is often also understood to include the downstream biochemical signaling from a mechanosensor. Furthermore, in the literature, "mechanotransduction" is also used to describe any cellular change in response to mechanical stimuli, for example, the alignment of endothelial cells in response to fluid shear stress, or the adaptation of bone density to microgravity or physical activity. In this chapter, we use the term "mechanotransduction signaling" to include both the initial mechanosensing event as well as the biochemical signaling downstream of the mechanosensor. Complementary to the intracellular propagation of biochemical signals is the process of propagating physical forces through the cell; we will call the structures involved in this process "mechanotransmitters." Importantly, based on the definition used here, a given protein could be involved in mechanotransduction signaling without itself being a mechanosensor.

Historically, research in mechanotransduction has attempted to find *the* cellular mechanosensor; it is now becoming apparent that there are numerous mechanosensors in the cell, ranging from stretch-sensitive channels in the plasma membrane to cytoplasmic proteins that undergo conformational changes in response to force.[2] Several recent studies support the idea of the nucleus being one such cellular mechanosensor, as discussed in detail below. At the same time, it is important to recognize that even if the nucleus may not directly sense mechanical stimuli, it certainly has a key role in regulating the cellular mechanoresponse via both physical force transmission and biochemical signal processing.

Although the specific function of the nucleus in cellular mechanotransduction is still incompletely understood, it is well established that mutations in numerous nuclear envelope proteins cause both defects in mechanotransduction signaling and force transmission.[3,4] These mutations can cause muscular dystrophy, dilated cardiomyopathy, partial familial lipodystrophy, cancer, and the accelerated aging disease Hutchinson–Gilford progeria syndrome (HGPS), among others. Many of these diseases are caused by mutations in a nuclear envelope protein, lamin A/C, which is encoded by the *LMNA* gene. To date, more than 450 disease-causing mutations have been discovered in the *LMNA* gene alone, with the vast majority of mutations affecting striated muscle, i.e., mechanically stressed tissues (http://www.umd.be/LMNA). In the case of the *LMNA* gene, the specific effects of subtle differences between these mutations on the resulting disease are fascinating. For instance, changing a single amino acid in lamin A/C at position 528 from threonine to lysine causes muscular dystrophy, while changing the identical amino acid position to methionine results in lipodystrophy symptoms.[5,6] Equally interesting is the fact that similar disease phenotypes can often be caused by mutations in one of multiple proteins (e.g., mutations in either lamins, emerin, nesprins, or the cytoskeletal protein dystrophin all cause muscular dystrophy). This suggests that these proteins are all involved in similar cellular functions, e.g., force transmission, mechanical stability, or mechanotransduction, and highlights the importance of intact force transmission and mechanotransduction pathways in cellular function. An improved understanding of the role of the nucleus in mechanotransduction would not only lead to better insights into normal cell biology but may also pave the way for novel therapies for the many diseases caused by mutations in nuclear (envelope) proteins.

2. OVERVIEW OF NUCLEAR STRUCTURE AND ORGANIZATION

As the age old maxim goes, structure imparts function. Like a mechanic fixing a car without understanding how the engine is built and connected to the rest of the car, trying to decipher the role of the nucleus in mechanotransduction and disease necessitates an understanding of nuclear structure and its connection to the cytoskeleton. Given the relevance to human disease, we restrict our discussion to mammalian cells. In eukaryotic cells, the nucleus houses not only the genome but also the transcriptional machinery, thus allowing it to act as the central processing center of incoming signals. The nucleus is typically the largest cellular organelle. It is separated from the surrounding cytoplasm by two lipid membranes and the underlying nuclear lamina meshwork which provides structural support. Together the membranes, lamina, and associated proteins make up the nuclear envelope, which also mechanically connects the cytoskeleton to the nuclear interior.[7] As the nucleus is substantially stiffer than the surrounding cytoplasm, the mechanical properties of the nucleus significantly contribute to the overall cell deformability and the transmission of forces across the cell. In the following, we provide a brief description of the structural and mechanical components of the cell nucleus, from the nuclear interior to the outer nuclear membrane and the proteins linking the nucleus to the cytoskeleton. These sections will illustrate that the nucleus is connected to the cellular environment by a continuous physical chain via the cytoskeleton, focal adhesions, and cell–cell junctions (Fig. 7.1). Consequently, forces from the extracellular matrix, neighboring cells, and the cytoskeleton can be transmitted directly to the nucleus, where they can induce substantial structural changes.[8]

2.1. Chromatin and chromosome territories

Long, contiguous DNA polymers are packaged into discrete units called chromosomes, which are made of chromatin, a fibrous combination of DNA and DNA-binding proteins. The most prominent of these proteins are histones; histone polypeptides facilitate the efficient packing of two meters of human DNA into a $\approx$10-μm-diameter nucleus.[9] This stunning packaging is achieved by winding the DNA-like thread around a histone spool. Histones H2A and H2B, and H3 and H4, unite as heterodimers to form octamer "beads" that are—together with the wound DNA—called nucleosomes. The odd histone out, histone H1, helps link nucleosomes together and stabilize higher-order structures. Histones are well suited to transmit signals directly to the genome through a plethora of covalent biochemical modifications such as acetylation, phosphorylation, methylation, poly-ADP-ribosylation, ubiquination, and SUMOylation which affect the structure and transcriptional activity of chromatin without altering the DNA sequence.[10,11] Broadly speaking, chromatin exists in two states: (1) the more transcriptionally active, more loosely organized euchromatin, which typically localizes to central regions of the nucleus, and (2) the transcriptionally inactive, more densely packed heterochromatin, which inhabits the nuclear periphery and the area surrounding nucleoli.[12] DNA-binding

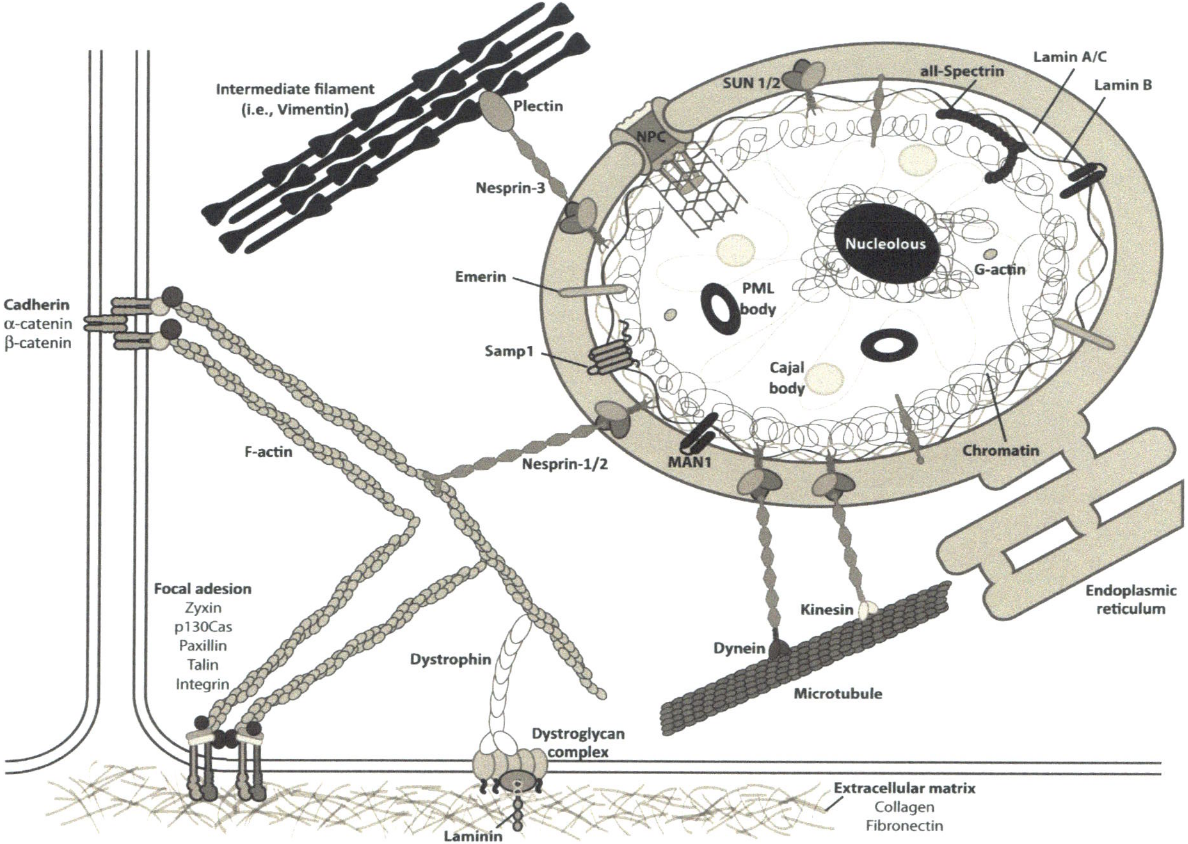

Figure 7.1 Cellular connectivity. The genome is connected to the extracellular environment through an array of proteins. The nucleoplasm houses the chromatin and subnuclear bodies (nucleolus, Cajal bodies, and PML bodies). The chromatin associates with the nuclear lamina along the inner nuclear membrane, where it also interacts with LEM-domain proteins. Reaching across the luminal space, SUN proteins on the inner nuclear membrane bind to the KASH domains of nesprins to form the LINC complexes that link the nuclear interior to the cytoskeleton filaments (actin, intermediate filaments, and microtubules), often through intermediary linker proteins such as plectin, dynein, or kinesin. The cytoskeletal filaments connect to cell–cell and cell–matrix adhesions, completing the physical connection from genome to extracellular environment. (See the color plate.)

dyes such as DAPI or Hoechst stain compact heterochromatin darker than its euchromatic counterpart. Euchromatin and heterochromatin can also be easily distinguished in transmission electron micrographs.[9]

The differences in chromatin compaction have important implications for deformation under force. Studies on isolated nuclei have shown that the nuclear interior, chromatin in particular, substantially contributes to the mechanical property of the nucleus.[13] Both condensed and noncondensed chromatin behave as complex viscoelastic fluids but demonstrate plastic changes after ≈10 s of stress in micropipette aspiration experiments.[14] In addition, condensed chromatin is substantially less deformable than noncondensed chromatin,[14] suggesting that the higher packing density limits further compaction. Importantly, the nucleoplasm is ≈2–10 times stiffer than the cytoplasm and resists further compression after being compacted more than 60%.[15–17] The "compression limit" is one contributor to chromatin's, and especially heterochromatin's, ability to bear up to 50% of the load applied to the nucleus[14,18] and may become highly relevant when cells must deform during migration through tight interstitial spaces.[19] These findings illustrate the interrelation between chromatin and nuclear mechanics and may hint at mechanisms by which force-induced changes in chromatin structure could contribute to nuclear mechanosensing.

Recent technological advancements have enabled stunning views of the nonrandom chromosome organization (chromosome territories) within the interphase nucleus. For example, fluorescence *in situ* hybridization (FISH) techniques with multiple fluorescent probes have been used to construct a probability model of chromosome territories. Using FISH, Cremer and Cremer showed that chromosomes have preferential locations that do not become intermingled.[20] Intriguingly, these territories appear to be relatively stable throughout mitosis.[21] This nonrandom orientation could enable specific directional forces, such as shear stress, to consistently impact the same chromatin region. For instance, it has been shown that cytoskeletal filaments in endothelial cells stiffen in regions exposed to high levels of localized shear.[22] It is conceivable that these forces could then propagate to specific nuclear territories and potentially influence gene expression.

While the nuclear interior is primarily occupied by chromatin, it also contains other distinct structures such as nucleoli and Cajal bodies (CBs), as well as structural proteins such as lamins, actin, myosin, titin, and spectrin[23,24] which will be discussed in the following.

2.2. Subnuclear structures and nuclear bodies

Interest in subnuclear organization stems from the desire to uncover links between structural changes and alterations in gene expression and cell signaling. The most prominent subnuclear structures are nucleoli, which appear as dark spots in the nuclear interior by phase-contrast light microscopy. Nucleoli are dynamic nuclear bodies primarily responsible for ribosome biogenesis. They are the largest subnuclear compartments and are functionally separate from the rest of the nucleus. When the nucleus is subjected to mechanical stress, nucleoli also undergo plastic deformations.[14]

In addition to nucleoli, the majority of mammalian nuclei contain 5–30 protein-based compartments called promyelocytic leukemia nuclear bodies (PML-NBs), which play a role in transcription, DNA damage repair, and nuclear response to viral infection.[25–27] PML-NBs show dynamic changes in number and size when cells are exposed to various stresses, such as heat shock and heavy metal exposure. However, their response to mechanical stress has not been well characterized.[28,29] This could yield interesting results given the specific interactions between PML-NBs and certain chromosomal loci.[26] In addition, a highly dynamic class of PML-NBs moves in an ATP-dependent fashion and may utilize nuclear actin and myosin in order to rapidly traverse the nucleus, further linking PML-NBs to mechanoresponsive elements.[27,30]

CBs are 0.5–1-μm-diameter subnuclear domains which feature two major multiprotein–mRNA complexes characterized by survival of motor neuron (SMN) and coilin.[31,32] Their main function is to assemble and modify ribonucleoproteins important in pre-mRNA splicing, telomere formation, and histone mRNA processing.[33] The number of these structures is a function of both cell type and cell cycle. CBs are more frequently seen in cells with high transcriptional demands (e.g., neurons, skeletal muscle cells, cancer cells), while in other cell types they may not be visible at all (skin epidermal cells, cardiac and smooth muscle cells).[32,34,35] Like PML-NBs, CBs are highly stress responsive to viral infection and radiation, among others. Linking to the genome, CBs associate dynamically with chromatin and they are mobile structures with their interactions and localization being influenced by phosphorylation.[33,36] As discussed in detail later, recent findings suggest that mechanical forces transmitted across the nuclear envelope may also affect CB structure and function.[31]

2.3. The nucleoskeleton

For decades, researchers have sought to characterize the putative nucleoskeleton as either an active framework, a passive framework, or an artifact.[23,37] The existence of simple, cell-free technologies to replicate and transcribe DNA proves that many of the quintessential functions of the nucleus can be carried out without higher-order structures (i.e., nucleoskeletal elements). At the same time, a nucleoskeletal framework or scaffold could help organize chromosome territories as well as position the transcriptional machinery and regulatory complexes. Relating back to mechanosensing, a dynamic, well-networked nucleoskeleton has the potential to directly couple extracellular forces to the genome. A prominent class of nucleoskeletal proteins are lamins, which exist as both a solid-like meshwork at the periphery and may form stable structures with transcription factors and other molecules in the interior.[38] Lamins and their interaction partners will be discussed in detail later.

In addition to specialized nuclear proteins, recent findings suggest that many proteins traditionally found in the cytoskeleton, such as spectrin II-alpha, titin, 4.1R, myosin, and actin, have roles in the nucleus as well.[39,40] The organization of these proteins in the nucleus remains uncertain—they may form a structural scaffold throughout the nuclear interior or they may form a peripheral network supporting the nuclear envelope.[24] Spectrin II-alpha and titin are cross-linking proteins known for their large size and elasticity. Consistent with these characteristics, nuclei that lack spectrin II-alpha have difficulty recovering from swelling and stretching.[40] Beyond its mechanical function, spectrin II-alpha is also important for DNA repair,[41] demonstrating how structural and regulatory roles often overlap in the nucleus.[12,40] Titin, normally found in muscle sarcomeres, has been shown to also interact with lamins both *in vitro* and *in vivo*. Structurally, loss of titin results in large nuclear blebs.[42]

One important player that is beginning to emerge in this context is nuclear actin. The presence of actin in the nucleus, a once hotly contested subject, is now widely accepted.[30,43] Actin exists as either monomeric G-actin or polymeric F-actin and is ubiquitously expressed in virtually all eukaryotic cells. In the nucleus, monomeric actin is highly active throughout transcription and has many interaction partners, including RNA polymerase I–III and certain transcription factors.[30] Exciting new fluorescent probes can now distinguish between monomeric (RPEL1 motif probes) and filamentous (Lifeact-NLS and Utr230-based probes) nuclear actin in living cells, showing distinct localization patterns and kinetics of these molecules.[30,44] Immunofluorescence studies indicate that Utr230-actin reporters distribute throughout the interchromatin space and neither colocalize with chromatin remodelers nor with traditional actin-associated proteins such as myosin I.[44] The main purpose of filamentous nuclear actin could then be to contribute to the viscoelasticity, or a stabilizing meshwork, to of the nuclear interior, similar to what is seen in *Xenopus laevis* oocytes.[45] Regardless of its structural role, F-actin assembly by formins or other actin nucleators[30] can control gene expression by freeing G-actin from the transcription factor MKL1 and enabling it to activate serum response factor (SRF) target genes, or alternatively, through modulating histone-acetyltransferases.[30] The role of nuclear actin in signaling will be discussed in more detail later. In general, nuclear actin and the nucleoskeleton are no longer considered artifacts; however, the ways by which mechanical force affects nucleoskeletal proteins and their functions remain incompletely understood.[12,23]

2.4. Lamins, the nuclear lamina, and nuclear mechanics

Lamins and their associated proteins form a dense meshwork (nuclear lamina) along the inner nuclear membrane. The lamina interacts with inner nuclear membrane proteins, nuclear pore complexes, and the nuclear interior. Lamins are type V intermediate filaments that can be grouped into two classes: (1) A-type lamins, which are generated by alternatively splicing of the *LMNA* gene into lamin A and C and some less abundant isoforms, and (2) B-type lamins, which are encoded by the *LMNB1* and *LMNB2* genes, which produce lamin B1 and B2/B3, respectively.[38] While A-type lamins and lamins B1 and B2 are expressed in almost all somatic cells, lamin B3 expression is restricted to germ cells. Lamin A and B-type lamins undergo extensive post-translational processing at the C-terminus, including farnesylation and endoproteolytic cleavage. B-type lamins remain permanently farnesylated and thus attached to the inner nuclear membrane, even during mitosis.[46] In contrast, lamin A undergoes an additional modification, where the protein Zmpste24 removes the farnesylated tail, resulting in mature lamin A. Lamin C, which has a distinct C-terminus, does not undergo the same processing and is not farnesylated.[10] Mature lamin A and lamin C, which lack the hydrophobic farnesyl tail, can be found both in the nucleoplasm and the nuclear lamina.[47]

Lamins, which have a half-life of $\approx$ 13 h, assemble into stable filaments.[48] They form parallel dimers through coiled-coil interaction of their central rod domains.[38] The dimers associate head to tail and then laterally assemble in an antiparallel fashion into nonpolar filaments with a final diameter of about 10 nm. In transmission electron micrographs of mammalian cells, the nuclear lamina is visible as a 25–50-nm thick dense protein layer underneath the inner nuclear membrane.[7,17] The higher-order structure of lamins in somatic cells is not completely understood due to the tight association of the lamina with chromatin, which makes high-resolution imaging challenging.[49] However, *Xenopus* oocytes do not pose the same challenges; electron micrographs in these cells show a lamin structure composed of a square lattice of ≈10-nm thick cross-linked filaments.[49,50] Because of this, lateral interactions between dimers and protofilaments are thought to be critical for maintaining the correct higher-order structure. Based on mathematical modeling, correct coiling direction of the heptads appears to be important to allow for "unzipping" and subsequent attachment to adjacent strand.[51] Mutations could result in increased or decreased stability due to incorrect assembly and/or binding.[52] It is important to note that these ideas await experimental confirmation. Intriguingly, although different lamin isoforms can all interact and form heteropolymers *in vitro*, they typically segregate into homopolymers and form distinct, but overlapping, networks *in vivo*.[53–56]

Although there are still some questions about the filament and structural assembly of the lamina *in vivo*, the importance of nuclear lamins in contributing to nuclear stiffness and stability has been unequivocally established. Based on micropipette aspiration experiments on isolated *Xenopus* oocyte nuclei, which can be osmotically swollen to separate the chromatin from the nuclear lamin, the lamin network has an elastic modulus of ≈25 mN/m.[57] For comparison, the plasma membrane of neutrophils has an elastic modulus of ≈0.03 mN/m and chondrocyte and endothelial cell membranes have a modulus of ≈0.5 mN/m.[58] Using a variety of experimental approaches, the stiffness of the nucleus has been determined to be 2–10 times stiffer than the surrounding cytoplasm, depending on the particular cell type and measurement method.[16,59,60] When comparing the lysis strain of the nuclear envelope (i.e., the nuclear lamina and the nuclear membranes) with that of a simple double lipid membrane to distinguish the contribution of the nuclear lamina, the lysis strain of the nuclear envelope was 12-fold higher than that of the standard double membrane system, highlighting the stabilizing impact of the nuclear lamina.[57] Similarly, when a fluorescent dye is injected into the nucleus of living cells, cells that lack lamins A/C show dramatically increased rates of nuclear rupture compared to wild-type cells.[61]

Given this important role of lamins in conferring structural integrity to the nucleus, what is the contribution of the different types of lamins to nuclear mechanics? While B-type lamins are nearly ubiquitously and uniformly expressed among different cell types and tissues, lamin A/C expression is highly tissue specific. For instance, muscle cells and other mesenchymal cells typically are among the highest in A-type lamin expression levels.[62,63] A recent study found that the ratios of A-type to B-type lamins in different tissues closely correlates with the tissue stiffness suggesting a mechanosensitive regulation of lamin levels,[62] which could help protect the nucleus from mechanical stress by increasing mechanical stability.[61] In cells that express both A-type and B-type lamins, lamins A and C are the major contributors to nuclear stability, with B-type lamins having a smaller role in overall nuclear stiffness.[64] Nonetheless, there may be some functional redundancy between lamins regarding mechanical properties. For instance, introducing lamin B into lamin A-null cells can partially rescue mechanical defects.[54,65] Furthermore, B-type lamins are important for nuclear anchoring to the cytoskeleton, particularly during neuronal migration/development in the brain, as these cells lack A-type lamins.[66–69]

Similarly, embryonic stem cells do not express A-type lamins until they begin to differentiate. Once they decrease their stemness, their nuclear stiffness increases up to sixfold compared to the undifferentiated state. This is most likely due to the increased levels of lamins A/C in the new lineage and possibly changes in chromatin configuration.[14,63] A few specialized differentiated cells, notably neutrophils and neurons, hardly express any A-type lamins even after differentiation.[68,70] The lack of A-type lamins in embryonic stem cells, neutrophils, and neurons may facilitate migration, enabling these cells to travel through dense tissues and interstitial spaces during development and inflammation.[71] For example, the decrease in lamin A/C levels along with the concomitant increase in expression of lamin B receptor (LBR) during granulopoiesis promotes the distinct highly lobulated nuclear shape of mature neutrophils.[15] In addition, the low levels of lamin A result in a highly deformable nucleus allowing neutrophils to easily squeeze through small spaces.[15] Similarly, regulation of lamin A/C levels may also regulate trafficking and lineage maturation of other hematopoietic cells types.[72]

In addition to changes in lamin expression, posttranslational modifications of lamins may further affect nuclear mechanics. Lamins are phosphorylated during mitosis, which causes them to become soluble and disperse into the cytoplasm.[47,73] Because farnesylation and phosphorylation of lamins changes their solubility, interaction, and localization, these posttranslational modification may also offer cells a way to dynamically adjust their nuclear stiffness in response to mechanical stimuli.[62]

2.5. Lamin-binding proteins

Beyond their direct structural role, lamins bind many different proteins, including transcriptional regulator, histones, and several inner nuclear membrane proteins (Fig. 7.2). Lamin-binding partners include LBR, nesprins, SUN (Sad-1/UNC-84) proteins, and members of the LEM-domain protein family, which comprise lamina-associated protein (LAP2), emerin, and MAN1 proteins.[74,75] In the absence of lamins, many of the lamin-binding proteins become mislocalized. This has severe consequences for force propagation across the nuclear envelope (see below) and mechanotransduction signaling. The LEM-domain proteins share a conserved ~40-amino acid sequence that allows them to bind BAF (barrier to autointegration factor), a protein with roles in nuclear organization and DNA bridging.[76,77] BAF helps to recruit lamin A and stabilize emerin at the nuclear envelope in addition to maintaining heterochromatin.[78,79] The LEM-domain proteins have additional binding partners that could affect mechanotransduction signaling.

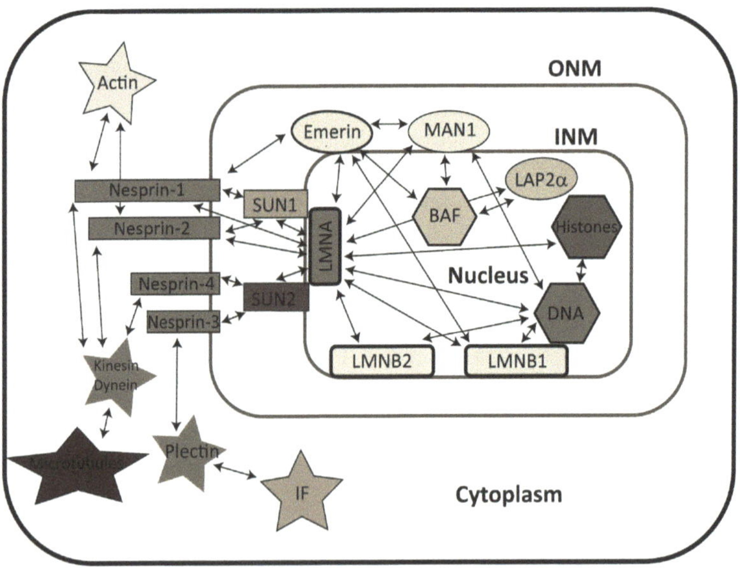

Figure 7.2 Nuclear envelope protein binding partners. Lamins associate with numerous proteins/structures in the nuclear interior and the nuclear membranes, thereby modulating intracellular force propagation and mechanotransduction signaling. Many of the lamin-binding proteins also bind to each other, creating a complex network of interactions. Please note that additional interactions may exist. In addition, many of the nuclear envelope proteins also bind a variety of transcriptional regulators not included in the figure. Stars = cytoskeletal elements; rectangles = LINC complex proteins; rounded rectangles = lamins; ellipses = LEM-domain proteins; hexagons = chromatin-associated elements. IF: Intermediate Filaments.

Broadly, LAP2, emerin, and MAN1 all interact with GCL, a transcriptional repressor. Emerin is particularly interesting in this regard because it competitively binds BAF and GCL.[80] Mutations in LEM-domain proteins can cause changes in downstream BAF signaling, but can also disrupt additional signaling pathways such as transforming growth factor β (TGF-β) and Wnt signaling, as will be discussed in detail later.[81,82]

Lamins can also directly connect to the genome by binding DNA and histones.[75,83] The areas of the genome that are bound to the nuclear lamina are called lamina-associated domains (LADs) and are often transcriptionally repressed heterochromatin. In addition, lamins also directly bind to numerous transcriptional factors such as extracellular-signal-regulated kinases 1 and 2 (ERK1/2), retinoblastoma protein Rb, Fos, sterol regulatory element-binding protein-1 (SREBP1), and MOK2, among others.[83–86] Of note is that phosphorylated Rb, one proposed regulator of lamin expression, can also directly bind lamins (together with LAP2α), suggesting a potential feedback loop. Some of these factors, such as ERK1/2 and c-Fos, have well established roles in mechanotransduction signaling and will be discussed in more detail later. For a detailed discussion of lamin-binding partners and their function, we refer to a recent review by Simon and Wilson.[10]

2.6. Nuclear membranes and nuclear pore complexes

Adjacent to the nuclear lamina and forming a barrier to the surrounding cytoplasm are the inner nuclear membrane and the outer nuclear membrane. The latter is continuous with the endoplasmic reticulum (ER), a large intracellular membrane system. The space enclosed by the inner and outer nuclear membranes is the ≈30–50 nm wide luminal or perinuclear space. Proteins in the inner nuclear membrane can interact with, and be retained, by the underlying lamina and nuclear interior. In the absence of these interactions, as in the case of lamin A/C-deficient cells, the inner nuclear membrane proteins can diffuse into the outer nuclear membrane at the nuclear pores, where the membranes converge, and then further into the ER. Similarly, proteins in the outer nuclear membrane are often retained there by their interaction with inner nuclear membrane proteins and can be mislocalized to the ER if this interaction is lost.[87] At least 74 membrane proteins specific to the nuclear envelope have been identified to date, conferring a variety of often tissue-specific functions from import and export, chromatin anchoring and regulation, and nucleocytoskeletal coupling.[88,89]

Nuclear pores help to regulate the active transport of large molecules (larger than ≈40 kDa) into and out of the nucleus. Ions and small molecules can passively diffuse through the membranes or the nuclear pores. Nuclear pore complexes are highly structured, symmetric channels that are regularly spaced along the nuclear envelope. This spatial regulation is likely due to interactions with the nuclear lamina, because loss of lamins can result in clustering of nuclear pore complexes.[90] Nuclear pore complex components interact with a larger number of proteins at the nuclear envelope (e.g., lamins, SUN proteins) and the nuclear interior.[91,92] Recent evidence suggests that proteins typically associated with the nuclear pore complex proteins can also be found deep inside the nuclear interior, where they may play a role in transcriptional regulation.[11,12]

2.7. LINC complexes

The lamina is connected to the cytoskeleton by the linker of nucleoskeleton and cytoskeleton (LINC) complex, which spans both the inner and outer nuclear membrane to transmit forces across the nuclear envelope (Fig. 7.1). LINC complex consists of SUN proteins at the inner nuclear membrane and KASH-domain proteins at the outer nuclear membrane. The LINC complex proteins are highly conserved across organisms and their vertebrate names are a combination of the names originally given in *C. elegans*, *D. melanogaster*, and *S. pombe*. The SUN proteins bind to the lamina and the SUN domain occupies the perinuclear space. There are two SUN proteins in somatic cells, SUN1 and SUN2, which have distinct but overlapping functions.[93] Three additional SUN proteins, SUN3–5, are only expressed in male germline cells.[94] All SUN proteins are composed of an N-terminal nucleoplasmic domain, a transmembrane domain, and a luminal coiled-coil rod domain that ends in a trimeric, SUN-domain head.[95] The main difference between SUN1 and SUN2 and the germline SUN proteins is the length of the rod domain.[95]

The trimeric C-terminal of the SUN proteins binds up to three KASH (Klarsicht/ANC-1/Syne homology) peptides, which make up the luminal C-terminus (and the transmembrane domain) of the aptly named KASH-domain proteins. The first four KASH-domain family members are referred to as nesprins (nuclear envelope spectrin repeat proteins), each consisting of several isoforms resulting from alternative splicing or initiation.[96] Historically, nesprin-1 and -2 were named Enaptin and NUANCE, respectively. Some nesprins were also referred to as SYNEs (synaptic nuclear envelope proteins) because of their presence in myonuclei at neuromuscular junctions.[97] A fifth family member, KASH5, is germ cell specific and lacks the name-sake spectrin repeats of nesprins.[98] Nesprins-1, -2, and -3 are generally widely expressed, and at least one nesprin isoform is found in all cell types.[99] In contrast, nesprin-4 expression is specific to secretory epithelial cells. On the cytoplasmic side, nesprins can interact with all major cytoskeletal structures. The largest isoforms of nesprin-1 and -2 directly bind to actin and nesprins-1 and -2 also bind to kinesin and dynein; nesprin-3 engages intermediate filaments via plectin; nesprin-4 connects to the microtubule network via kinesin-1 (Fig. 7.1).[100–103] The large, "giant," nesprin isoforms are found on the outer nuclear membrane, while smaller nesprin isoforms like nesprin-1α and -2β also localize to the inner nuclear membrane and bind lamins.[102,104,105]

The integrity of the LINC complex and cytoskeletal filaments is needed to support intracellular force transmission to the nucleus. LINC complex disruption causes changes in both the nuclear and cytoskeletal morphology, including loss of the actin and vimentin network surrounding the nucleus.[106,107] Depletion of LINC complex components, or expression of dominant-negative nesprin constructs, results in altered nuclear cross-sectional area and an increase in nuclear height. This is presumably due to loss of cytoskeletal tension or compression acting on the nucleus.[108] In addition to changes in nuclear morphology, downregulation of nesprin-1 also increases focal adhesion strength, potentially in an attempt to balance the forces normally distributed to the nucleus. In addition to these findings, many other groups have demonstrated the effect of the LINC complex on nuclear morphology based on substrate stiffness, 2-dimensional (2-D) versus 3-dimensional (3-D) microenvironment, and mechanical strain.[109–113]

In many ways, the LINC complexes appear analogous to the focal adhesion complexes at the plasma membrane, which help transmit forces from the cellular environment to the cytoskeleton and are also involved in mechanosensing.[114] Comparing the current view of the simple LINC complex structure with the complex structure and interaction of focal adhesion complexes and the many associated proteins, it becomes apparent that there is still much work needed to better understand nucleocytoskeletal coupling. As of now, there are a few other proteins that are known to help control the interaction between nesprins and SUN proteins. Samp1 was recently identified as a LINC complex-associated protein that interacts with SUN1 and aids in nuclear positioning.[115,116] Torsin-A appears to be another potential regulator of SUN–nesprin interaction,[117] but much remains to be uncovered about how binding between SUN proteins and nesprins is controlled. Furthermore, there are still many questions regarding the formation, maintenance, and regulation of LINC complexes, especially in response to mechanical stimuli. For example, does the same principle of concurrent inside-out and outside-in signaling that governs cell adhesion also apply to the LINC complex? And which other proteins could be involved in such regulation or potential mechanosensing?

3. MECHANICALLY INDUCED CHANGES IN NUCLEAR STRUCTURE

While the above section paints a picture of the static nuclear structure, mechanosensing and mechanotransduction are active processes. It is now well established that forces from the extracellular environment and the cytoskeleton are transmitted to the nucleus, where they can result in (intranuclear) deformations. Physiologically, such forces can result from fluid shear stress, tissue compression, cell migration, and tissue strain. Cell generated traction forces, responding to the stiffness of the underlying substrate, also affects nuclear morphology. To probe specific interactions and intracellular force transmission, more targeted, but also more invasive, experimental techniques have been developed and applied to a variety of cellular systems (Fig. 7.3). These experiments firmly illustrate how the nuclear structure and morphology reflect changes in the extracellular and intracellular mechanical environment, and how changing the structure and composition of the nuclear envelope can significantly affect the intracellular force propagation.

3.1. Nuclear deformation during cell migration

As previously discussed, the nucleus is the largest organelle and resists deformation to greater extent than the surrounding cytoplasm. Because of this, the rate of nuclear deformation may constitute a limiting factor in cell migration through narrow constrictions.[15,19,118] In cells migrating through dense collagen matrices, large nuclear deformations and protrusions can be observed as the cell/nucleus squeezes through openings as small as $\approx 3\ \mu m$ in diameter.[19,71] The limit of nuclear compressibility appears to be $\approx 10\%$ of the initial nuclear diameter as migration halts when trying to navigate smaller constrictions.[19] Some specialized cells, such as neutrophils, may have evolved highly lobulated nuclei with very low levels of lamins to mitigate this limitation.[15] Ongoing research is aimed at further establishing the role of nuclear mechanics in 3-D cell migration, including the contributions of lamin levels, which are altered in many cancers. However, changes in the nuclear envelope composition not only affect nuclear

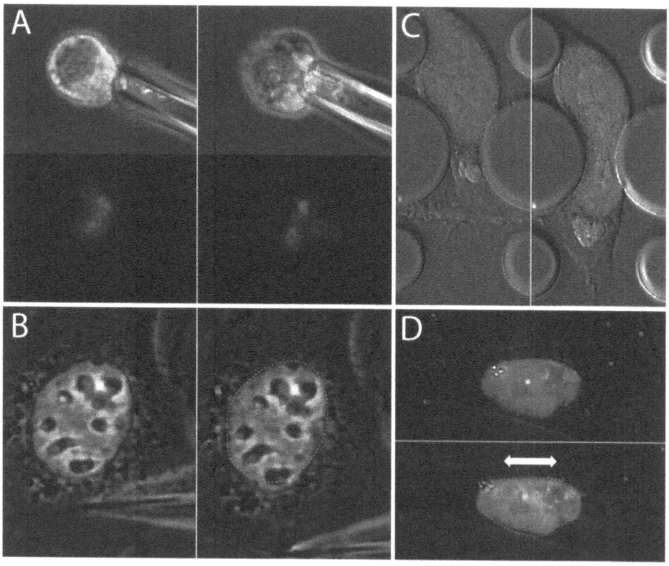

Figure 7.3 Mechanically induced deformation of the nucleus. Examples of biophysical assays to probe nuclear mechanics and nucleocytoskeletal coupling. Images show overlay of phase-contrast images with fluorescent signal (blue) from Hoechst 33342-labeled nuclei. (A) Micropipette aspiration assay to probe nuclear deformability. Images of mouse embryo fibroblast with fluorescently labeled nuclei before (left) and during (right) micropipette aspiration. Bottom: Corresponding fluorescence images alone, illustrating large nuclear deformation during aspiration into the micropipette tip. (B) Microneedle manipulation of a mouse embryo fibroblast to assess force transmission between the cytoskeleton and nucleus. A microneedle is inserted into the cytoplasm and moved away from the nucleus, resulting in large nuclear deformations. Left, initial position; right, induced nuclear and cytoskeletal deformation after microneedle movement. Dotted line indicates initial nuclear outline. (C) Migration of MDA-MB-231 breast cancer cell through microfluidic constrictions channels. The nucleus must undergo substantial deformation as the cell passes through the ≈ 3 μm × 5 μm opening. Left, nucleus before entering constriction; right, nucleus inside the constriction. (D) Nuclear deformation of human skin fibroblast subjected to $\approx 20\%$ uniaxial substrate strain. Top, cell before strain application; bottom, cell and nuclear deformation during strain application. Dotted line indicates nuclear outline before strain application. (See the color plate.)

deformability but can also disrupt force transmission between the nucleus and cytoskeleton (and possibly signaling pathways). This manifests in decreased overall cell migration speeds and lack of persistence in lamin A/C-null and -mutant cells.[111,119] Nuclear deformation and positioning has been extensively studied on cells migrating on 2-D substrates, particularly in response to stiffness. At the nuclear envelope, protein arrays consisting of SUN2 and nesprin-2G bind dorsal actin cables to form so-called transmembrane actin-associated nuclear lines (TAN lines), which allow the cell to use retrograde actin flow to reposition the nucleus for migration.[120] Highlighting the importance of intact connections between the cytoskeleton and the nucleus, depleting lamins, emerin, nesprin-1, nesprin-2, nesprin-3, SUN, or plectin, all results in an increased distance between the nucleus and the centrosome, along with inhibited cell polarization and migration.[119,121–124]

3.2. Nuclear deformation under shear stress

Vascular endothelial cells experience constant fluid shear stress. This consistent force results in the alignment of cells in the flow direction.[22,125] Importantly, fluid shear stress application for 24 h also causes endothelial cell nuclei to flatten, lengthen, and increase up to 50% in stiffness.[126] What is fascinating is that these changes persist even after nuclei are isolated, suggesting that stable nuclear reorganization or remodeling occurred.[126] Some of these changes may be facilitated by reorganization of the actin cytoskeleton, as related to the magnitude of shear stress. The cytoskeletal reorganization and resulting irreversible deformations require the presence of nesprins-2 and/or -3, highlighting again the importance of LINC complexes.[122,127] Beyond the gross nuclear level, shear stress application also affects subnuclear structures; after a half hour of force application, rDNA upstream of polymerase I showed directionally enhanced diffusion, on the same time scale as cytoskeletal remodeling.[128]

3.3. Nuclear deformation under compression

Cells of mesenchymal origin frequently experience compressive forces, for example, in cartilage and during muscle contraction. Compression of chondrocytes results in large changes in nuclear shape and height[129]; associated changes in cellular osmolarity could further affect nuclear structure and chromatin organization.[130] In response to compressive forces, mouse embryonic fibroblasts lacking lamin A/C deform isotropically, while wild-type cells exhibit anisotropic deformations.[131] Because isolated nuclei deform isotropically, it is hypothesized that the anisotropic deformation observed in the wild-type cells is due to cytoskeleton linkages, which are disrupted in the lamin A/C-deficient cells. Importantly, lamin A/C-deficient cells show less resistance to compression and more frequent nuclear rupture,[132] again illustrating the importance of A-type lamins in stabilizing the nucleus.

3.4. Nuclear deformation under strain application

In cells subjected to strain, e.g., by stretching the underlying substrate or an entire tissue, the nucleus undergoes substantial deformations. In human and mouse embryo fibroblasts plated on flexible silicone membrane, nuclear deformation accounts for up to 30% of the applied membrane strain, with depletion of lamins A/C resulting in significant increases in nuclear strain.[61,64] Similar results can be observed in intact muscle tissue subjected to uniaxial stretch.[52] Elongated and fragmented nuclei have been observed in muscle tissues of lamin A/C-deficient mice and in patients with lamin A/C and emerin mutations suggesting that the mechanical stress experienced in muscle cells can cause dramatic changes and defects in nuclear structure.[52,61,105] Importantly, disruption of the LINC complex using either dominant-negative nesprin, SUN–protein mutations, or nesprin knock down/out can almost completely abolish strain-induced nuclear deformation or alignment,[107,108,112] stressing the importance of intact nucleocytoskeletal coupling to transmit forces from the cytoskeleton to the nucleus.

3.5. Substrate stiffness and patterning

Cells fine-tune their cytoskeletal tension to match the stiffness of the underlying substrate, a process which may have a profound effect on the forces experienced by the nucleus. On stiffer substrates, the nucleus flattens due to increased actomyosin-generated contractile forces that are transmitted to the nucleus via the LINC complex.[109] In lamin A/C-mutant cells, growing the cells on softer hydrogels can help minimize nuclear rupture and distortion by reducing the forces acting on the nucleus.[110] The cellular details by which the actin cytoskeleton dictates changes in nuclear shape is still incompletely understood, with competing ideas of whether apical actin cables or basal actin stress fibers have a stronger influence on nuclear morphology.[110,133] Substrate patterning also effects cell spreading and the cytoskeletal organization.

3.6. Micromanipulation

Before the implementation of powerful micromanipulation tools, such as microneedle manipulation,[134] it was unknown whether forces at the cell membrane could lead to subnuclear changes. By harpooning cells with a glass microneedle and exerting a pulling force on their cytoskeleton or on beads attached to the plasma membrane, Maniotis *et al.*[134] demonstrated the structural coupling between the cell surface and the nucleus via the actin cytoskeleton. Using a refined version of the technique coupled with detailed nuclear and cytoskeletal displacement mapping, Lombardi and colleagues proved that an intact LINC complex is required for this intracellular force transmission.[92]

4. POTENTIAL MECHANISMS FOR DIRECT NUCLEAR MECHANOSENSING

The way the nucleus mechanically responds to various physical stimuli has been well characterized (see above section), yet few studies have provided molecular details directly linking force-induced deformations to specific transcriptional changes. Thus, we return to our original question: can the nucleus act as a mechanosensor? So far, the pursuit within the field of mechanobiology to identify force transducers has primarily focused on stretch-activated ion channels, cell adhesion complexes, and the cortical cytoskeleton. Nonetheless, a number of exciting recent findings have called attention to the nuclear interior and nuclear envelope as potential sites of cellular mechanosensing.

Distinguishing the exclusive role of the nucleus in mechanosensing can be difficult given the myriad of biochemical signals originating from the cytoplasm that are processed in the nucleus. Researchers have been applying two alternative approaches to address this problem, each with its own advantages and limitations. One way to overcome this challenge is by experimenting on isolated nuclei, which allows direct control over the forces exerted on the nucleus and eliminates confounding effects from cytoplasmic signaling, but may not accurately recapitulate the physiological forces and biochemical environment of the nucleus in living cells. The other approach is to conduct experiments on intact cells at high temporal resolution, with the assumption that mechanical forces are transmitted significantly faster (milliseconds vs. seconds and minutes) to the nucleus than biochemical signals,[8] so that the earliest observed changes represent direct forced-induced events in the nucleus rather than downstream responses to cytoplasmic signals.

Following the former experimental approach, Swift *et al.*[62] employed a cysteine-shotgun mass spectrometry (CS-MS) approach to assess structural changes in isolated nuclei exposed to controlled shear stress. In CS-MS, fluorescently or otherwise labeled tags can bind to cryptic cysteine residues that are buried deep within the protein structure and become exposed when the protein partially unfolds under the applied force. The CS-MS technique is a high-throughput means to characterize and quantify the modifications (e.g., phosphorylation) of endogenous proteins by comparing their mass-to-charge ratios under different conditions. The CS-MS shear rheometer assay identified a cryptic site (Cys522) in the Ig-domain of lamin A that is responsive to shear stress on the order of 1 Pa. This structural change might alter the association of the Ig-domain with its interaction partners, which include actin, DNA, LAP2α, and nesprin-2,[38] thus acting as a molecular mechanosensor. In an attempt to recapitulate the Ig-domain unfolding in living cells, mesenchymal stem cells were grown on substrates of varying stiffness, which induces increasing cytoskeletal tension, and then analyzed by CS-MS. However, under these conditions, there was no change in Cys522 labeling, suggesting that intracellular forces were either too weak to induce unfolding of the Ig-domain under physiological conditions, or that the cells adapted to the altered mechanical environment by strengthening the nuclear lamina and preventing protein further unfolding.[62] Interestingly, cells grown on stiffer substrates showed reduced phosphorylation and altered nuclear distribution of lamin[62]; however, the underlying molecular mechanism and whether this phenomenon can be directly induced mechanically remain unclear.

Another recent report of direct nuclear mechanosensing probed deeper into the heart of the nucleus to study nuclear body dynamics in response to oscillatory stress exerted at the cell membrane. Wang and colleagues[31] measured fluorescence resonance energy transfer (FRET) ratio changes between two fluorescently conjugated CB proteins as an indicator of physical association between the two proteins. Force application in the range of $\approx$14–17 Pa caused rapid and irreversible dissociation between SMN and coilin within the CB of HeLa nuclei.[31] The effect required an intact F-actin–LINC complex–lamin A/C network and was sensitive to changes in substrate stiffness, suggesting that a certain cytoskeletal tension is required to transmit forces to the nucleus and induce nuclear deformations.[110] As the observed changes occurred rapidly (<1 s), the SMN–coilin dissociation was proposed to occur independent of intermediate biochemical cascades. Since CBs interact directly with chromatin, the findings support the idea that force-induced dissociation of nuclear proteins can alter gene expression, but follow-up studies will be required to detect functional consequences and whether long-term transcriptional changes occur.

Following a similar strategy, Iyer *et al.* visualized force-induced changes in chromatin structure in HeLa cells subjected to mechanical forces via magnetic particles on the plasma membrane.[135] They observed a decrease in the anisotropy of the fluorescence signal of green fluorescent protein (GFP)-labeled histone, interpreted as chromatin decondensation, that occurred on the order of seconds, followed by slower changes in the nuclear cross-sectional area (tens of seconds) and translocation of the transcription factor MKL1 to the nucleus (minutes).

Further evidence for mechanically induced changes in the nuclear interior was provided by the Dahl group, which studied nuclear body dynamics in the context of shear and compression forces. Using fluorescent-tagged upstream binding factor-1 (UBF-1) and fibrillarin to track subnuclear movements, they discovered force magnitude- and duration-dependent changes in UBF1-GFP and fibrillarin-GFP displacement.[128] While the increased subnuclear movement could be indicative of mechanically induced genome reorganization, the fact that many of the changes only occurred after 30 min may indicate that these changes primarily reflect a response to cytoskeletal reorganization, prompting further investigation.

Although the above findings support a direct mechanism whereby chromatin and other subnuclear structures can sense and respond to mechanical forces exerted at the cell membrane, another recent study makes it clear that nuclear deformation is not required for cellular mechanotransduction. Using dominant-negative nesprin and SUN–protein constructs to disrupt force transmission between the cytoskeleton and the nucleus, Lombardi and colleagues found that despite abolishing mechanically induced nuclear deformation, expression of the mechanosensitive genes *Egr-1*, *Iex-1*, *Pai-1*, and *vinculin* was indistinguishable from control cells.[107] These findings suggest that at least for the investigated genes, alternative mechanosensing mechanisms in the cytoplasm that are then biochemically transmitted to the nucleus may be sufficient for transcriptional activation.

5. MECHANOTRANSDUCTION SIGNALING IN THE NUCLEUS

When force is exerted on a cell, it is propagated through the various intracellular structural scaffolds. Along the way, the force causes mechanosensitive proteins to change conformation, resulting in exposure of cryptic binding sites, phosphorylation of specific residues, and/or altered binding affinities and interaction partners. Ultimately, many of the mechanically induced signaling pathways converge on the nucleus and involve the import/export of transcription factors, their interaction with cofactors, and changes in chromatin organization (Fig. 7.4).[136] Recent findings suggest

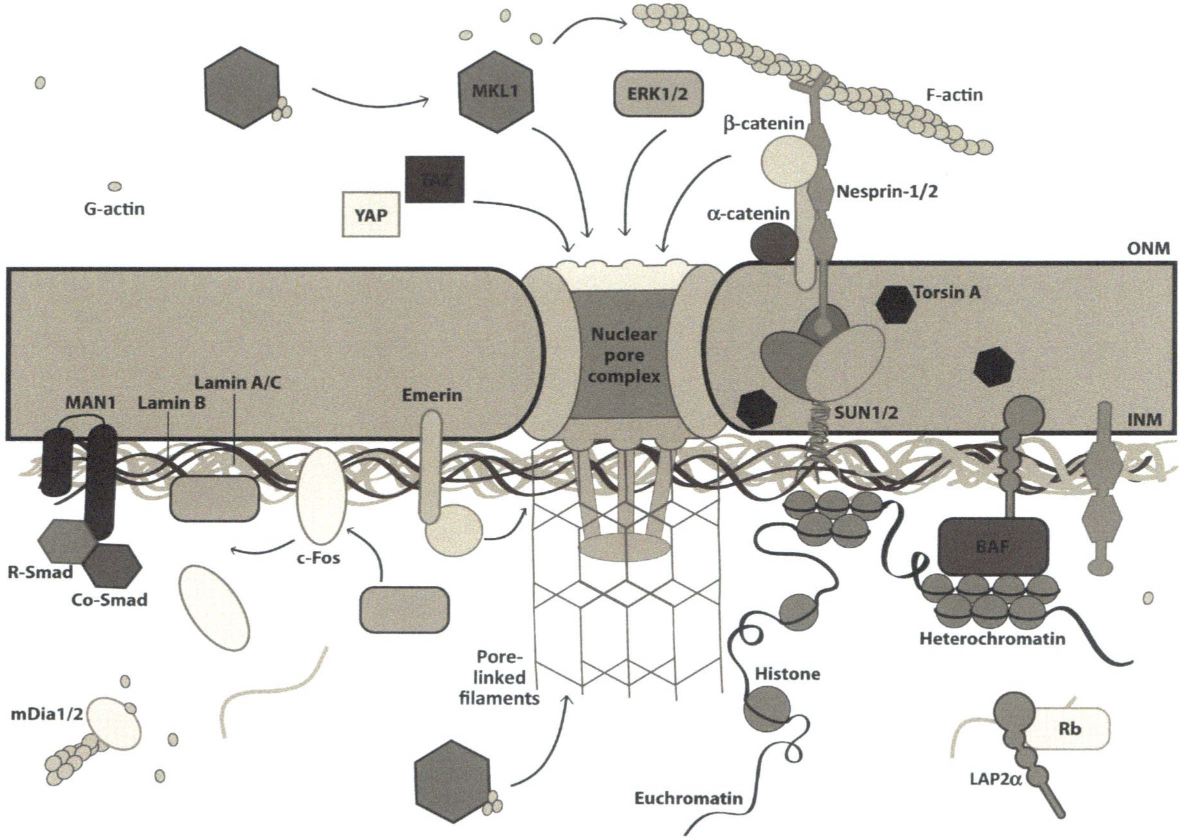

Figure 7.4 Nuclear envelope connectivity and signaling. Schematic overview of signaling pathways (MKL1/SRF, MAPK and ERK1/2, YAP/TAZ, TGF-β, and Wnt) associated with nuclear envelope proteins. (See the color plate.)

that many of the involved pathways directly or indirectly interact with nuclear (envelope) proteins, as discussed in detail below. Consequently, the role of the nucleus in mechanotransduction is not limited to direct mechanosensing; it also includes the modulation of biochemical signals arriving at the nucleus.

5.1. MAPK/ERK/Fos

One of the most prevalent and best-characterized responses to mechanical stress is the phosphorylation of proteins, which leads to changes in their binding affinities and often triggers further downstream signaling. The mitogen-activated protein kinase (MAPK) pathway mediates cellular reactions to stress and growth factors by regulating differentiation, apoptosis, and inflammation.[137] There are multiple distinct MAPKs, including ERK1/2, p38, JNK, and ERK3-5. The first three constitute particularly important branches of the MAPK pathway in cellular mechanotransduction.[2] Intriguingly, recent findings have implicated lamins A/C and emerin in causing impaired MAPK signaling resulting in cardiac and skeletal muscle disease. Patient samples carrying different *LMNA* mutations responsible for dilated cardiomyopathy have an increase in phosphorylated ERK and JNK in the nucleus.[138] Cardiomyocytes from mouse models that either have a lamin A/C mutation (*Lmna* H222P) or lack emerin also have enhanced ERK1/2 signaling.[139,140] Surprisingly, emerin and lamin mutations appear to have somewhat different effects. While the Lmna H222P mutation affects the JNK branch, loss of emerin does not.[88,89] Both, however, have similar effects on the ERK1/2 branch. In addition, p38 phosphorylation is shown to be increased in cardiac tissue samples from laminopathy patients.[141,142] Lastly, lamin A/C- and emerin-deficient fibroblasts subjected to cyclic strain respond with decreased expression of the mechanosensitive genes *egr-1* and *iex-1*, which are downstream targets of the MAPK pathway.[61,143]

At least some of the effects of lamin mutations on MAPK signaling may be attributed to the direct interaction of lamin A/C with c-Fos; when ERK1/2 enters the nucleus, it displaces c-Fos by binding to lamins and becomes transcriptionally active (Fig. 7.4).[144] Although this mechanism has not been definitively confirmed, there are multiple reports supporting it. For instance, cells with mutated lamins have normal activation of ERK1/2 in the cytoplasm but decreased expression of downstream genes.[61,145,146]

5.2. Wnt signaling

Wnt signaling is important for multiple developmental processes such as axis patterning, proliferation, and migration[147] and is modulated by mechanical forces in bone cell differentiation.[148] There are three Wnt pathways, the canonical Wnt pathway, which involves translocation of the transcriptional coactivator β-catenin from the cytoplasm and cell–cell adhesions to the nucleus, and two noncanonical pathways. Importantly, β-catenins can directly interact with several nuclear envelope proteins. At the nuclear envelope, emerin is responsible for nuclear export of β-catenin, so cells lacking emerin or lamins A/C show enhanced growth and proliferation due to β-catenin accumulation in the nucleus (Fig. 7.4).[82] Furthermore, nesprin-2 can interact with α-catenin, together with emerin and β-catenin.[149] Recently, it has been proposed that nesprin-2, α-catenin, emerin, and β-catenin form a complex that recruits β-catenin to the perinuclear region and facilitates its nuclear import, as β-catenin lacks a nuclear localization sequence.[150] This mechanism could explain why depletion of nesprin-2 reduces levels of nucleoplasmic β-catenin.[149] Other potential effects of altered Wnt signaling include impaired fibronectin and collagen Col11A1 synthesis, which could affect cellular mechanosensing at the plasma membrane.[151]

5.3. TGF-β and Smad signaling

TGF-β stimulates cell proliferation, extracellular matrix synthesis, and degradation, as well as aspects of embryogenesis and wound healing.[150,152] TGF-β expression levels have been shown to directly correlate with mechanical loading in tendons and to be activated downstream of integrin-mediated contraction in myofibroblasts.[153,154] Binding of TGF-β to its receptor on the plasma membrane induces formation of a receptor ligand complex and activation of the cytoplasmic receptor kinase domain, which results in phosphorylation of Smads, transcription factors that translocate from the cytoplasm into the nucleus, where they activate transcription of TGF-β target genes. There are three different kinds of Smads; regulating Smads (R-Smads), Co-Smads, and inhibitory Smads. The nuclear envelope LEM-domain protein MAN1 binds R-Smads in a manner that also regulates BMP and activin signaling in addition to TGF-β.[155] R-Smads bound to Co-Smads are transcriptionally active and enter the nucleus. MAN1 breaks this complex apart by selectively binding to R-Smads, promoting nuclear export of the Smads and resulting in attenuation of TGF-β signaling

(Fig. 7.4).[156] In addition to MAN1, lamin A/C regulates TGF-β signaling by modulating the phosphorylation of Smads in a mechanism that involves lamin-bound PP2, a nuclear phosphatase.[157,158] The role of lamins A/C and emerin in modulating Smad activity may be particularly relevant for muscle tissue, because TGF-β1 stimulates satellite cell differentiation. Disruption of this pathway by lamin or emerin mutations may contribute to impaired muscle regeneration.

5.4. MKL1/SRF signaling

The SRF transcription factor and its cofactor, megakaryocytic acute leukemia (MKL1, also known as MAL or myocardin-related transcription factor-A, MRTF-A), control a large number of genes involved in growth factor response, cytoskeletal remodeling (e.g., vinculin, actin, and actin binding proteins), and muscle-specific functions such as myoblast fusion and differentiation.[159] The MKL1/SRF pathway is highly sensitive to changes in actin dynamics.[160] MKL1 is normally sequestered in the cytoplasm through interactions with G-actin's three RPEL motifs. This prevents importin dimers from binding with a bipartite NLS signal in the regulatory domain of MKL1.[159,161] Serum- or mechanically induced polymerization of actin releases G-actin from MKL1 and allows the translocation of MKL1 to the nucleus, where it, together with SRF, activates SRF/MKL1 target genes (Fig. 7.4). Recent studies illustrate the importance of this pathway in diseases affecting mechanically stressed tissues, particularly cardiac and skeletal muscle. For example, MKL regulates the fibrotic repair gene program in injured cardiac tissue following myocardial infarction,[162] and SRF-null mice develop severe cardiac defects.[163]

A recent study revealed a novel mechanism by which nuclear envelope proteins can affect actin dynamics and MKL1/SRF signaling.[164] Lamin A/C mutations are shown to displace emerin from the inner nuclear membrane, thereby leading to disturbed actin polymerization and impaired nuclear import and increased nuclear export of MKL1. This causes decreased nuclear localization of MKL1 and reduced expression of MKL1/target genes *in vitro* and *in vivo*.[164] This effect was attributed to the ability of emerin to bind and polymerize actin,[165] but it remains unclear to what extent emerin at the inner nuclear membrane and its effect on nuclear actin, versus emerin on the outer nuclear membrane and its effect on cytoskeletal actin,[166] contributes to this mechanism.

Taken together, these findings shine a growing spotlight on the largely uncharted territory of nuclear actin structure and function. A recent study demonstrated the ability of the formins mDia1 and mDia2 to drive nuclear actin polymerization and to induce MKL1/SRF activity.[167] While formins were previously known to aid actin filament assembly in the cytoplasm, this study first documented a role of formins in the nucleus, suggesting that formins (and emerin) could play an important role in controlling nuclear actin polymerization and in modulating MKL1/SRF activity. Considering the complex feedback loop whereby changes in actin dynamics and activation of MKL1/SRF signaling results in increased expression of SRF, actin, and actin-modulating proteins, it becomes challenging to conclude whether cytoplasmic events drive nuclear signaling or the other way around. Furthermore, several other signaling pathways converge with MKL1/SRF signaling. For example, Smad3 can bind MKL1 and prevent it from binding to SRF. On the other hand, β-catenin binds both Smad3 and MKL1, but without disrupting the MKL1/SRF complex.[168] Regardless of the many remaining questions, the SRF pathway has already immerged as an important link between gene regulation and nuclear and cytoskeletal structure. Disturbed MKL1/SRF signaling may in part explain the altered mechanosensitivity as well as the tissue-specific phenotypes in diseases caused by mutations in lamins A/C and emerin.

5.5. YAP/TAZ signaling

Recent findings indicate that the YAP/TAZ pathway, traditionally considered to be a component of the Hippo pathway, can operate independently to mediate the cellular response to mechanical cues.[169,170] YAP/TAZ activation is particularly sensitive to substrate stiffness and cytoskeletal tension, resulting in an increased presence of YAP/TAZ in the nucleus when cells are grown on stiffer substrates that promote cell spreading.[169,170] Unlike the SRF pathway discussed above, YAP/TAZ shows nuclear accumulation in response to cell spreading and does not appear to be sensitive to changes in F-/G-actin ratios.[169] While lamin A/C overexpression decreases YAP1 levels and nuclear localization, the exact interplay between lamins and YAP/TAZ signaling remains unresolved. Interestingly, expression of YAP1 and its binding partners did not scale with the stiffness of the corresponding tissue.[62]

5.6. Interaction of retinoblastoma protein, lamins A/C, and LAP2α

Retinoblastoma protein (Rb) is part of the E2F pathway and helps to regulate cell cycle progression and differentiation.[171] Nuclear accumulation of Rb is present in proliferating cells. LAP2α and lamin A/C bind to Rb and halt cell cycle progression; when LAP2α is depleted, cells continue to proliferate, preventing differentiation of stem cells.[172,173] This defect is rescued by LAP2α knockdown.[174] Recently, Rb has been proposed to act as a potential regulator of lamin expression levels, but the mechanistic details remain to be determined.[62]

5.7. Phosphorylation of nuclear envelope proteins

In addition to modulating localization and activity of many transcriptional regulators, several nuclear envelope proteins themselves may be substrates of signaling pathways. Emerin has at least six known phosphorylation sites, four of which are regulated by the Src, Abl, and Her2 kinases, which in turn regulate emerin's ability to bind BAF.[175,176] The Her2 pathway is particularly interesting, as it has important implications for breast cancer and cardiac muscle function. Approximately a quarter of human breast tumors have upregulated Her2 expression,[177] but the role of emerin signaling has yet to be studied. Illustrating the role of Her signaling on muscle function, mice without Her2 develop cardiomyopathy and other symptoms such as lack of coordination and muscle wasting similar to muscular dystrophies.[175] Lmo7, another emerin-interacting protein, can regulate breast cancer cell migration by modulating GTPase/RhoA activity.[178] Lmo7 shuttles between focal adhesions and the nuclear envelope, where emerin binding inhibits its activity.[178] At the focal adhesions, Lmo7 specifically binds to p130Cas, a known mechanosensitive protein.[179] Of note, emerin also interacts with proteins in the MAPK and Wnt pathways. With so many interactions, one question that emerges is how emerin regulates its myriad of binding partners. It has been hypothesized that emerin exists in multiple types of protein complexes at once, allowing it to have a designated complex for various signaling pathways; many of these interactions may be regulated by emerin phosphorylation.[180] Other LEM-domain proteins similarly show altered interaction upon phosphorylation. In MAN1, phosphorylation of three different residues interrupts binding to BAF[136,181]; changing the phosphorylation state of LAP2α affects its binding to chromatin but not to the nuclear envelope.[182]

For lamins and other nuclear envelope proteins, which can also undergo various posttranslational modifications, the functional consequences and prevalence of such modifications outside of their role in mitosis are only beginning to emerge and are likely to produce exciting new insights.[10,62]

6. FUNCTIONAL CONSEQUENCES OF IMPAIRED MECHANOTRANSDUCTION AND DISEASE

As illustrated by the above sections, proper mechanosensing and mechanotransduction signaling is essential for numerous cellular functions. Both the external mechanical input and the integrity of the internal connectivity can affect nuclear morphology, mechanosensing, and transcription. Mutations in nuclear envelope proteins and disturbed expression of nuclear proteins can thus have severe functional consequences, resulting in a variety of human diseases. Therefore, it is crucial to understand how mechanical forces affect the cell, both the structurally and by rerouting its signaling pathways. This is especially true for mechanically active tissue such as skeletal and cardiac muscle which is primarily affected in many of the pathologies resulting from defects in nuclear (envelope) proteins. Beyond the immediate effects of disturbed protein function, the mechanical microenvironment is implicated in several other diseases, such as cancer, where changes in the mechanical environment are thought to contribute to the disease progression, or vascular disease, where plaques often form at sites of perturbed fluid shear stress.[2] An improved understanding of the underlying disease mechanisms is critical for developing therapeutic approaches. In particular, it is important to distinguish which of the following scenarios constitutes the primary disease driver: (1) changes in the mechanical environment, in which cellular mechanotransduction signaling functions normally but responds to an altered input; (2) physical defects in cellular structure, resulting in altered force transmission, mechanosensing, and/or direct physical damage in response to mechanical stress; or (3) altered mechanotransduction signaling. Only the third scenario is readily amenable to pharmaceutical intervention with small molecules to restore normal signaling.

In the following, we highlight some of the diseases that can be caused by mutations in nuclear envelope proteins such as lamins A/C and emerin as examples of how defects in nuclear envelope proteins can affect cellular mechanics and mechanotransduction. What is particularly perplexing about the large number of diseases resulting from mutations in nuclear envelope proteins is that although the implicated proteins are almost ubiquitously expressed throughout the body, many of the diseases primarily affect skeletal and cardiac muscle and tendons. Several hypotheses have been proposed to explain these tissue-specific effects. One potential explanation is the "structural hypothesis"—loss of protein function causes nuclear fragility, rupture, and death. This mechanism helps to explain the progressive nature of many of the laminopathies, especially in mechanically stressed tissues. Another hypothesis is that of disturbed gene regulation—mutations in nuclear envelope proteins affect their (tissue-specific) binding partners and lead to altered gene expression. A third hypothesis suggests that the mutations alter the ability for stem cells to differentiate and decrease their self-renewal capacity, thereby causing defects in the maintenance and repair of tissues.[183] Importantly, these hypotheses are not mutually exclusive, and the specific disease mechanism probably lies at an intersection of these propositions.

6.1. Muscular dystrophy

The discovery that mutations in the nuclear envelope protein emerin cause Emery–Dreifuss muscular dystrophy (EDMD) was one of the main motivators for studying nuclear structure and mechanics.[184] To date, the nuclear envelope proteins lamin A/C, emerin, nesprin-1/2, and SUN1 have all been linked to EDMD.[5,105,184] EDMD presents with early contractures, humeroperoneal weakness, and cardiac conduction defects, although overall the disease is less severe than Duchenne muscular dystrophy, which is caused by mutations in the dystrophin gene.[185] Muscle biopsies from patients with EDMD often show highly fragmented nuclei.[82] Another signature of muscular dystrophy is the mislocalization of the nucleus in muscle fibers, with cells lacking lamins or nesprins showing loss of myonuclei from the neuromuscular junction, suggestive of defects in anchoring of the nucleus to the cytoskeleton.[175,186,187] Impaired nucleocytoskeletal coupling and disrupted force transmission between the cytoskeleton and nucleus have also been reported in fibroblasts expressing lamin A/C mutations responsible for muscular dystrophies.[52,188] A detailed analysis comparing a large set of lamin A/C mutations found that mutations that cause striated muscle disease often lead to impaired lamin assembly, reduced nuclear stability, and impaired nucleocytoskeletal coupling, while mutations that cause familial partial lipodystrophy have no effect on nuclear stability.[52] Importantly, cells lacking lamins A/C or emerin have severe defects in several mechanotransduction pathways *in vitro* and *in vivo*, including failure to adequately activate expression of the mechanosensitive genes *Egr-1* and *Iex-1*,[61,64,189] and in the translocation of MKL1 to the nucleus.[164,165] However, it remains to be established whether the defects in mechanotransduction signaling are caused by altered nuclear mechanics and impaired force transmission to the nucleus or disturbed interactions between lamins/emerin and binding partners involved in transcriptional regulation. Regardless of the mechanism, the impaired mechanotransduction signaling may not only weaken muscle adaptation to mechanical stress[189] but could also affect appropriate stem cell differentiation, as lineage commitment is strongly modulated by cells sensing their mechanical environment.[190]

6.2. Dilated cardiomyopathy

While dilated cardiomyopathy is often associated with muscular dystrophy, some lamin A/C mutations also cause cardiomyopathies without affecting skeletal muscle.[3] Dilated cardiomyopathy is characterized by progressive weakening of the heart muscles, thinning of the left ventricle wall, and insufficient pumping that typically leads to heart failure.[191] Dilated cardiomyopathy is the most prevalent disease caused by mutations in *LMNA* but can also be caused by mutations in emerin, nesprin-1 and -2, LUMA, an inner nuclear membrane protein that binds emerin, and LAP2α.[3,192] Many of the disease mechanisms may mirror those for muscular dystrophy described above, and it remains unclear why/how some mutations specifically affect cardiac but not skeletal muscle. Possible explanations include cell-type-specific differences in (mechanotransduction) signaling pathways or the fact that skeletal muscle has a higher potential for repair and regeneration compared to cardiac tissue. Recently, animal studies have produced promising evidence that blocking hyperactive MAPK signaling can significantly delay the development of cardiomyopathy and prolong survival in mice expressing lamin or emerin mutations.[141,142,193–195]

6.3. Hutchinson–Gilford progeria syndrome

HGPS is a premature aging disease that results from the buildup of progerin, a mutant form of lamin A with a 50-amino acid deletion in the tail domain that leads to permanent farnesylation and membrane accumulation.[196] Children with HGPS appear normal at birth, but soon begin to present with stunted growth, thinned skin, hair loss, atherosclerosis, and cardiac defects; patients typically die from heart attacks and strokes in their early teens.[197,198] HGPS cells have stiffer nuclei and are more prone to mechanically induced damage when stretched or squeezed through narrow constrictions.[128,199,200] In addition to increased sensitivity to mechanical stress, cells from HGPS patients display impaired proliferation in response to strain, altered extracellular matrix synthesis, and disturbed Wnt signaling, which may provide a mechanism explaining the predominance of vascular disease in HGPS.[151,199] Ongoing clinical trials are aiming to block accumulation of progerin at the nuclear envelope by treatment with farnesyltransferase inhibitors, statins, and bisphosphonates and have provided some encouraging findings, but it remains to be seen whether the beneficial results can be attributed to the drugs' effects on lamin farnesylation or other molecular targets.[201]

7. OPEN QUESTIONS AND FUTURE RESEARCH DIRECTIONS

Moving forward, there are several key areas that need to be addressed regarding the role of the nucleus in mechanotransduction. First, a more thorough understanding of LINC complex proteins is needed to better understand intracellular force transmission. Further work is needed in identifying LINC complex-associated proteins as well as in determining interaction partners with structural and signaling molecules at the nuclear envelope. For example, how is the SUN/nesprin interaction dynamically regulated to allow controlled engagement of nucleocytoskeletal connections without "locking up" the nucleus? The recent discoveries of Samp1 and Torsin-A as potential mediators[115,117] are likely just the tip of the iceberg. Considering that more than 100 unique nuclear membrane proteins have been identified to date and only a minuscule subset of these proteins have been functionally characterized, it is likely that additional proteins contribute to mechanical coupling and/or signal processing at the nuclear envelope.[202] Combining this research with advanced mass spectrometry methods, which labels mechanically sensitive cysteine residues in proteins, will have the potential to reveal novel nuclear mechanosensitive proteins. A better understanding of the mechanical coupling between the nucleus and the cytoskeleton may also explain which cytoskeletal structures and connections are primarily responsible for mediating nuclear shape and morphology, for example, during cell migration or when cells are subjected to mechanical force or grown on substrates with varying shapes or stiffness.

The elephant in the room, and probably one of the biggest questions in mechanobiology, is whether mechanically induced nuclear deformations alone can result in a consistent, dose-responsive change in gene expression, i.e., whether the nucleus (or part of it) can serve as a cellular mechanosensor. To answer this question, single-cell assays will be useful in matching precise measurements of (sub-)nuclear deformation with reliable readouts for mechanically induced gene expression. These may include both DNA and RNA FISH technologies to assess intranuclear localization of specific genes and could be complemented by emerging technologies to visualize transcriptional activity in living cells.[9] To avoid confounding factors from cytoplasmic mechanotransduction signals, these experiments should be complemented by assays involving isolated nuclei subjected to mechanical stimulation. However, finding suitable conditions that recapitulate both the mechanical and biochemical environment of the nucleus in isolation may prove challenging.

Lastly, another important factor for understanding nuclear mechanosensing and transcription at large is determining the biophysical aspects of transcription in the 3-D context inside the nucleus. We have already learned about chromosome territories and higher-order organization of the genome. But it remains unclear whether active genes move to specialized transcription sites or if transcription factories are recruited to where they are needed, and what role the nucleoskeleton plays in these processes. The answers to each of these questions will help to determine whether the nucleus is a true mechanosensor or mostly serves as a central processor for a variety of mechanotransduction signals converging from all parts of the cell.

8. CONCLUSIONS

Since the discovery of the first disease-causing mutations in nuclear envelope proteins 20 years ago, research on nuclear mechanics and nuclear mechanotransduction has been a rapidly growing field (Fig. 7.5). In the last decade alone, many details have emerged on the role of lamins in conferring nuclear stability, the 3-D organization of chromatin, and the molecular details involved in transmitting forces between the cytoskeleton and nuclear interior, not to mention the growing number of mutations, diseases, and binding partners associated with nuclear envelope proteins. Nonetheless, the opening question—what exactly is the role of the nucleus in mechanosensing and mechanotransduction signaling—has not yet been completely answered. Increasing evidence suggests that external force application can lead to direct changes in chromatin organization and dissociation of protein complexes in subnuclear structures, but insights into mechanistic

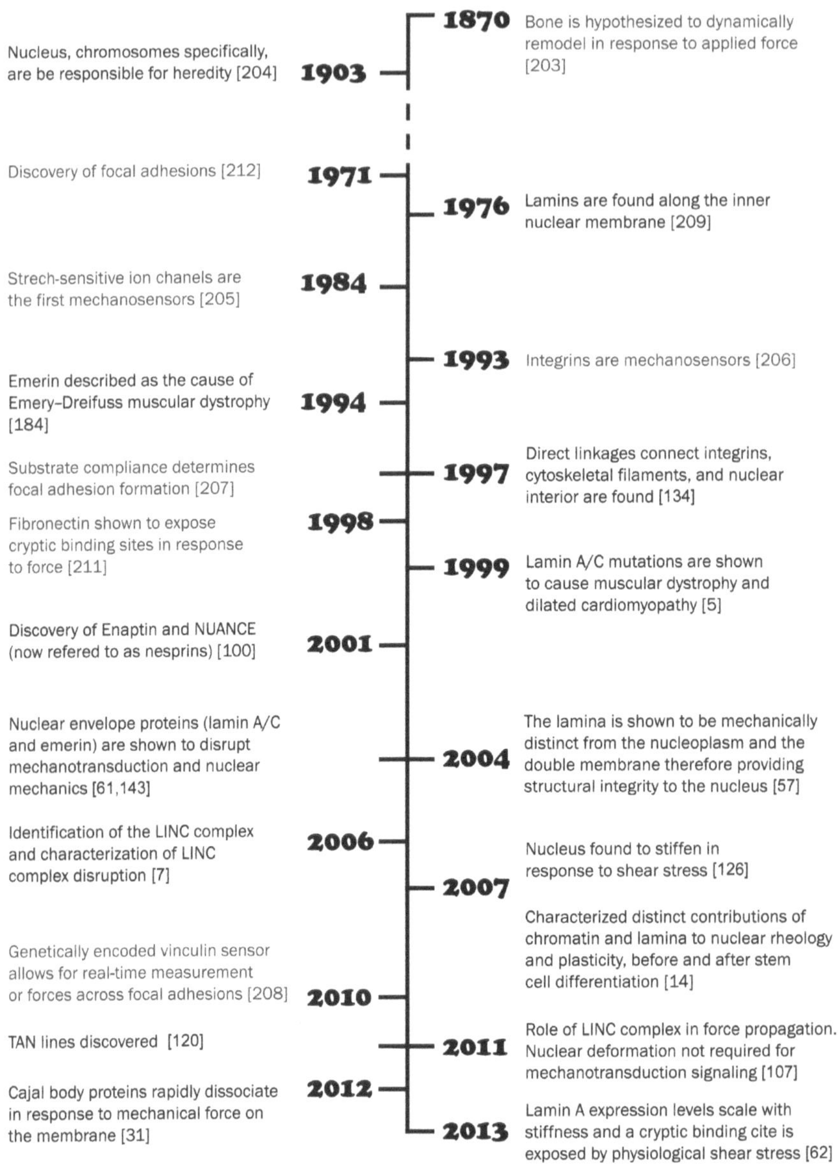

Figure 7.5 Timeline of (nuclear) mechanotransduction discoveries. Roadmap of selected milestones in the field of mechanobiology that have paved the way for our current understanding of the role of the nucleus in mechanotransduction.[203–212] Discoveries most relevant to nuclear mechanotransduction are printed in bold.

details and functional consequences are still lacking. Undoubtedly, the direct physical connection of the nucleus to the cytoskeleton and the cellular exterior provides a means for physical forces to propagate through the cell and results in nuclear deformation, which may ultimately cause changes in chromatin organization and mediate transcriptional regulation. Concurrently, many nuclear envelope proteins play a crucial role in correctly interpreting mechanically induced biochemical signals that originated in the cytoplasm. Mutations in these proteins, including lamins, emerin, and nesprin, therefore not only affect force transmission to the nucleus and potential nuclear mechanosensing, but also disturb biochemical pathways independent of immediate structural changes to the nucleus, which may further affect gene expression, differentiation, and proliferation. Lastly, the example of impaired MKL1 activity in lamin A/C-deficient cells due to emerin mislocalization illustrates how tightly some of these ideas are interconnected. Going forward, new experimental techniques to probe deeper into associations between the nuclear envelope, the nucleoskeleton, and the genome are expected to elicit many of the mechanistic details currently lacking. Ultimately, insights gained from these studies will not only enhance our understanding of normal cellular processes and mechanotransduction but may also provide important novel therapeutic targets for diseases ranging from deafness to muscular dystrophy, heart disease, and cancer.

ACKNOWLEDGMENTS

We apologize all authors whose work could not be cited due to space constraints. This work was supported by the National Institutes of Health awards (R01 HL082792), the National Science Foundation CAREER award (CBET-1254846), the Department of Defense Breast Cancer Idea Award (BC102152), NSF Graduate Research Fellowships to A. K. (2013160437) and G.F (2014163403), and the Pilot Project Award by the Cornell Center on the Microenvironment & Metastasis through Award Number U54CA143876 from the National Cancer Institute. The content is solely the responsibility of the authors and does not necessarily represent the official views of the National Cancer Institute or the National Institutes of Health.

REFERENCES

1. Yang L, Munck M, Swaminathan K, Kapinos LE, Noegel AA, Neumann S. Mutations in LMNA modulate the lamin a—nesprin-2 interaction and cause LINC complex alterations. *PLoS One*. 2013;8(8):e71850.
2. Jaalouk DE, Lammerding J. Mechanotransduction gone awry. *Nat Rev Mol Cell Biol*. 2009;10(1):63–73.
3. Schreiber KH, Kennedy BK. When lamins go bad: nuclear structure and disease. *Cell*. 2013;152(6):1365–1375.
4. Zwerger M, Ho CY, Lammerding J. Nuclear mechanics in disease. *Annu Rev Biomed Eng*. 2011;13:397–428.

5. Bonne G, Di Barletta MR, Varnous S, et al. Mutations in the gene encoding lamin A/C cause autosomal dominant Emery-Dreifuss muscular dystrophy. *Nat Genet.* 1999;21(3): 285–288.
6. Savage DB, Soos MA, Powlson A, et al. Familial partial lipodystrophy associated with compound heterozygosity for novel mutations in the LMNA gene. *Diabetologia.* 2004;47(4):753–756.
7. Crisp M, Liu Q, Roux K, et al. Coupling of the nucleus and cytoplasm role of the LINC complex. *J Cell Biol.* 2006;172(1):41–53.
8. Wang N, Tytell JD, Ingber DE. Mechanotransduction at a distance: mechanically coupling the extracellular matrix with the nucleus. *Nat Rev Mol Cell Biol.* 2009;10(1):75–82.
9. Martins RP, Finan JD, Guilak F, Lee DA. Mechanical regulation of nuclear structure and function. *Annu Rev Biomed Eng.* 2012;14:431.
10. Simon DN, Wilson KL. Partners and post-translational modifications of nuclear lamins. *Chromosoma.* 2013;122(1–2):13–31.
11. Kouzarides T. Chromatin modifications and their function. *Cell.* 2007;128(4):693–705.
12. Simon DN, Wilson KL. The nucleoskeleton as a genome-associated dynamic 'network of networks'. *Nat Rev Mol Cell Biol.* 2011;12(11):695–708.
13. Dahl KN, Engler AJ, Pajerowski JD, Discher DE. Power-law rheology of isolated nuclei with deformation mapping of nuclear substructures. *Biophys J.* 2005;89(4):2855–2864.
14. Pajerowski JD, Dahl KN, Zhong FL, Sammak PJ, Discher DE. Physical plasticity of the nucleus in stem cell differentiation. *Proc Natl Acad Sci USA.* 2007;104(40):15619–15624.
15. Rowat AC, Jaalouk DE, Zwerger M, et al. Nuclear envelope composition determines the ability of neutrophil-type cells to passage through micron-scale constrictions. *J Biol Chem.* 2013;288(12):8610–8618.
16. Guilak F, Mow VC. The mechanical environment of the chondrocyte: a biphasic finite element model of cell-matrix interactions in articular cartilage. *J Biomech.* 2000;33(12): 1663–1673.
17. Versaevel M, Riaz M, Grevesse T, Gabriele S. Cell confinement: putting the squeeze on the nucleus. *Soft Matter.* 2013;9:6665–6676.
18. Mazumder A, Shivashankar GV. Emergence of a prestressed eukaryotic nucleus during cellular differentiation and development. *J R Soc Interface.* 2010;7:S321–S330.
19. Wolf K, te Lindert M, Krause M, et al. Physical limits of cell migration: control by ECM space and nuclear deformation and tuning by proteolysis and traction force. *J Cell Biol.* 2013;201(7):1069–1084.
20. Cremer T, Cremer M. Chromosome territories. *Cold Spring Harb Perspect Biol.* 2010;2(3):22.
21. Walter J, Schermelleh L, Cremer M, Tashiro S, Cremer T. Chromosome order in HeLa cells changes during mitosis and early G1, but is stably maintained during subsequent interphase stages. *J Cell Biol.* 2003;160(5):685–697.
22. Sato M, Ohshima N, Nerem RM. Viscoelastic properties of cultured porcine aortic endothelial cells exposed to shear stress. *J Biomech.* 1996;29(4):461–467.
23. Dahl KN, Kalinowski A. Nucleoskeleton mechanics at a glance. *J Cell Sci.* 2011;124(5): 675–678.
24. Dahl KN, Ribeiro AJ, Lammerding J. Nuclear shape, mechanics, and mechanotransduction. *Circ Res.* 2008;102(11):1307–1318.
25. Zhong S, Salomoni P, Pandolfi PP. The transcriptional role of PML and the nuclear body. *Nat Cell Biol.* 2000;2(5):E85–E90.
26. Lallemand-Breitenbach V, de The H. PML nuclear bodies. *Cold Spring Harb Perspect Biol.* 2010;2(5):17.
27. Muratani M, Gerlich D, Janicki SM, Gebhard M, Eils R, Spector DL. Metabolic-energy-dependent movement of PML bodies within the mammalian cell nucleus. *Nat Cell Biol.* 2002;4(2):106–110.

28. Zimber A, Nguyen QD, Gespach C. Nuclear bodies and compartments: functional roles and cellular signalling in health and disease. *Cell Signal.* 2004;16(10):1085–1104.
29. Bernardi R, Pandolfi PP. Structure, dynamics and functions of promyelocytic leukaemia nuclear bodies. *Nat Rev Mol Cell Biol.* 2007;8(12):1006–1016.
30. Grosse R, Vartiainen MK. To be or not to be assembled: progressing into nuclear actin filaments. *Nat Rev Mol Cell Biol.* 2013;14(11):693–697.
31. Poh Y-C, Shevtsov SP, Chowdhury F, et al. Dynamic force-induced direct dissociation of protein complexes in a nuclear body in living cells. *Nat Commun.* 2012;3:866.
32. Machyna M, Heyn P, Neugebauer KM. Cajal bodies: where form meets function. *Wiley Interdiscip Rev RNA.* 2013;4(1):17–34.
33. Hebert MD. Signals controlling Cajal body assembly and function. *Int J Biochem Cell Biol.* 2013;45(7):1314–1317.
34. Cioce M, Lamond AI. Cajal bodies: a long history of discovery. *Annu Rev Cell Dev Biol.* 2005;21:105–131.
35. Morris GE. The Cajal body. *Biochim Biophys Acta.* 2008;1783(11):2108–2115.
36. Platani M, Goldberg I, Lamond AI, Swedlow JR. Cajal body dynamics and association with chromatin are ATP-dependent. *Nat Cell Biol.* 2002;4(7):502–508.
37. Cook PR. The nucleoskeleton—artifact, passive framework or active-site. *J Cell Sci.* 1988;90:1–6.
38. Ho CY, Lammerding J. Lamins at a glance. *J Cell Sci.* 2012;125(9):2087–2093.
39. Castano E, Philimonenko VV, Kahle M, et al. Actin complexes in the cell nucleus: new stones in an old field. *Histochem Cell Biol.* 2010;133(6):607–626.
40. Zhong ZX, Wilson KL, Dahl KN. Beyond lamins: other structural components of the nucleoskeleton. In: Shivashankar GV, ed. *Nuclear Mechanics and Genome Regulation.* Vol. 98:San Diego: Elsevier Academic Press Inc.; 2010:97–119.
41. McMahon LW, Sangerman J, Goodman SR, Kumaresan K, Lambert MW. Human alpha spectrin II and the FANCA, FANCC, and FANCG proteins bind to DNA containing psoralen interstrand cross-links. *Biochemistry.* 2001;40(24):7025–7034.
42. Zastrow MS, Flaherty DB, Benian GM, Wilson KL. Nuclear titin interacts with A- and B-type lamins in vitro and in vivo. *J Cell Sci.* 2006;119(2):239–249.
43. Nürnberg A, Kitzing T, Grosse R. Nucleating actin for invasion. *Nat Rev Cancer.* 2011;11(3):177–187.
44. Belin BJ, Cimini BA, Blackburn EH, Mullins RD. Visualization of actin filaments and monomers in somatic cell nuclei. *Mol Biol Cell.* 2013;24(7):982–994.
45. Bohnsack MT, Stuven T, Kuhn C, Cordes VC, Gorlich D. A selective block of nuclear actin export stabilizes the giant nuclei of Xenopus oocytes. *Nat Cell Biol.* 2006;8(3): 257–263.
46. Rober RA, Weber K, Osborn M. Differential timing of nuclear lamin a/c expression in the various organs of the mouse embryo and the young animal—a developmental-study. *Development.* 1989;105(2):365–378.
47. Ottaviano Y, Gerace L. Phosphorylation of the nuclear lamins during interphase and mitosis. *J Biol Chem.* 1985;260(1):624–632.
48. Eden E, Geva-Zatorsky N, Issaeva I, et al. Proteome half-life dynamics in living human cells. *Science.* 2011;331(6018):764–768.
49. Aebi U, Cohn J, Buhle L, Gerace L. The nuclear lamina is a meshwork of intermediate-type filaments. *Nature.* 1986;323(6088):560–564.
50. Goldberg MW, Huttenlauch I, Hutchison CJ, Stick R. Filaments made from A- and B-type lamins differ in structure and organization. *J Cell Sci.* 2008;121(2): 215–225.
51. Gangemi F, Degano M. Disease-associated mutations in the coil 2B domain of human lamin A/C affect structural properties that mediate dimerization and intermediate filament formation. *J Struct Biol.* 2013;181(1):17–28.

52. Zwerger M, Jaalouk DE, Lombardi ML, et al. Myopathic lamin mutations impair nuclear stability in cells and tissue and disrupt nucleo-cytoskeletal coupling. *Hum Mol Genet.* 2013;22(12):2335–2349.
53. Kapinos LE, Schumacher J, Mücke N, et al. Characterization of the head-to-tail overlap complexes formed by human lamin A, B1 and B2 "half-minilamin" dimers. *J Mol Biol.* 2010;396(3):719–731.
54. Zuela N, Bar DZ, Gruenbaum Y. Lamins in development, tissue maintenance and stress. *EMBO Rep.* 2012;13(12):1070–1078.
55. Shimi T, Pfleghaar K, Kojima S-I, et al. The A- and B-type nuclear lamin networks: microdomains involved in chromatin organization and transcription. *Genes Dev.* 2008; 22(24):3409–3421.
56. Kolb T, Maass K, Hergt M, Aebi U, Herrmann H. Lamin A and lamin C form homodimers and coexist in higher complex forms both in the nucleoplasmic fraction and in the lamina of cultured human cells. *Nucleus.* 2011;2(5):425–433.
57. Dahl KN, Kahn SM, Wilson KL, Discher DE. The nuclear envelope lamina network has elasticity and a compressibility limit suggestive of a molecular shock absorber. *J Cell Sci.* 2004;117(20):4779–4786.
58. Hochmuth RM. Micropipette aspiration of living cells. *J Biomech.* 2000;33(1):15–22.
59. Caille N, Thoumine O, Tardy Y, Meister JJ. Contribution of the nucleus to the mechanical properties of endothelial cells. *J Biomech.* 2002;35(2):177–187.
60. Kha HN, Chen BK, Clark GM, Jones R. Stiffness properties for nucleus standard straight and contour electrode arrays. *Med Eng Phys.* 2004;26(8):677–685.
61. Lammerding J, Schulze PC, Takahashi T, et al. Lamin A/C deficiency causes defective nuclear mechanics and mechanotransduction. *J Clin Invest.* 2004;113(3):370–378.
62. Swift J, Ivanovska IL, Buxboim A, et al. Nuclear lamin-A scales with tissue stiffness and enhances matrix-directed differentiation. *Science.* 2013;341(6149):1240104.
63. Stewart C, Burke B. Teratocarcinoma stem-cells and early mouse embryos contain only a single major lamin polypeptide closely resembling lamin-B. *Cell.* 1987;51(3):383–392.
64. Lammerding J, Fong LG, Ji JY, et al. Lamins A and C but not lamin B1 regulate nuclear mechanics. *J Biol Chem.* 2006;281(35):25768–25780.
65. Chen H, Chen X, Zheng Y. The nuclear lamina regulates germline stem cell niche organization via modulation of EGFR signaling. *Cell Stem Cell.* 2013;13(1):73–86.
66. Ji JY, Lee RT, Vergnes L, et al. Cell nuclei spin in the absence of lamin B1. *J Biol Chem.* 2007;282(27):20015–20026.
67. Jung HJ, Nobumori C, Goulbourne CN, et al. Farnesylation of lamin B1 is important for retention of nuclear chromatin during neuronal migration. *Proc Natl Acad Sci USA.* 2013;110(21):E1923–E1932.
68. Jung HJ, Lee JM, Yang SH, Young SG, Fong LG. Nuclear lamins in the brain—new insights into function and regulation. *Mol Neurobiol.* 2013;47(1):290–301.
69. Coffinier C, Fong LG, Young SG. LINCing lamin B2 to neuronal migration growing evidence for cell-specific roles of B-type lamins. *Nucleus.* 2010;1(5):407–411.
70. Olins AL, Herrmann H, Lichter P, Kratzmeier M, Doenecke D, Olins DE. Nuclear envelope and chromatin compositional differences comparing undifferentiated and retinoic acid- and phorbol ester-treated HL-60 cells. *Exp Cell Res.* 2001;268(2):115–127.
71. Friedl P, Wolf K, Lammerding J. Nuclear mechanics during cell migration. *Curr Opin Cell Biol.* 2011;23(1):55–64.
72. Shin J-W, Spinler KR, Swift J, Chasis JA, Mohandas N, Discher DE. Lamins regulate cell trafficking and lineage maturation of adult human hematopoietic cells. *Proc Natl Acad Sci USA.* 2013;110(47):18892–18897.
73. Kuga T, Nozaki N, Matsushita K, Nomura F, Tomonaga T. Phosphorylation statuses at different residues of lamin B2, B1, and A/C dynamically and independently change throughout the cell cycle. *Exp Cell Res.* 2010;316(14):2301–2312.

74. Wilson KL, Berk JM. The nuclear envelope at a glance. *J Cell Sci.* 2010;123(12): 1973–1978.
75. Lin F, Blake DL, Callebaut I, et al. MAN1, an inner nuclear membrane protein that shares the LEM domain with lamina-associated polypeptide 2 and emerin. *J Biol Chem.* 2000;275(7):4840–4847.
76. Lee KK, Wilson KL. All in the family: evidence for four new LEM-domain proteins Lem2 (NET-25), Lem3, Lem4 and Lem5 in the human genome. In: *Paper Presented at Symposia of the Society for Experimental Biology*; 2003.
77. Margalit A, Brachner A, Gotzmann J, Foisner R, Gruenbaum Y. Barrier-to-autointegration factor—a BAFfling little protein. *Trends Cell Biol.* 2007;17(4):202–208.
78. Haraguchi T, Koujin T, Segura-Totten M, et al. BAF is required for emerin assembly into the reforming nuclear envelope. *J Cell Sci.* 2001;114(24):4575–4585.
79. Margalit A, Segura-Totten M, Gruenbaum Y, Wilson KL. Barrier-to-autointegration factor is required to segregate and enclose chromosomes within the nuclear envelope and assemble the nuclear lamina. *Proc Natl Acad Sci USA.* 2005;102(9):3290–3295.
80. Dorner D, Gotzmann J, Foisner R. Nucleoplasmic lamins and their interaction partners, LAP2 alpha, Rb, and BAF, in transcriptional regulation. *FEBS J.* 2007;274(6): 1362–1373.
81. Lin F, Morrison JM, Wu W, Worman HJ. MAN1, an integral protein of the inner nuclear membrane, binds Smad2 and Smad3 and antagonizes transforming growth factor-beta signaling. *Hum Mol Genet.* 2005;14(3):437–445.
82. Markiewicz E, Tilgner K, Barker N, et al. The inner nuclear membrane protein Emerin regulates beta-catenin activity by restricting its accumulation in the nucleus. *EMBO J.* 2006;25(14):3275–3285.
83. Wilson KL, Foisner R. Lamin-binding proteins. *Cold Spring Harb Perspect Biol.* 2010; 2(4):17.
84. Rodriguez J, Calvo F, Gonzalez JM, Casar B, Andres V, Crespo P. ERK1/2 MAP kinases promote cell cycle entry by rapid, kinase-independent disruption of retinoblastoma-lamin A complexes. *J Cell Biol.* 2010;191(5):967–979.
85. Moiseeva O, Bourdeau V, Vernier M, Dabauvalle MC, Ferbeyre G. Retinoblastoma-independent regulation of cell proliferation and senescence by the p53-p21 axis in lamin A/C-depleted cells. *Aging Cell.* 2011;10(5):789–797.
86. Lloyd DJ, Trembath RC, Shackleton S. A novel interaction between lamin A and SREBP1: implications for partial lipodystrophy and other laminopathies. *Hum Mol Genet.* 2002;11(7):769–777.
87. Zuleger N, Kerr ARW, Schirmer EC. Many mechanisms, one entrance: membrane protein translocation into the nucleus. *Cell Mol Life Sci.* 2012;69(13):2205–2216.
88. Schirmer EC, Florens L, Guan TL, Yates JR, Gerace L. Nuclear membrane proteins with potential disease links found by subtractive proteomics. *Science.* 2003;301(5638): 1380–1382.
89. de las Heras JI, Batrakou DG, Schirmer EC. Cancer biology and the nuclear envelope: a convoluted relationship. *Semin Cancer Biol.* 2013;23(2):125–137.
90. Sullivan T, Escalante-Alcalde D, Bhatt H, et al. Loss of A-type lamin expression compromises nuclear envelope integrity leading to muscular dystrophy. *J Cell Biol.* 1999;147(5):913–919.
91. Lu WS, Gotzmann J, Sironi L, et al. Sun1 forms immobile macromolecular assemblies at the nuclear envelope. *Biochim Biophys Acta.* 2008;1783(12):2415–2426.
92. Lombardi ML, Lammerding J. Keeping the LINC: the importance of nucleocytoskeletal coupling in intracellular force transmission and cellular function. *Biochem Soc Trans.* 2011;39:1729–1734.
93. Rothballer A, Schwartz TU, Kutay U. LINCing complex functions at the nuclear envelope: what the molecular architecture of the LINC complex can reveal about

its function. *Nucleus*. 2013;4(1):29–36. http://dx.doi.org/10.4161/nucl.23387, PMID: 23324460.
94. Fridkin A, Penkner A, Jantsch V, Gruenbaum Y. SUN-domain and KASH-domain proteins during development, meiosis and disease. *Cell Mol Life Sci*. 2009;66(9):1518–1533.
95. Sosa BA, Kutay U, Schwartz TU. Structural insights into LINC complexes. *Curr Opin Struct Biol*. 2013;23(2):285–291.
96. Mellad JA, Warren DT, Shanahan CM. Nesprins LINC the nucleus and cytoskeleton. *Curr Opin Cell Biol*. 2011;23(1):47–54.
97. Apel ED, Lewis RM, Grady RM, Sanes JR. Syne-1, a dystrophin- and Klarsicht-related protein associated with synaptic nuclei at the neuromuscular junction. *J Biol Chem*. 2000;275(41):31986–31995.
98. Morimoto A, Shibuya H, Zhu XQ, et al. A conserved KASH domain protein associates with telomeres, SUN1, and dynactin during mammalian meiosis. *J Cell Biol*. 2012; 198(2):165–172.
99. Rajgor D, Shanahan CM. Nesprins: from the nuclear envelope and beyond. *Expert Rev Mol Med*. 2013;15:e5.
100. Zhang QP, Skepper JN, Yang FT, et al. Nesprins: a novel family of spectrin-repeat-containing proteins that localize to the nuclear membrane in multiple tissues. *J Cell Sci*. 2001;114(24):4485–4498.
101. Wilhelmsen K, Litjens SH, Kuikman I, et al. Nesprin-3, a novel outer nuclear membrane protein, associates with the cytoskeletal linker protein plectin. *J Cell Biol*. 2005;171(5):799–810.
102. Zhang QP, Ragnauth C, Greener MJ, Shanahan CM, Roberts RG. The nesprins are giant actin-binding proteins, orthologous to Drosophila melanogaster muscle protein MSP-300. *Genomics*. 2002;80(5):473–481.
103. Roux KJ, Crisp ML, Liu Q, et al. Nesprin 4 is an outer nuclear membrane protein that can induce kinesin-mediated cell polarization. *Proc Natl Acad Sci USA*. 2009;106(7): 2194–2199.
104. Wheeler MA, Davies JD, Zhanq QP, et al. Distinct functional domains in nesprin-1 alpha and nesprin-2 beta bind directly to emerin and both interactions are disrupted in X-linked Emery-Dreifuss muscular dystrophy. *Exp Cell Res*. 2007;313(13): 2845–2857.
105. Zhang QP, Bethmann C, Worth NF, et al. Nesprin-1 and -2 are involved in the pathogenesis of Emery-Dreifuss muscular dystrophy and are critical for nuclear envelope integrity. *Hum Mol Genet*. 2007;16(23):2816–2833.
106. Lu W, Schneider M, Neumann S, et al. Nesprin interchain associations control nuclear size. *Cell Mol Life Sci*. 2012;69(20):3493–3509.
107. Lombardi ML, Jaalouk DE, Shanahan CM, Burke B, Roux KJ, Lammerding J. The interaction between nesprins and sun proteins at the nuclear envelope is critical for force transmission between the nucleus and cytoskeleton. *J Biol Chem*. 2011;286(30): 26743–26753.
108. Chancellor T, Lee J, Thodeti CK, Lele T. Actomyosin tension exerted on the nucleus through nesprin-1 connections influences endothelial cell adhesion, migration, and cyclic strain-induced reorientation. *Biophys J*. 2010;99(1):115–123.
109. Lovett DB, Shekhar N, Nickerson JA, Roux KJ, Lele TP. Modulation of nuclear shape by substrate rigidity. *Cell Mol Bioeng*. 2013;6(2):230–238.
110. Tamiello C, Kamps MA, van den Wijngaard A, et al. Soft substrates normalize nuclear morphology and prevent nuclear rupture in fibroblasts from a laminopathy patient with compound heterozygous LMNA mutations. *Nucleus*. 2013;4(1):61–73.
111. Khatau SB, Bloom RJ, Bajpai S, et al. The distinct roles of the nucleus and nucleus-cytoskeleton connections in three-dimensional cell migration. *Sci Rep*. 2012;2:488.

112. Anno T, Sakamoto N, Sato M. Role of nesprin-1 in nuclear deformation in endothelial cells under static and uniaxial stretching conditions. *Biochem Biophys Res Commun.* 2012;424(1):94–99.
113. Houben F, Ramaekers F, Snoeckx L, Broers J. Role of nuclear lamina-cytoskeleton interactions in the maintenance of cellular strength. *Biochim Biophys Acta.* 2007;1773(5): 675–686.
114. Geiger B, Bershadsky A, Pankov R, Yamada KM. Transmembrane extracellular matrix-cytoskeleton crosstalk. *Nat Rev Mol Cell Biol.* 2001;2(11):793–805.
115. Gudise S, Figueroa RA, Lindberg R, Larsson V, Hallberg E. Samp1 is functionally associated with the LINC complex and A-type lamina networks. *J Cell Sci.* 2011;124(12): 2077–2085.
116. Gomes ER, Jani S, Gundersen GG. Nuclear movement regulated by Cdc42, MRCK, myosin, and actin flow establishes MTOC polarization in migrating cells. *Cell.* 2005; 121(3):451–463.
117. Nery FC, Zeng J, Niland BP, et al. TorsinA binds the KASH domain of nesprins and participates in linkage between nuclear envelope and cytoskeleton. *J Cell Sci.* 2008;121(20):3476–3486.
118. Krause M, te Riet J, Wolf K. Probing the compressibility of tumor cell nuclei by combined atomic force-confocal microscopy. *Phys Biol.* 2013;10(6):13.
119. Lee JS, Hale CM, Panorchan P, et al. Nuclear lamin A/C deficiency induces defects in cell mechanics, polarization, and migration. *Biophys J.* 2007;93(7):2542–2552.
120. Luxton GWG, Gomes ER, Folker ES, Vintinner E, Gundersen GG. Linear arrays of nuclear envelope proteins harness retrograde actin flow for nuclear movement. *Science.* 2010;329(5994):956–959.
121. Hale CM, Shrestha AL, Khatau SB, et al. Dysfunctional connections between the nucleus and the actin and microtubule networks in laminopathic models. *Biophys J.* 2008;95(11):5462–5475.
122. Morgan JT, Pfeiffer ER, Thirkill TL, et al. Nesprin-3 regulates endothelial cell morphology, perinuclear cytoskeletal architecture, and flow-induced polarization. *Mol Biol Cell.* 2011;22(22):4324–4334.
123. Salpingidou G, Smertenko A, Hausmanowa-Petrucewicz I, Hussey PJ, Hutchison CJ. A novel role for the nuclear membrane protein emerin in association of the centrosome to the outer nuclear membrane. *J Cell Biol.* 2007;178(6):897–904.
124. Niwa T, Saito H, Imajoh-ohmi S, et al. BRCA2 interacts with the cytoskeletal linker protein plectin to form a complex controlling centrosome localization. *Cancer Sci.* 2009;100(11):2115–2125.
125. Sato M, Levesque MJ, Nerem RM. Micropipette aspiration of cultured bovine aortic endothelial-cells exposed to shear-stress. *Arteriosclerosis.* 1987;7(3):276–286.
126. Deguchi S, Maeda K, Ohashi T, Sato M. Flow-induced hardening of endothelial nucleus as an intracellular stress-bearing organelle. *J Biomech.* 2005;38(9):1751–1759.
127. Chambliss AB, Khatau SB, Erdenberger N, et al. The LINC-anchored actin cap connects the extracellular milieu to the nucleus for ultrafast mechanotransduction. *Sci Rep.* 2013;3:1087.
128. Booth-Gauthier EA, Alcoser TA, Yang G, Dahl KN. Force-induced changes in subnuclear movement and rheology. *Biophys J.* 2012;103(12):2423–2431.
129. Guilak F. Compression-induced changes in the shape and volume of the chondrocyte nucleus. *J Biomech.* 1995;28(12):1529–1541.
130. Finan JD, Leddy HA, Guilak F. Osmotic stress alters chromatin condensation and nucleocytoplasmic transport. *Biochem Biophys Res Commun.* 2011;408(2):230–235.
131. Peeters EAG, Bouten CVC, Oomens CWJ, Bader DL, Snoeckx L, Baaijens FPT. Anisotropic, three-dimensional deformation of single attached cells under compression. *Ann Biomed Eng.* 2004;32(10):1443–1452.

132. Broers JLV, Peeters EAG, Kuijpers HJH, et al. Decreased mechanical stiffness in LMNA−/− cells is caused by defective nucleo-cytoskeletal integrity: implications for the development of laminopathies. *Hum Mol Genet.* 2004;13(21): 2567–2580.
133. Nagayama K, Yahiro Y, Matsumoto T. Apical and basal stress fibers have different roles in mechanical regulation of the nucleus in smooth muscle cells cultured on a substrate. *Cell Mol Bioeng.* 2013;6(4):473–481.
134. Maniotis AJ, Chen CS, Ingber DE. Demonstration of mechanical connections between integrins, cytoskeletal filaments, and nucleoplasm that stabilize nuclear structure. *Proc Natl Acad Sci USA.* 1997;94(3):849–854.
135. Iyer KV, Pulford S, Mogilner A, Shivashankar G. Mechanical activation of cells induces chromatin remodeling preceding MKL nuclear transport. *Biophys J.* 2012;103(7): 1416–1428.
136. Maraldi NM, Capanni C, Cenni V, Fini M, Lattanzi G. Laminopathies and lamin-associated signaling pathways. *J Cell Biochem.* 2011;112(4):979–992.
137. Peti W, Page R. Molecular basis of MAP kinase regulation. *Protein Sci.* 2013;22(12): 1698–1710.
138. Emerson LJ, Holt MR, Wheeler MA, Wehnert M, Parsons M, Ellis JA. Defects in cell spreading and ERK1/2 activation in fibroblasts with lamin A/C mutations. *Biochim Biophys Acta.* 2009;1792(8):810–821.
139. Muchir A, Pavlidis P, Bonne G, Hayashi YK, Worman HJ. Activation of MAPK in hearts of EMD null mice: similarities between mouse models of X-linked and autosomal dominant Emery-Dreifuss muscular dystrophy. *Hum Mol Genet.* 2007;16(15): 1884–1895.
140. Muchir A, Pavlidis P, Decostre V, et al. Activation of MAPK pathways links LMNA mutations to cardiomyopathy in Emery-Dreifuss muscular dystrophy. *J Clin Invest.* 2007;117(5):1282–1293.
141. Muchir A, Wu W, Choi JC, et al. Abnormal p38 mitogen-activated protein kinase signaling in dilated cardiomyopathy caused by lamin A/C gene mutation. *Hum Mol Genet.* 2012;21(19):4325–4333.
142. Muchir A, Reilly SA, Wu W, et al. Treatment with selumetinib preserves cardiac function and improves survival in cardiomyopathy caused by mutation in the lamin A/C gene. *Cardiovasc Res.* 2012;93(2):311–319.
143. Lammerding J, Hsiao J, Schulze PC, Kozlov S, Stewart CL, Lee RT. Abnormal nuclear shape and impaired mechanotransduction in emerin-deficient cells. *J Cell Biol.* 2005;170(5):781–791.
144. Gonzalez JM, Navarro-Puche A, Casar B, Crespo P, Andres V. Fast regulation of AP-1 activity through interaction of lamin A/C, ERK1/2, and c-Fos at the nuclear envelope. *J Cell Biol.* 2008;183(4):653–666.
145. Muchir A, Wu W, Worman HJ. Reduced expression of A-type lamins and emerin activates extracellular signal-regulated kinase in cultured cells. *Biochim Biophys Acta.* 2009; 1792(1):75–81.
146. Ivorra C, Kubicek M, Gonzalez JM, et al. A mechanism of AP-1 suppression through interaction of c-Fos with lamin A/C. *Genes Dev.* 2006;20(3):307–320.
147. Gordon MD, Nusse R. Wnt signaling: multiple pathways, multiple receptors, and multiple transcription factors. *J Biol Chem.* 2006;281(32):22429–22433.
148. Robinson JA, Chatterjee-Kishore M, Yaworsky PJ, et al. Wnt/beta-catenin signaling is a normal physiological response to mechanical loading in bone. *J Biol Chem.* 2006; 281(42):31720–31728.
149. Neumann S, Schneider M, Daugherty RL, et al. Nesprin-2 interacts with α-catenin and regulates Wnt signaling at the nuclear envelope. *J Biol Chem.* 2010;285(45): 34932–34938.

150. Fagotto F, Gluck U, Gumbiner BM. Nuclear localization signal-independent and importin/karyopherin-independent nuclear import of beta-catenin. *Curr Biol.* 1998;8(4):181–190.
151. Hernandez L, Roux KJ, Wong ESM, et al. Functional coupling between the extracellular matrix and nuclear lamina by Wnt signaling in progeria. *Dev Cell.* 2010;19(3): 413–425.
152. Bengtsson L. What MAN1 does to the smads—TGF beta/BMP signaling and the nuclear envelope. *FEBS J.* 2007;274(6):1374–1382.
153. Maeda T, Sakabe T, Sunaga A, et al. Conversion of mechanical force into TGF-beta-mediated biochemical signals. *Curr Biol.* 2011;21(11):933–941.
154. Wipff PJ, Rifkin DB, Meister JJ, Hinz B. Myofibroblast contraction activates latent TGF-beta 1 from the extracellular matrix. *J Cell Biol.* 2007;179(6):1311–1323.
155. Pan D, Estevez-Salmeron LD, Stroschein SL, et al. The integral inner nuclear membrane protein MAN1 physically interacts with the R-Smad proteins to repress signaling by the transforming growth factor-beta superfamily of cytokines. *J Biol Chem.* 2005;280(16):15992–16001.
156. Osada SI, Ohmori S, Taira M. XMAN1, an inner nuclear membrane protein, antagonizes BMP signaling by interacting with Smad1 in Xenopus embryos. *Development.* 2003;130(9):1783–1794.
157. Van Berlo JH, Voncken JW, Broers JLV, et al. A-type lamins are essential for TGF-beta 1 induced PP2A to dephosphorylate transcription factors. *Hum Mol Genet.* 2005;14(19): 2839–2849.
158. Worman HJ. Inner nuclear membrane and regulation of Smad-mediated signaling. *Biochim Biophys Acta.* 2006;1761(5–6):626–631.
159. Mouilleron S, Langer CA, Guettler S, McDonald NQ, Treisman R. Structure of a pentavalent G-actin*MRTF-A complex reveals how G-actin controls nucleocytoplasmic shuttling of a transcriptional coactivator. *Sci Signal.* 2011;4(177):ra40.
160. Miralles F, Posern G, Zaromytidou AI, Treisman R. Actin dynamics control SRF activity by regulation of its coactivator MAL. *Cell.* 2003;113(3):329–342.
161. Pawlowski R, Rajakyla EK, Vartiainen MK, Treisman R. An actin-regulated importin alpha/beta-dependent extended bipartite NLS directs nuclear import of MRTF-A. *EMBO J.* 2010;29(20):3448–3458.
162. Small EM, Thatcher JE, Sutherland LB, et al. Myocardin-related transcription factor-a controls myofibroblast activation and fibrosis in response to myocardial infarction. *Circ Res.* 2010;107(2):294–304.
163. Parlakian A, Tuil D, Hamard G, et al. Targeted inactivation of serum response factor in the developing heart results in myocardial defects and embryonic lethality. *Mol Cell Biol.* 2004;24(12):5281–5289.
164. Ho CY, Jaalouk DE, Lammerding J. Novel insights into the disease etiology of laminopathies. *Rare Dis.* 2013;1(1):e27002.
165. Ho CY, Jaalouk DE, Vartiainen MK, Lammerding J. Lamin A/C and emerin regulate MKL1-SRF activity by modulating actin dynamics. *Nature.* 2013;497(7450):507–511.
166. Chang W, Folker ES, Worman HJ, Gundersen GG. Emerin organizes actin flow for nuclear movement and centrosome orientation in migrating fibroblasts. *Mol Biol Cell.* 2013;24(24):3869–3880. http://dx.doi.org/10.1091/mbc.E13-06-0307.
167. Baarlink C, Wang H, Grosse R. Nuclear actin network assembly by formins regulates the SRF coactivator MAL. *Science.* 2013;340(6134):864–867.
168. Charbonney E, Speight P, Masszi A, Nakano H, Kapus A. β-Catenin and Smad3 regulate the activity and stability of myocardin-related transcription factor during epithelial–myofibroblast transition. *Mol Biol Cell.* 2011;22(23):4472–4485.
169. Halder G, Dupont S, Piccolo S. Transduction of mechanical and cytoskeletal cues by YAP and TAZ. *Nat Rev Mol Cell Biol.* 2012;13(9):591–600.

170. Dupont S, Morsut L, Aragona M, et al. Role of YAP/TAZ in mechanotransduction. *Nature*. 2011;474(7350):179–183.
171. Andres V, Gonzalez JM. Role of A-type lamins in signaling, transcription, and chromatin organization. *J Cell Biol*. 2009;187(7):945–957.
172. Dorner D, Vlcek S, Foeger N, et al. Lamina-associated polypeptide 2 alpha regulates cell cycle progression and differentiation via the retinoblastoma-E2F pathway. *J Cell Biol*. 2006;173(1):83–93.
173. Markiewicz E, Dechat T, Foisner R, Quinlan RA, Hutchison CJ. Lamin A/C binding protein LAP2 alpha is required for nuclear anchorage of retinoblastoma protein. *Mol Biol Cell*. 2002;13(12):4401–4413.
174. Naetar N, Korbei B, Kozlov S, et al. Loss of nucleoplasmic LAP2α–lamin A complexes causes erythroid and epidermal progenitor hyperproliferation. *Nat Cell Biol*. 2008; 10(11):1341–1348.
175. Tifft KE, Bradbury KA, Wilson KL. Tyrosine phosphorylation of nuclear-membrane protein emerin by Src, Abl and other kinases. *J Cell Sci*. 2009;122(20):3780–3790.
176. Bose R, Molina H, Patterson AS, et al. Phosphoproteomic analysis of Her2/neu signaling and inhibition. *Proc Natl Acad Sci USA*. 2006;103(26):9773–9778.
177. Olayioye MA, Neve RM, Lane HA, Hynes NE. The ErbB signaling network: receptor heterodimerization in development and cancer. *EMBO J*. 2000;19(13):3159–3167.
178. Wozniak MA, Baker BM, Chen CS, Wilson KL. The emerin-binding transcription factor Lmo7 is regulated by association with p130Cas at focal adhesions. *PeerJ*. 2013;1:e134.
179. Sawada Y, Tamada M, Dubin-Thaler BJ, et al. Force sensing by mechanical extension of the Src family kinase substrate p130Cas. *Cell*. 2006;127(5):1015–1026.
180. Bengtsson L, Wilson KL. Multiple and surprising new functions for emerin, a nuclear membrane protein. *Curr Opin Cell Biol*. 2004;16(1):73–79.
181. Hirano Y, Iwase Y, Ishii K, Kumeta M, Horigome T, Takeyasu K. Cell cycle-dependent phosphorylation of MAN1. *Biochemistry*. 2009;48(7):1636–1643.
182. Gajewski A, Csaszar E, Foisner R. A phosphorylation cluster in the chromatin-binding region regulates chromosome association of LAP2 alpha. *J Biol Chem*. 2004;279(34): 35813–35821.
183. Scaffidi P, Misteli T. Lamin A-dependent misregulation of adult stem cells associated with accelerated ageing. *Nat Cell Biol*. 2008;10(4):452–459.
184. Bione S, Maestrini E, Rivella S, et al. Identification of a novel x-linked gene responsible for Emery-Dreifuss muscular-dystrophy. *Nat Genet*. 1994;8(4):323–327.
185. Emery AEH. Emery-Dreifuss muscular dystrophy—a 40 year retrospective. *Neuromuscul Disord*. 2000;10(4–5):228–232.
186. Mejat A, Decostre V, Li J, et al. Lamin A/C-mediated neuromuscular junction defects in Emery-Dreifuss muscular dystrophy. *J Cell Biol*. 2009;184(1):31–44.
187. Zhang XC, Xu RN, Zhu BG, et al. Syne-1 and Syne-2 play crucial roles in myonuclear anchorage and motor neuron innervation. *Development*. 2007;134(5):901–908.
188. Folker ES, Ostlund C, Luxton GWG, Worman HJ, Gundersen GG. Lamin A variants that cause striated muscle disease are defective in anchoring transmembrane actin-associated nuclear lines for nuclear movement. *Proc Natl Acad Sci USA*. 2011;108(1): 131–136.
189. Cupesi M, Yoshioka J, Gannon J, Kudinova A, Stewart CL, Lammerding J. Attenuated hypertrophic response to pressure overload in a lamin A/C haploinsufficiency mouse. *J Mol Cell Cardiol*. 2010;48(6):1290–1297.
190. Engler AJ, Sen S, Sweeney HL, Discher DE. Matrix elasticity directs stem cell lineage specification. *Cell*. 2006;126(4):677–689.
191. Nikolova V, Leimena C, McMahon AC, et al. Defects in nuclear structure and function promote dilated cardiomyopathy in lamin A/C-deficient mice. *J Clin Invest*. 2004;113(3):357–369.

192. Bengtsson L, Otto H. LUMA interacts with emerin and influences its distribution at the inner nuclear membrane. *J Cell Sci*. 2008;121(4):536–548.
193. Muchir A, Shan J, Bonne G, Lehnart SE, Worman HJ. Inhibition of extracellular signal-regulated kinase signaling to prevent cardiomyopathy caused by mutation in the gene encoding A-type lamins. *Hum Mol Genet*. 2009;18(2):241–247.
194. Wu W, Iwata S, Homma S, Worman HJ, Muchir A. Depletion of extracellular signal-regulated kinase 1 in mice with cardiomyopathy caused by lamin A/C gene mutation partially prevents pathology before isoenzyme activation. *Hum Mol Genet*. 2014;23(1): 1–11.
195. Wu W, Muchir A, Shan JA, Bonne G, Worman HJ. Mitogen-activated protein kinase inhibitors improve heart function and prevent fibrosis in cardiomyopathy caused by mutation in lamin A/C gene. *Circulation*. 2011;123(1):53–61.
196. Qin Z, Kalinowski A, Dahl KN, Buehler MJ. Structure and stability of the lamin A tail domain and HGPS mutant. *J Struct Biol*. 2011;175(3):425–433.
197. Merideth MA, Gordon LB, Clauss S, et al. Phenotype and course of Hutchinson-Gilford progeria syndrome. *N Engl J Med*. 2008;358(6):592–604.
198. Al-Shali KZ, Hegele RA. Laminopathies and atherosclerosis. *Arterioscler Thromb Vasc Biol*. 2004;24(9):1591–1595.
199. Verstraeten V, Ji JY, Cummings KS, Lee RT, Lammerding J. Increased mechanosensitivity and nuclear stiffness in Hutchinson-Gilford progeria cells: effects of farnesyltransferase inhibitors. *Aging Cell*. 2008;7(3):383–393.
200. Dahl KN, Scaffidi P, Islam MF, Yodh AG, Wilson KL, Misteli T. Distinct structural and mechanical properties of the nuclear lamina in Hutchinson-Gilford progeria syndrome. *Proc Natl Acad Sci USA*. 2006;103(27):10271–10276.
201. Young SG, Yang SH, Davies BSJ, Jung HJ, Fong LG. Targeting protein prenylation in progeria. *Sci Transl Med*. 2013;5(171):6.
202. Schirmer EC, Foisner R. Proteins that associate with lamins: many faces, many functions. *Exp Cell Res*. 2007;313(10):2167–2179.
203. Wolff J. The classic: on the significance of the architecture of the spongy substance for the question of bone growth: a preliminary publication. *Clin Orthop Relat Res*. 2011;469(11):3077–3078.
204. Satzinger H. Series on historical profiles—timeline—Theodor and Marcella Boveri: chromosomes and cytoplasm in heredity and development. *Nat Rev Genet*. 2008;9(3): 231–238.
205. Guharay F, Sachs F. Stretch-activated single ion channel currents in tissue-cultured embryonic chick skeletal-muscle. *J Physiol*. 1984;352:685–701.
206. Wang N, Butler JP, Ingber DE. Mechanotransduction across the cell-surface and through the cytoskeleton. *Science*. 1993;260(5111):1124–1127.
207. Pelham RJ, Wang YL. Cell locomotion and focal adhesions are regulated by substrate flexibility. *Proc Natl Acad Sci USA*. 1997;94(25):13661–13665.
208. Grashoff C, Hoffman BD, Brenner MD, et al. Measuring mechanical tension across vinculin reveals regulation of focal adhesion dynamics. *Nature*. 2010;466(7303): 263–266.
209. Dwyer N, Blobel G. Modified procedure for isolation of a pore complex lamina fraction from rat-liver nuclei. *J Cell Biol*. 1976;70(3):581–591.
210. Humphries JD, Wang P, Streuli C, Geiger B, Humphries MJ, Ballestrem C. Vinculin controls focal adhesion formation by direct interactions with talin and actin. *J Cell Biol*. 2007;179(5):1043–1057.
211. Zhong CL, Chrzanowska-Wodnicka M, Brown J, Shaub A, Belkin AM, Burridge K. Rho-mediated contractility exposes a cryptic site in fibronectin and induces fibronectin matrix assembly. *J Cell Biol*. 1998;141(2):539–551.
212. Abercrom M, Heaysman JE, Pegrum SM. Locomotion of fibroblasts in culture. 4. Electron microscopy of leading lamella. *Exp Cell Res*. 1971;67(2):359–367.

CHAPTER EIGHT

Nuclear Forces and Cell Mechanosensing

Samer Alam*, David B. Lovett*, Richard B. Dickinson*, Kyle J. Roux[†], Tanmay P. Lele*

*Department of Chemical Engineering, University of Florida, Gainesville, Florida, USA
[†]Sanford Children's Health Research Center, University of South Dakota, Sioux Falls, South Dakota, USA

Contents

Abstract

Cells respond to mechanical signals, but the subcellular mechanisms are not well understood. The nucleus has recently emerged as an important mechanosensory organelle in the cell, as it is intimately connected to the cytoskeleton. Mechanical forces applied to cells that act on membrane-embedded receptors are transmitted through the cytoskeleton to the nuclear surface. Interfering with linkers of the nucleus to the cytoskeleton causes defects in cell mechanosensing and cell function. In this chapter, we discuss recent work in this area, highlighting the role that the nuclear linkages with the cytoskeleton play in cellular mechanotransduction.

1. INTRODUCTION

Cells in the body are exposed to both "active" and "passive" mechanical stimuli. For example, endothelial cells that line the blood vessels are constantly exposed to shear stresses and cyclic stretching imposed by the pulsatile flow of blood, actively remodeling their cytoskeleton and overall morphology. Adherent cells are also exposed to widely varying mechanical cues from the extracellular matrix (ECM) depending on the organ they inhabit neurons, for example, are surrounded by much softer tissue than smooth muscle

Progress in Molecular Biology and Translational Science, Volume 126
ISSN 1877-1173
http://dx.doi.org/10.1016/B978-0-12-394624-9.00008-7

cells or osteoblasts. These passive cues also elicit a response from cells. The mechanisms by which cells respond to mechanical stimuli are of strong current interest in the emerging area of mechanobiology.

Cells can transduce mechanical signals into a biochemical response. This process is known as mechanotransduction; however, there is no singular mechanism by which this happens. Mechanical forces applied to cells, which transmit signals ~40-fold faster than diffusion of some chemical signals,[1] can cause conformational changes in heterodimeric integrin proteins in cell-matrix adhesion sites. This can ultimately alter signaling pathways and gene expression. Mechanical stimuli can open ion channels (see review by Morris[2]), alter binding of proteins in focal adhesions,[3] and cause changes in overall cell morphology[4] (also see review by Ingber[5]). In recent years, it has become recognized that the nucleus of the cell can act as a mechanosensory organelle (see review by Wang *et al.*[6]), which not only experiences and transmits forces directly but also influences cell mechanosensing, through mechanisms that are beginning to be understood.

In this chapter, we review the evidence that supports the concept that the nucleus mediates mechanosensing. We discuss how force propagation occurs to the nuclear surface, how cytoskeletal coupling to the nucleus is necessary for mechanosensing, and how the nucleus should be seen as an integrated component with the cytoskeleton in models for cell mechanosensing.

2. CYTOSKELETAL FORCES ARE EXERTED ON THE NUCLEUS

Early evidence that the nucleus is under tension came from Ingber and coworkers in 1992, who showed that perturbing actomyosin forces altered cell and nuclear shape.[7] In a landmark paper in 1997, they showed that tugging on integrin receptors in the cell membrane causes nuclear distortion and motion.[8] This established the concept that forces applied externally to the cell are propagated to the nuclear surface, consistent with mechanical models of the cell cytoskeleton that are 'hardwired' to the nuclear envelope.[9–14] These external forces have now been shown to induce clearly detectable nuclear deformation.[8,15–22] The F-actin cytoskeleton plays a major role in propagating the mechanical forces from integrin receptors to the nuclear surface, although the molecules which connect the nucleus to the cytoskeleton have only recently been identified.

In recent years, members of the so-called LINC complex (for *Li*nker of *N*ucleoskeleton to the *C*ytoskeleton) have been discovered in the nuclear envelope.[23–26] The LINC complex is comprised of two protein families that span the nuclear envelope and physically connect the cytoskeleton to the nucleoskeleton. The SUN (*S*ad1p, *UN*C-84) domain proteins span the inner nuclear membrane (INM) and translumenally bind the KASH (*K*larsicht/*A*NC-1/*S*yne *H*omology) domain proteins that span the outer nuclear membrane (ONM) (Fig. 8.1). In this way, the KASH and SUN domain proteins create a mechanical tether that connects both membranes of the nuclear envelope. The KASH domain proteins bind to various cytoskeletal constituents, whereas the SUN domain proteins associate with the nuclear lamina. Thus, the mechanical connections created by the LINC complex can integrate the forces of the cytoskeleton and the nucleus.

The LINC complex is functionally well conserved in eukaryotes, including single-celled organisms such as yeasts; however, the number and nature

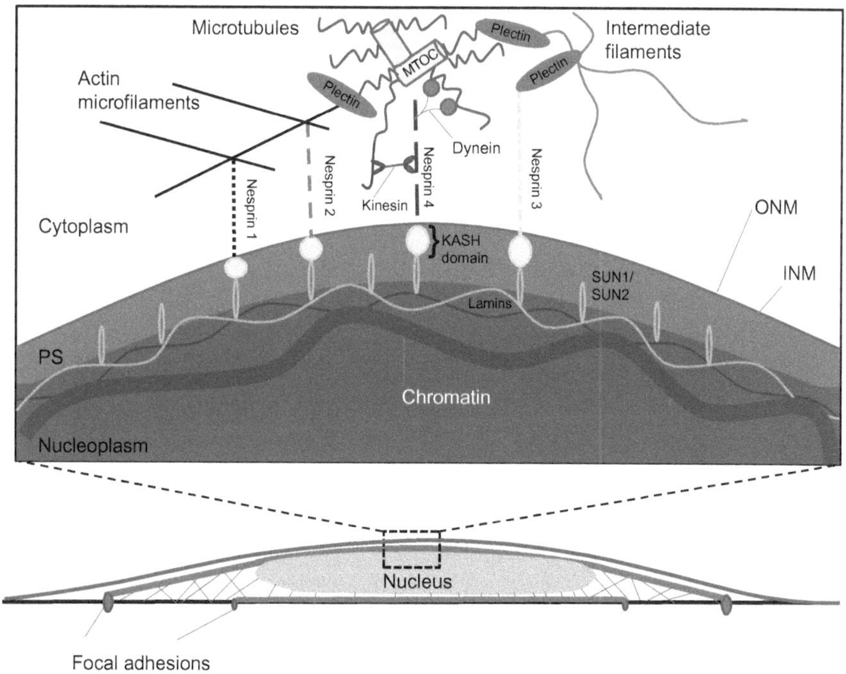

Figure 8.1 The LINC complex is formed by the SUN domain proteins spanning the inner nuclear membrane (INM) and translumenally binding the KASH domain proteins that span the outer nuclear membrane (ONM). The SUN domain proteins associate with the nuclear lamina, whereas the KASH domain proteins bind to various cytoskeletal constituents through the nesprin protein family (nesprins 1–4). Nesprin 1 and nesprin 2 bind to actin, and to dynein and kinesin, whereas nesprin 3 binds plectin, an intermediate filament-associated cytolinker. (See the color plate.)

of the KASH and SUN domain constituents varies between divergent species. In mammals, there are five SUN domain proteins (SUNs 1–5), but only SUN1 and SUN2 appear widely expressed. SUNs 3–5 are predominantly, if not exclusively expressed in the testis. SUN1 and SUN2 have been shown to associate with the nuclear lamina; however, the role of any additional nucleoplasmic associations of SUN domain proteins in LINC complex function remains unclear. The crystal structure of the large SUN2 lumenal domain has revealed that the protein forms a trimeric[27,28] oligomer, mediated by a lumenal coiled-coil domain. Each of the three SUN domains in this trimer binds to a unique KASH domain. This oligomeric association likely enhances the physical strength the LINC complex to transfer forces between the cytoskeleton and nucleus.

There are six mammalian KASH domain proteins (nesprins 1–4, LRMP, and KASH5), only three of which (nesprins 1–3) are generally ubiquitous. Nesprin 1 and nesprin 2 bind to actin, and to dynein and kinesin,[29,30] whereas nesprin 3 binds plectin, an intermediate filament-associated cytolinker. Nesprin 4 is predominately expressed in highly polarized epithelial cells and its loss leads to hearing defects in mice and humans associated with perturbation of nuclear positioning.[31,32] KASH5 expression appears limited to meiotic cells where it binds dynein to move chromosomes during homologous recombination.[33,34] Predominantly expressed in lymphocytes and taste cells, LRMP does not appear to interact with the cytoskeleton but instead binds to the calcium channel IP3 receptors.[33,35] There are multiple splice isoforms of nesprin 1 and nesprin 2 that do not contain the ONM-targeting KASH domain. Many of these KASH-less isoforms likely have functions other than formation of the LINC complex.[36–40]

The LINC complex is the only known structure by which cytoskeletal stresses can be directly transferred to the nuclear surface (Fig. 8.1). Since the cytoskeleton ultimately connects to focal adhesions[41,42] (also see review by Geiger *et al.*[43]), the LINC complex enables a mechanical linkage between the nucleus, the cytoskeleton, and the extracellular matrix.[20,25,26]

3. THE LINC COMPLEX TRANSMITS CYTOSKELETAL FORCES TO THE NUCLEAR SURFACE

The nucleus has been suggested to be a cellular mechanosensor,[6] and besides being a key player in the physical signaling pathway, it is possible that the LINC complex is involved in chemical signal transduction pathways as well.[44–46] The LINC complex is essential for efficient migration,[17,47,48] normal structure,[48,49] function,[17,48] and maintaining nuclear shape and position.[17,48,50–52] For a review on disrupted LINC complexes causing defects in mechanotransduction, see the recent review by Jaalouk and Lammerding.[53] The LINC complex has been found to be a key player in force transmission between the nucleus and cytoskeleton,[17,48] and an intact LINC complex is required for nuclear positioning, cell polarization, and normal propagation of cytoskeletal forces.[17]

It has been demonstrated that external forces applied to the apical surface of a living cell propagate through the cytoskeleton and all the way to the nucleus[8] and that forces applied to integrins can cause motion of intranuclear organelles.[22] Chancellor *et al.* recently showed that nesprin 1 knockdown significantly increased the nuclear height, suggesting an essential role of nesprin 1 in flattening the nucleus in endothelial cells.[54,55] Inhibiting myosin activity similarly produced a vertically rounded nucleus. Chancellor *et al.* proposed a model in which actomyosin force pulls laterally on the nucleus and flattens it to the shape of a disk. In this model, the nucleus acts as a scaffold that balances actomyosin forces internally, while the substrate balances them externally. In the absence of nesprin 1, the pulling force on the nucleus is substantially reduced and the nucleus is free to relax vertically into a rounded shape,[54] while the excess force is now balanced at an increased number of adhesion sites with the substratum.

In a series of recent papers, Gundersen and coworkers showed that dorsal actin bundles (on top of the nuclear surface) are directly linked to the nucleus via TAN (for *T*ransmembrane *A*ctin-associated *N*uclear) lines assembled from nesprin 2 giant and SUN2 proteins.[47,56] The authors suggest that the TAN lines across the nuclear membrane function in a manner similar to focal adhesions across the cellular membrane in that both assemblies are linked to actin cables and transmit mechanical force.[47,57] They also showed in fibroblasts that emerin and myosin IIB function to polarize nuclear movement and flow of actin, which suggests a new role of the nuclear envelope in establishing cytoskeletal polarity and directional actin flow.[58]

Recently, Khatau *et al.* suggested that apical stress fibers on top of the nucleus shape it by squeezing from the top as the fibers contract.[59] The authors call these distinct bundles the "actin cap." These bundles terminate at a small distinct subset of focal adhesions, which are proposed to regulate mechanosensing via the actin cap. The actin cap associate focal adhesions have been shown to be larger in size than conventional focal adhesions, are located only at the periphery, and experience fast turnover dynamics.[60]

Wirtz and coworkers have proposed that the actin cap and its associated focal adhesions play a key role in the fast and efficient physical pathway for mechanotransduction by providing a continuous mechanical linkage from the ECM to the nucleus.[60–62] Coupling of the actin cap to the nuclear lamina occurs via the LINC complex[59,63,64] and its associated nesprins.

Like actin bundles, microtubule motor proteins can also generate tension that is transmitted to the nuclear surface through the LINC complex. For example, Splinter *et al.* showed that by binding to the nuclear membrane at their cargo end, dynein, and kinesin-1 can pull the nucleus in opposite directions as they walk. Kinesin-1 pulls the nucleus away from the centrosome, while dynein pulls it toward the centrosome.[65] The combined activities of these two processive motor proteins control nuclear translation and rotation.[66–68] Wu *et al.* showed that dynein walking on microtubules in the vicinity of the nucleus can produce nuclear rotations.[69] They also modeled the rotation of the nucleus computationally to show how this could mechanically occur in the cell. As mentioned previously, nesprin 4, and likely nesprins 1 and 2, function to position the nucleus within the cell through the action of microtubule motors. KASH5 (and its orthologs) functions to transmit microtubule motor forces through the nuclear envelope directly to meiotic chromosomes.[33,34,70–72]

4. THE ROLE OF THE NUCLEUS IN CELL MECHANOSENSING

Although the exact mechanisms are not completely understood, cells have been shown to sense and respond to different mechanical cues from their surroundings such as shear stress,[73] substrate strain,[74] and substrate rigidity.[75] Evidence that the nuclear force balance is important in this cell mechanosensing has come from studies in which the LINC complex was disrupted. For example, endothelial cells were found to lose their ability to reorient in response to uniaxial cyclic strain upon knockdown of nesprin 1.[54] Because the nucleus no longer acts as an internal scaffold to balance actomyosin tension, the nesprin 1-deficient cells apply increased traction on the substratum, resulting in stronger adhesion and less propensity to reorient in response to mechanical strain.

There is evidence that nuclear forces might regulate gene expression. Changing nuclear shape by controlling the degree of cell spreading alters protein synthesis.[76] In response to cyclic strain, lamin A/C-deficient and emerin-deficient mouse embryonic fibroblasts have impaired expression of genes lex-1 and Egr-1.[20,55] Philip and Dahl showed that in response to shear stress, lamins are upregulated and reorganized.[77] More recently, defects in lamin A/C have been demonstrated to impair nuclear translocation of MKL-1, a transcription factor.[78]

In NIH/3T3 fibroblasts, the nuclear shape has been shown to depend on the rigidity of the underlying substrate.[79] The nucleus takes a spherical shape on soft substrates, likely due to the smaller actomyosin tension and a flattened ellipsoid on rigid substrates where higher forces are generated in cells.[79] Tuning the actomyosin tension by changing the substrate rigidity enables control of nuclear shape. Disrupting the LINC complex via KASH overexpression or inhibiting myosin II eliminates this shape dependence of nucleus on substrate rigidity, suggesting that the coupling of the nucleus to the cytoskeleton is essential for mechanosensing.[79] The cell motility and the cell spreading area both correlate with the underlying substrate rigidity in a manner that depends on the nuclear linkages to the cytoskeleton; on softer substrates, cells have lower speed and spreading area, whereas they have higher speed and spreading area on stiffer substrates, and KASH overexpression ablated this trend.[79] Separately, Swift and coworkers have demonstrated that the lamin A levels are proportional to the tissue rigidity,[80] indicating that lamin A stabilizes the nucleus under stress. Consistent with this, LINC complex disruption decreases cellular mechanical stiffness.[49]

5. CONCLUSIONS

In conclusion, physical connectivity from the nucleus to the cytoskeleton and cell membrane is required for normal cell mechanosensing. Force transmission from external receptors to the nuclear surface requires an intact nuclear lamina and physical connectivity between the nucleus and the cytoskeleton by the LINC complex. Ultimately, how the integrated nuclear–cytoskeleton integrates mechanical stimuli with complex intracellular signaling pathways and gene regulatory networks will require a systems biology approach that seeks to understand cell mechanosensing in an integrated manner.

ACKNOWLEDGMENTS

This work is supported by the National Science Foundation under award CMMI 0954302 (T. P. L.) and by the National Institutes of Health under awards R01GM102486-01 (T. P. L., K. R., and R. B. D.) and 5R01EB014869-02 (T. P. L. and K. R.).

REFERENCES

1. Na S, Collin O, Chowdhury F, et al. Rapid signal transduction in living cells is a unique feature of mechanotransduction. *Proc Natl Acad Sci USA.* 2008;105(18):6626–6631.
2. Morris CE. Mechanosensitive ion channels. *J Membr Biol.* 1990;113(2):93–107.
3. Lele TP, Pendse J, Kumar S, Salanga M, Karavitis J, Ingber DE. Mechanical forces alter zyxin unbinding kinetics within focal adhesions of living cells. *J Cell Physiol.* 2006;207(1):187–194.
4. Yeung T, Georges PC, Flanagan LA, et al. Effects of substrate stiffness on cell morphology, cytoskeletal structure, and adhesion. *Cell Motil Cytoskeleton.* 2005;60(1):24–34.
5. Ingber D. Cellular mechanotransduction: putting all the pieces together again. *FASEB J.* 2006;20(7):811–827.
6. Wang N, Tytell JD, Ingber DE. Mechanotransduction at a distance: mechanically coupling the extracellular matrix with the nucleus. *Nat Rev Mol Cell Biol.* 2009;10(1):75–82.
7. Sims JR, Karp S, Ingber DE. Altering the cellular mechanical force balance results in integrated changes in cell, cytoskeletal and nuclear shape. *J Cell Sci.* 1992;103 (Pt. 4):1215–1222.
8. Maniotis AJ, Chen CS, Ingber DE. Demonstration of mechanical connections between integrins, cytoskeletal filaments, and nucleoplasm that stabilize nuclear structure. *Proc Natl Acad Sci USA.* 1997;94(3):849–854.
9. Ingber DE. Tensegrity: the architectural basis of cellular mechanotransduction. *Annu Rev Physiol.* 1997;59:575–599.
10. Chen CS, Ingber DE. Tensegrity and mechanoregulation: from skeleton to cytoskeleton. *Osteoarthr Cartil.* 1999;7(1):81–94.
11. Ingber DE, Tensegrity I. Cell structure and hierarchical systems biology. *J Cell Sci.* 2003;116(Pt 7):1157–1173.
12. Wang N, Naruse K, Stamenovic D, et al. Mechanical behavior in living cells consistent with the tensegrity model. *Proc Natl Acad Sci USA.* 2001;98(14):7765–7770.
13. Stamenovic D, Mijailovich SM, Tolic-Norrelykke IM, Chen J, Wang N. Cell prestress. II. Contribution of microtubules. *Am J Physiol Cell Physiol.* 2002;282(3): C617–C624.
14. Wang N, Tolic-Norrelykke IM, Chen J, et al. Cell prestress. I. Stiffness and prestress are closely associated in adherent contractile cells. *Am J Physiol Cell Physiol.* 2002;282(3): C606–C616.
15. Guilak F, Tedrow JR, Burgkart R. Viscoelastic properties of the cell nucleus. *Biochem Biophys Res Commun.* 2000;269(3):781–786.
16. Guilak F. Compression-induced changes in the shape and volume of the chondrocyte nucleus. *J Biomech.* 1995;28(12):1529–1541.
17. Lombardi ML, Jaalouk DE, Shanahan CM, Burke B, Roux KJ, Lammerding J. The interaction between nesprins and sun proteins at the nuclear envelope is critical for force transmission between the nucleus and cytoskeleton. *J Biol Chem.* 2011;286(30): 26743–26753.
18. Lammerding J, Lee RT. Mechanical properties of interphase nuclei probed by cellular strain application. *Methods Mol Biol.* 2009;464:13–26.
19. Lammerding J, Fong LG, Ji JY, et al. Lamins A and C but not lamin B1 regulate nuclear mechanics. *J Biol Chem.* 2006;281(35):25768–25780.
20. Lammerding J, Schulze PC, Takahashi T, et al. Lamin A/C deficiency causes defective nuclear mechanics and mechanotransduction. *J Clin Invest.* 2004;113(3):370–378.
21. Pajerowski JD, Dahl KN, Zhong FL, Sammak PJ, Discher DE. Physical plasticity of the nucleus in stem cell differentiation. *Proc Natl Acad Sci USA.* 2007;104(40):15619–15624.
22. Poh YC, Shevtsov SP, Chowdhury F, et al. Dynamic force-induced direct dissociation of protein complexes in a nuclear body in living cells. *Nat Commun.* 2012;3:866.

23. Burke B, Roux KJ. Nuclei take a position: managing nuclear location. *Dev Cell.* 2009;17(5):587–597.
24. Crisp M, Liu Q, Roux K, et al. Coupling of the nucleus and cytoplasm: role of the LINC complex. *J Cell Biol.* 2006;172(1):41–53.
25. Starr DA, Fridolfsson HN. Interactions between nuclei and the cytoskeleton are mediated by SUN-KASH nuclear-envelope bridges. *Annu Rev Cell Dev Biol.* 2010;26:421–444.
26. Simon DN, Wilson KL. The nucleoskeleton as a genome-associated dynamic 'network of networks'. *Nat Rev Mol Cell Biol.* 2011;12(11):695–708.
27. Wang W, Shi Z, Jiao S, et al. Structural insights into SUN-KASH complexes across the nuclear envelope. *Cell Res.* 2012;22(10):1440–1452.
28. Sosa BA, Rothballer A, Kutay U, Schwartz TU. LINC complexes form by binding of three KASH peptides to domain interfaces of trimeric SUN proteins. *Cell.* 2012;149(5):1035–1047.
29. Yu J, Lei K, Zhou M, et al. KASH protein Syne-2/Nesprin-2 and SUN proteins SUN1/2 mediate nuclear migration during mammalian retinal development. *Hum Mol Genet.* 2011;20(6):1061–1073.
30. Zhang X, Lei K, Yuan X, et al. SUN1/2 and Syne/Nesprin-1/2 complexes connect centrosome to the nucleus during neurogenesis and neuronal migration in mice. *Neuron.* 2009;64(2):173–187.
31. Horn HF, Brownstein Z, Lenz DR, et al. The LINC complex is essential for hearing. *J Clin Invest.* 2013;123(2):740–750.
32. Roux KJ, Crisp ML, Liu Q, et al. Nesprin 4 is an outer nuclear membrane protein that can induce kinesin-mediated cell polarization. *Proc Natl Acad Sci USA.* 2009;106(7):2194–2199.
33. Horn HF, Kim DI, Wright GD, et al. A mammalian KASH domain protein coupling meiotic chromosomes to the cytoskeleton. *J Cell Biol.* 2013;202(7):1023–1039.
34. Morimoto A, Shibuya H, Zhu X, et al. A conserved KASH domain protein associates with telomeres, SUN1, and dynactin during mammalian meiosis. *J Cell Biol.* 2012;198(2):165–172.
35. Shindo Y, Kim MR, Miura H, et al. Lrmp/Jaw1 is expressed in sweet, bitter, and umami receptor-expressing cells. *Chem Senses.* 2010;35(2):171–177.
36. Kobayashi Y, Katanosaka Y, Iwata Y, Matsuoka M, Shigekawa M, Wakabayashi S. Identification and characterization of GSRP-56, a novel Golgi-localized spectrin repeat-containing protein. *Exp Cell Res.* 2006;312(16):3152–3164.
37. Rajgor D, Mellad JA, Autore F, Zhang Q, Shanahan CM. Multiple novel nesprin-1 and nesprin-2 variants act as versatile tissue-specific intracellular scaffolds. *PLoS One.* 2012;7(7):e40098.
38. Cottrell JR, Borok E, Horvath TL, Nedivi E. CPG2: a brain- and synapse-specific protein that regulates the endocytosis of glutamate receptors. *Neuron.* 2004;44(4):677–690.
39. Marme A, Zimmermann HP, Moldenhauer G, et al. Loss of Drop1 expression already at early tumor stages in a wide range of human carcinomas. *Int J Cancer.* 2008;123(9):2048–2056.
40. Warren DT, Tajsic T, Mellad JA, Searles R, Zhang Q, Shanahan CM. Novel nuclear nesprin-2 variants tether active extracellular signal-regulated MAPK1 and MAPK2 at promyelocytic leukemia protein nuclear bodies and act to regulate smooth muscle cell proliferation. *J Biol Chem.* 2010;285(2):1311–1320.
41. Burridge K, Fath K, Kelly T, Nuckolls G, Turner C. Focal adhesions: transmembrane junctions between the extracellular matrix and the cytoskeleton. *Annu Rev Cell Biol.* 1988;4:487–525.
42. Wang N, Butler JP, Ingber DE. Mechanotransduction across the cell surface and through the cytoskeleton. *Science.* 1993;260(5111):1124–1127.

43. Geiger B, Spatz JP, Bershadsky AD. Environmental sensing through focal adhesions. *Nat Rev Mol Cell Biol.* 2009;10(1):21–33.
44. Bissell MJ, Aggeler J. Dynamic reciprocity: how do extracellular matrix and hormones direct gene expression? *Prog Clin Biol Res.* 1987;249:251–262.
45. Bissell MJ, Barcellos-Hoff MH. The influence of extracellular matrix on gene expression: is structure the message? *J Cell Sci Suppl.* 1987;8:327–343.
46. Bissell MJ, Hall HG, Parry G. How does the extracellular matrix direct gene expression? *J Theor Biol.* 1982;99(1):31–68.
47. Luxton GW, Gomes ER, Folker ES, Vintinner E, Gundersen GG. Linear arrays of nuclear envelope proteins harness retrograde actin flow for nuclear movement. *Science.* 2010;329(5994):956–959.
48. Lombardi ML, Lammerding J. Keeping the LINC: the importance of nucleocytoskeletal coupling in intracellular force transmission and cellular function. *Biochem Soc Trans.* 2011;39(6):1729–1734.
49. Stewart-Hutchinson PJ, Hale CM, Wirtz D, Hodzic D. Structural requirements for the assembly of LINC complexes and their function in cellular mechanical stiffness. *Exp Cell Res.* 2008;314(8):1892–1905.
50. Fischer-Vize JA, Mosley KL. Marbles mutants: uncoupling cell determination and nuclear migration in the developing Drosophila eye. *Development.* 1994;120(9): 2609–2618.
51. Malone CJ, Fixsen WD, Horvitz HR, Han M. UNC-84 localizes to the nuclear envelope and is required for nuclear migration and anchoring during C. elegans development. *Development.* 1999;126(14):3171–3181.
52. Mosley-Bishop KL, Li Q, Patterson L, Fischer JA. Molecular analysis of the klarsicht gene and its role in nuclear migration within differentiating cells of the Drosophila eye. *Curr Biol.* 1999;9(21):1211–1220.
53. Jaalouk DE, Lammerding J. Mechanotransduction gone awry. *Nat Rev Mol Cell Biol.* 2009;10(1):63–73.
54. Chancellor TJ, Lee J, Thodeti CK, Lele T. Actomyosin tension exerted on the nucleus through nesprin-1 connections influences endothelial cell adhesion, migration, and cyclic strain-induced reorientation. *Biophys J.* 2010;99(1):115–123.
55. Lammerding J, Hsiao J, Schulze PC, Kozlov S, Stewart CL, Lee RT. Abnormal nuclear shape and impaired mechanotransduction in emerin-deficient cells. *J Cell Biol.* 2005;170(5):781–791.
56. Gundersen GG, Worman HJ. Nuclear positioning. *Cell.* 2013;152(6):1376–1389.
57. Luxton GW, Gomes ER, Folker ES, Worman HJ, Gundersen GG. TAN lines: a novel nuclear envelope structure involved in nuclear positioning. *Nucleus.* 2011;2(3):173–181.
58. Chang W, Folker ES, Worman HJ, Gundersen GG. Emerin organizes actin flow for nuclear movement and centrosome orientation in migrating fibroblasts. *Mol Biol Cell.* 2013;24(24):3869–3880.
59. Khatau SB, Hale CM, Stewart-Hutchinson PJ, et al. A perinuclear actin cap regulates nuclear shape. *Proc Natl Acad Sci USA.* 2009;106(45):19017–19022.
60. Kim DH, Khatau SB, Feng Y, et al. Actin cap associated focal adhesions and their distinct role in cellular mechanosensing. *Sci Rep.* 2012;2:555.
61. Chambliss AB, Khatau SB, Erdenberger N, et al. The LINC-anchored actin cap connects the extracellular milieu to the nucleus for ultrafast mechanotransduction. *Sci Rep.* 2013;3:1087.
62. Kim DH, Chambliss AB, Wirtz D. The multi-faceted role of the actin cap in cellular mechanosensation and mechanotransduction. *Soft Matter.* 2013;9(23):5516–5523.
63. Khatau SB, Kusuma S, Hanjaya-Putra D, et al. The differential formation of the LINC-mediated perinuclear actin cap in pluripotent and somatic cells. *PLoS One.* 2012;7(5): e36689.

64. Khatau SB, Bloom RJ, Bajpai S, et al. The distinct roles of the nucleus and nucleus-cytoskeleton connections in three-dimensional cell migration. *Sci Rep*. 2012;2:488.
65. Splinter D, Tanenbaum ME, Lindqvist A, Jaarsma D, Flotho A, Yu KL, Grigoriev I, Engelsma D, Haasdijk ED, Keijzer N, Demmers J, Fornerod M, Melchior F, Hoogenraad CC, Medema RH, Akhmanova A. Bicaudal D2, dynein, and kinesin-1 associate with nuclear pore complexes and regulate centrosome and nuclear positioning during mitotic entry. *PLoS Biol* 2010;8:e1000350.
66. Wilson MH, Holzbaur EL. Opposing microtubule motors drive robust nuclear dynamics in developing muscle cells. *J Cell Sci*. 2012;125(Pt 17):4158–4169.
67. Fridolfsson HN, Ly N, Meyerzon M, Starr DA. UNC-83 coordinates kinesin-1 and dynein activities at the nuclear envelope during nuclear migration. *Dev Biol*. 2009;338(2):237–250.
68. Fridolfsson HN, Starr DA. Kinesin-1 and dynein at the nuclear envelope mediate the bidirectional migrations of nuclei. *J Cell Biol*. 2010;191(1):115–128.
69. Wu J, Lee KC, Dickinson RB, Lele TP. How dynein and microtubules rotate the nucleus. *J Cell Physiol*. 2011;226(10):2666–2674. http://dx.doi.org/10.1002/jcp.22616.
70. Sato A, Isaac B, Phillips CM, et al. Cytoskeletal forces span the nuclear envelope to coordinate meiotic chromosome pairing and synapsis. *Cell*. 2009;139(5):907–919.
71. Miki F, Kurabayashi A, Tange Y, Okazaki K, Shimanuki M, Niwa O. Two-hybrid search for proteins that interact with Sad1 and Kms1, two membrane-bound components of the spindle pole body in fission yeast. *Mol Genet Genomics*. 2004;270(6):449–461.
72. Yoshida M, Katsuyama S, Tateho K, et al. Microtubule-organizing center formation at telomeres induces meiotic telomere clustering. *J Cell Biol*. 2013;200(4):385–395.
73. Chiu JJ, Usami S, Chien S. Vascular endothelial responses to altered shear stress: pathologic implications for atherosclerosis. *Ann Med*. 2009;41(1):19–28.
74. Seliktar D, Nerem RM, Galis ZS. Mechanical strain-stimulated remodeling of tissue-engineered blood vessel constructs. *Tissue Eng*. 2003;9(4):657–666.
75. Discher DE, Janmey P, Wang YL. Tissue cells feel and respond to the stiffness of their substrate. *Science*. 2005;310(5751):1139–1143.
76. Thomas CH, Collier JH, Sfeir CS, Healy KE. Engineering gene expression and protein synthesis by modulation of nuclear shape. *Proc Natl Acad Sci USA*. 2002;99(4): 1972–1977.
77. Philip JT, Dahl KN. Nuclear mechanotransduction: response of the lamina to extracellular stress with implications in aging. *J Biomech*. 2008;41(15):3164–3170.
78. Ho CY, Jaalouk DE, Vartiainen MK, Lammerding J. Lamin A/C and emerin regulate MKL1-SRF activity by modulating actin dynamics. *Nature*. 2013;497(7450):507–511.
79. Lovett DB, Shekhar N, Nickerson JA, Roux KJ, Lele TP. Modulation of nuclear shape by substrate rigidity. *Cell Mol Bioeng*. 2013;6(2):230–238.
80. Swift J, Ivanovska IL, Buxboim A, et al. Nuclear lamin-A scales with tissue stiffness and enhances matrix-directed differentiation. *Science*. 2013;341(6149):1240104.

PART FOUR

Mechano-Sensing in Stem Cells

CHAPTER NINE

From Stem Cells to Cardiomyocytes: The Role of Forces in Cardiac Maturation, Aging, and Disease

Gaurav Kaushik, Adam J. Engler
Department of Bioengineering, University of California, San Diego, La Jolla, California, USA

Contents

Abstract

Stem cell differentiation into a variety of lineages is known to involve signaling from the extracellular niche, including from the physical properties of that environment. What regulates stem cell responses to these cues is there ability to activate different mechanotransductive pathways. Here, we will review the structures and pathways that regulate stem cell commitment to a cardiomyocyte lineage, specifically examining proteins within muscle sarcomeres, costameres, and intercalated discs. Proteins within these structures stretch, inducing a change in their phosphorylated state or in their localization to initiate different signals. We will also put these changes in the context of stem cell differentiation into cardiomyocytes, their subsequent formation of the chambered heart, and explore negative signaling that occurs during disease.

Progress in Molecular Biology and Translational Science, Volume 126
ISSN 1877-1173
http://dx.doi.org/10.1016/B978-0-12-394624-9.00009-9

1. INTRODUCTION

Mechanics play an essential role in developing or maintaining function at every stage of the heart's lifespan, from differentiation and maturation to regulation of cardiac structure with advancing age. Cells not only respond to external stresses but are also capable of affecting them through internal and external restructuring. This remarkable ability is due to the abundance of mechanosensitive molecules and mechanisms populating the cardiac tissue, which form a closed feedback loop in which mechanics regulate mechanics.

Mechanotransduction, the process by which cells sense external forces and translate them into biochemical signals that can change cell function, is regulated in the heart by a diverse array of factors operating at different length scales. Externally, arterial blood pressure, valve compliance, passive stiffness and adhesivity of the cellular niche, and ventricular wall stress have all been shown to impact form and function of the heart. Through intracellular mechanosensitive pathways, cardiac cells can sense these changes and remodel themselves and their surroundings in order to achieve and maintain a level of function that meets physiological demand.[1] Evidence also suggests that "inside-out" mechanical signaling is crucial for tissue morphogenesis, maintenance of homeostasis, and prolonging function over decades of life.[2–4]

In this chapter, we will first describe the establishment of cardiac fate from stem cells and subsequent morphogenesis of the heart. Then, we will highlight several major mechanosensitive subcompartments of the heart, noting the way in which they engage in mechanical and biochemical cross talk. Throughout the chapter, we will also discuss how mechanical signaling helps establish cardiac fate, construct the contractile apparatus, shape cardiac morphogenesis, regulate force transmission between myocytes and their niche, and underline multiscale remodeling during aging and altered mechanical loads. We will also argue that establishment and long-term heart maintenance are highly dependent upon the cardiomyocyte's ability to remodel its intracellular structure in order to adapt to changing mechanical loads and physiological demand. By dissecting the effector and affected pathways of cardiac mechanotransduction, we hope that the reader will appreciate how mechanics regulates cardiac differentiation and how physical parameters help engineer the function of adult cardiac myocytes in addition to developing a better understanding of the pathophysiology of genetic and age-related cardiomyopathies.

2. CARDIAC MORPHOGENESIS DURING THE LIFESPAN OF THE HEART

2.1. Specification, differentiation, and heart morphogenesis

The cells that eventually become the myocardium are derived from the mesoderm within the primitive streak.[5,6] Early cardiogenesis is driven by time-dependent biochemical signaling, such as bone morphogenic protein (BMP) and suppression of wingless-related integration site (WNT) signaling.[7,8] At this stage, cardiac progenitors begin to migrate and form two populations of cells, one of which will eventually become the early, beating heart tube and the other the outflow tract and portions of the right heart.[5,9] It is shortly after formation of the heart tube that contractions begin and underline further growth and remodeling to loop and subdivide into a primitive four-chambered heart. Morphogenesis can continue in embryonic mice hearts *ex vivo*[10] as it is guided by an internal mechanism: the forces created by interactions of myosin and actin.[11]

2.2. Cell maturation and maintenance

For the remainder of embryogenesis, growth is caused by hyperplasia or proliferation of the early cardiomyocyte population.[12,13] Postnatal cardiomyocytes continue remodeling in a process dubbed maturation, which includes both hyperplasia and hypertrophy that are considered *adaptive* as the growth contributes to improved function.[13] Postmaturation myocardial remodeling, either through concentric or through eccentric hypertrophy, is underlined by the addition of sarcomeres, remodeling of cortical ultrastructure, protein expression, and altered cell morphology, and is associated with age-related dysfunction such as impaired fractional shortening.[1,14,15] While primarily composed of terminally differentiated adult cardiomyocytes, cardiac stem cells have also been recently identified[16–18]; despite the presence of these progenitor cells, however, the adult heart is still thought to have limited regenerative potential compared to other tissue systems given that the heart does not repair itself like other muscles. Therefore, adult cardiomyocytes must be extremely responsive to these changing mechanical environments (e.g., elevated arterial pressure, fibrosis) to maintain function over several decades, and mechanosensitive molecules provide a convenient feedback mechanism to maintain cardiac function.

3. MECHANOSENSITIVE COMPARTMENTS IN CARDIOMYOCYTES

Cardiomyocytes are composed of several subcompartments involved in mechanotransduction, including the contractile sarcomeres, the cytoskeletal filament networks, transmembrane cell–cell and cell–matrix junctions, stretch-sensitive membrane structures, and calcium-handling machinery as shown in Fig. 9.1. Each of these is involved in providing structural integrity as well as generation, sensation, transmission, and/or modulation of forces in the heart. In the following sections, we will summarize the basic structure and highlight known and hypothesized functions of each of these subcompartments.

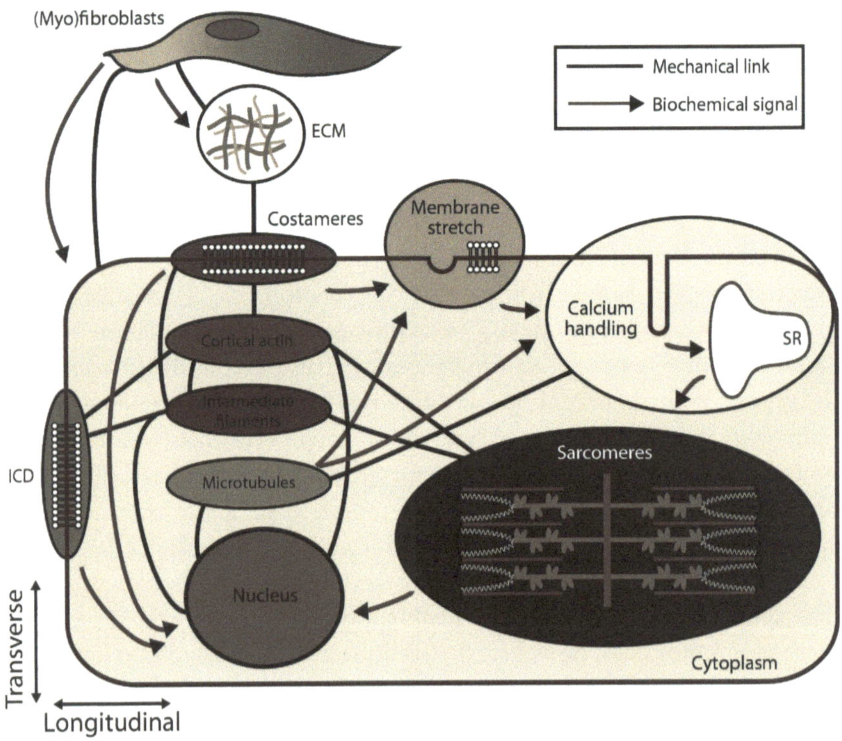

Figure 9.1 *Mechanosignaling between subcellular compartments.* Illustration highlights the mechanical and chemical interplay of indicated intracellular, extracellular, and membrane components. Black lines indicate mechanic and/or structural interaction between elements, whereas arrows indicate the direction of signaling pathways. Note the following abbreviations: intercalated disc, ICD; sarcoplasmic reticulum, SR; extracellular matrix, ECM. (See the color plate.)

4. THE SARCOMERE

The primary function of the heart is to pump blood throughout the circulatory system, dispersing oxygen, nutrients, and biochemical factors to both distal organs as well as itself.[19] This contraction is contingent upon the sarcomere, a dynamic protein complex that serves as the basic mechanical unit of muscle. A sarcomere is composed of actin-based thin filaments and myosin-based thick filaments as illustrated in Fig. 9.1; these filaments slide past one another to create contraction. These structures are cross-linked together by Z-disks. Each of these sarcomeric components is critical in regulating how muscle performs mechanical work, maintains passive tension,[20] undergoes biochemical signaling,[21] and facilitates mechanotransduction.[22] In vertebrates, the expression of these molecules during development is highly coordinated[23] and mutations or alternative splicing of genes encoding these proteins are associated with a variety of congenital defects and cardiomyopathies[24] that manifest in aberrant molecular, cellular, or organ-level structure.

4.1. Cardiac structure and mechanosignaling

Sarcomeric contraction and subsequent force production begins with an action potential. Depolarization of the cardiac membrane results in calcium influx from the extracellular space via L-type calcium channels located in structural membrane invaginations known as the transverse-tubules.[25] Local increases in calcium concentration are detected by proximal ryanodine receptors in the sarcoplasmic reticulum (SR), causing the latter to release additional calcium ions into the cytosolic space. This event is known as a "calcium spark"[26,27] and is part of a calcium-induced calcium-release mechanism central to cardiac excitation–contraction coupling. Though each spark is spatiotemporally constrained, it can result in a significant elevation of global cytoplasmic calcium concentration when summed ($\sim>10^4$ sparks). This elevation is transient, as sarco/endoplasmic reticulum ATP-ase pump and the sodium–calcium exchanger will transport calcium back into the SR and interstitium, respectively. This rise and fall of calcium is referred to as the calcium transient.

Myosin thick filaments bind to actin-based thin filaments transiently, and as they bind and unbind, they undergo a power stroke where the myosin head moves forward relative to the actin filament. Myosin's cyclic binding is known as cross-bridge cycling, and this process creates the net contraction of the sarcomere. Binding can be blocked in the absence of calcium by the troponin–

tropomyosin complex[28] where troponin C can cover up myosin's binding site on the thin filament. The probability of acto-myosin binding is therefore associated with increased intracellular calcium concentration.[29] Changes in the calcium transient directly affect contraction and relaxation dynamics and short- and long-term power output of the heart, making it a potent regulator of transient and long-term cardiac mechanotransduction.[30]

In addition to perturbations in calcium handling, sarcomeric function is also modulated by changes in absolute gene expression,[31] alternative splicing,[24,32] impaired protein homeostasis,[33,34] decreased protein quality,[35] and altered ultrastructure of its subcomponents. Biochemical and mechanical cell stress can affect the quantity, quality, and integrity of the sarcomeres. In kind, molecules in the sarcomere are also capable of acting as stress sensors that are capable of nuclear signaling, allowing the myocyte to respond.[36,37]

The Z-disc was once thought to be a static architectural support for the myofilaments, responsible for anchoring and transversely cross-linking adjacent thin filaments. However in recent years, an intricate complex of force sensing and signaling molecules have been identified within Z-discs, and many of these proteins have critical roles in development and disease.[30] To briefly outline the contents of the Z-disc, it is primarily composed of α-actinin, a spectrin-family protein required for actin-filament anchoring to the Z-disc.[38,39] Nonsarcomeric clustering of α-actinin is a marker of sarcomere degeneration and is associated age-related dilated cardiomyopathy.[40]

In addition to α-actinin, Z-discs also contain a significant amount of titin, which helps maintain resting passive tension and longitudinal stiffness in the cell. Titin also acts as a molecular spring which spans the length of the Z-disk to the M-line.[41–43] Embryonic titin is considerably shorter than adult titin, suggesting that transverse elasticity interplays with the changes in cardiac morphology and mechanics during development.[44] On the other hand, maladaptive cross-linking and differential splicing of titin have been shown to modulate resting sarcomere length and perturb sarcomere force production.[45] In addition to its mechanical properties, titin has several binding sites for ankyrin-repeat proteins (ARPs).[46] ARP signaling proteins are capable of translocating to the nucleus, where they presumably alter the transcriptome of the cell.[46] In addition to binding α-actinin at its N-terminus, titin forms a complex with muscle LIM protein (MLP), which has been hypothesized to act as a stress and/or length tensor.[36,47] MLP-null myocytes experience decreased longitudinal passive stiffness and muted response to stretch in the form of BMP expression[48,49] as well as decreased power output.[50] MLP mutations are also associated with diastolic dysfunction followed by dilated cardiomyopathy, suggesting time-dependent remodeling.[51] Other LIM-domain-containing proteins, including myopodin[52] and zyxin,[53] localize at the Z-disc and further suggest a critical sensory role for these types of proteins. During myoblast differentiation, myopodin is localized to the nucleus but translocates to the nascent Z-disk as sarcomerogenesis progresses.[54] Acute, short-term heat shock of differentiated myotubes results in reverse translocation of myopodin from mature Z-discs to the nucleus, suggesting that myopodin plays a role in stress-response signaling.[52,54] Nebulin, a hypothesized sarcomeric "ruler,"[55–57] is associated with nebulin-related anchoring protein (N-RAP), another well-known LIM protein and member of the Z-disk-associated mechanosensitive network.[58,59] Myomesin is another sarcomeric mechanosensor that is thought to bind titin near its C-terminal and act as a molecular spring.[60,61] In addition to its contributions to mature sarcomere stability, it is associated with myosin during the earliest stages of sarcomerogenesis, suggesting that it is crucial for performing mechanical work given the few proteins present in these early contractile structures.[62]

4.2. Sarcomere mutations, microenvironmental changes, and their impact

Mutations in vital myofilament proteins are associated with congenital defects and adult-onset cardiomyopathies and can arise due to increased or decreased sensitivity to calcium.[36,63–73] These changes reflect a potential role of the myofilaments in cardiogenesis and maintenance of structure. A missense mutation in α-MHC can result in maladaptive hypertrophy, myofibrillar disarray, and fibrosis due to impaired calcium homeostasis in the SR.[74,75] This process can be ameliorated by inhibition of L-type calcium channel activity upstream[76] and potentially other negative-inotropic agents, suggesting that the mutation results in a maladaptive gain of function. Certain mutations in cardiac troponins are associated with restrictive cardiomyopathy,[68,69,77] a condition in which diastolic and systolic dimensions dramatically decline. Recent studies show that such mutations in critical regions can expose the myosin-binding site on thin filaments without the need for calcium influx, resulting in cross-bridge formation independent of calcium release, elevated diastolic tension, cortical stiffening, and impaired fractional shortening.[68,71,78] A mutation in troponin C resulting in calcium-independent sarcomeric perturbation has been shown to be rescued by an engineered molecule.[79]

Sarcomeres are also known to respond to alterations in the mechanical forces being presented to the heart, i.e., "outside-in" mechanotransduction. For example, application of stretch on cardiomyocytes *in vitro* results in increased sarcomerogenesis and hypertrophic signaling.[80–82] The addition of new sarcomeres appears to occur at the intercalated disc, suggesting synergy between the two compartments in this phenomena.[83] Developing embryonic chick cardiomyocytes has increased sarcomerogenesis from when plated on time-dependent, stiffening hydrogels as compared to static elastic substrates.[84,85] The rate of MHC and thin filament actin turnover is stretch sensitive.[86,87] Thus, sarcomerogenesis and maintenance of sarcomeric protein quality may depend upon the ability of the sarcomeres to act as a stress/strain sensor; perturbations of this structure could impair the resultant signaling.

5. OTHER INTRACELLULAR MECHANOSENSITIVE STRUCTURES

In order for sarcomeric shortening to translate into cellular contraction, and for the contractile apparatus to sense changes in external loads, they require mechanical coupling to the membrane. This is achieved via the sarcolemma, which consists of highly ordered junctions and a cortical cytoskeletal network. Cardiomyocytes are coupled longitudinally by an electromechanically active junctional complex known as the intercalated disc (ICD) and transversely at the Z-disc by special focal adhesion plaques known as costameres.[12] Sensation and application of external loads by the sarcomeres are enabled by these transmembrane contacts and their associated cytoskeletal networks, making them the primary responders to forces leading to and from the contractile apparatus. Here, we will discuss these structures, their influence on downstream signaling, and its impact on contractile function.

5.1. Actin-associated intercalated disc and costameric proteins

Recent work has identified the intercalated disc and costameres, shown in Fig. 9.1 at the cell membrane facilitating cell–cell and cell–extracellular matrix (ECM) connections, as being remarkably mechanosensitive, i.e., capable of responding to and producing forces and enabling biochemical signaling through a variety of sensing molecules.[88–93] Most of these studies have focused on *in vitro* systems[89] or noncardiac cells,[88,92] which have the same components but are organized differently. For example, Le Duc and coworkers applied cyclic stress to E-cadherin-coated magnetic beads bonded to endothelial cells and measured changes in bead deflection as the cell cytoskeleton responded.[93] They found that cadherin-mediated binding is mechanosensitive and results in local stiffening, potentially driven by reinforcement actin of the cortical cytoskeleton. The adhesions eventually asymptote to a new stiffness, the rate and magnitude of which depends on the presence of the actin-binding, mechanosensitive molecule vinculin.[90,93] In this way, the actin cytoskeleton downstream of cadherin–cadherin bonds reacts similarly to those proximal to integrin–ECM bonds.[91,94] It is reasonable to assume that cadherin contacts perform similarly in cardiomyocytes as all the critical machinery exists and cell–cell contacts in cultured myocytes are known to remodel during cyclic and static stretch.[95] However, the mechanosensing machinery is not always present or highly organized. In early development, cadherins are expressed ubiquitously throughout the membrane.[12] It is only as sarcomerogenesis advances and cells elongate and hypertrophy that cadherins become polarized and constrained to the longitudinal ends. Once it is fully formed, the intercalated disc does not remain static; advancing age and dilation associates with convolution of the ICD ultrastructure and enrichment with mechanotransductive molecules.[96,97] Remodeling of actin-binding mechanosensors is hypothesized to result in cortical or transverse stiffening during aging,[38] though the ultimate impact of this event on cardiac function remains unclear.

Beyond normal changes with development and age, studies suggest that cortical remodeling from disease or mutation may play a functional role in modulating sarcomeric function. For example, Tangney and coworkers showed that vinculin-null neonatal mouse cardiomyocytes experience decreased cortical stiffness and increased interfilament spacing, the distance between thin and thick filaments in the myofilament lattice.[98] Interfilament spacing impacts the calcium sensitivity and power output of the sarcomeres,[99–101] perhaps due to a lack of cortical transverse compression upon the myofilaments. Cortical compression through osmotic loading *in vitro* has been shown to affect contractile function in rat myocardium.[101,102]

Lastly, it is important to consider mechanotransduction at focal adhesions, which play a leading role in contractility[103,104] and mechanical induction of myogenic differentiation.[88] In cardiomyocytes specifically, integrin expression and signaling, as well as focal adhesion and integrin-linked kinases, have been the focus of understanding age-related hypertrophic signaling.[105–111] Integrin expression is regulated by both passive and active mechanical forces as well as the ligands and ligand density presented to the cell.[112] Integrin clustering and isoform expression depend upon the

ECM proteins and their arrangement.[113–115] Another crucial component in lateral coupling to the membrane is the actin-binding dystrophin molecule, which is thought to stabilize the membrane and provide additional mechanical coupling to the cortical actin cytoskeleton.[116–118] Dystrophin knockout or expression of a mutated isoform results in impaired sarcomeric force transmission to the membrane. Furthermore, without dystrophin as a structural support, conformational changes in stretch-activated channels are thought to occur, resulting in pathological intracellular leakage of calcium and eventual cell death.[119–121]

5.2. Intermediate filament and microtubule networks

While actin plays a dominant role in regulating cardiomyocyte development and function, both the intermediate filament (IF) and microtubule networks play important supporting roles. For example, IFs are extremely deformable, capable of being stretched to several times their slack length.[122–124] One of these IFs, desmin, provides scaffolding around the Z-disk and links them laterally, binds to desmosomes at the intercalated disc, and bridges the nucleus to adjacent sarcomeres.[125] Desmin is also believed to localize to costameres, perhaps conferring additional coupling between sarcomeres and the ECM.[118] It is an early marker of cardiogenesis.[126,127] Desminopathies result from mutations in the desmin gene and subsequent cytoplasmic aggregation of desmin and impaired myofibrillar assembly.[128–130] Patients with arrhythmogenic right ventricular cardiomyopathy have been shown to have desmin mutations, suggesting that it may play a crucial role in EC coupling via longitudinal load bearing along the sarcomeres or by stabilizing gap junctions at the intercalated disc.[131–133]

Another IF protein prevalent in cardiomyocytes is lamin, which, as with desmin, is also extremely compliant. For example, during lineage specification of stem cells, lamin expression in stem cells scales with substrate stiffness[134] and is a determinant of nuclear cytoskeleton deformability.[135,136] Moreover, it may act as a universal transducer of mechanical signals into the nucleus to regulate gene expression.[137] Thus, defects in lamin A/C can adversely affect signal transduction, making them less efficient. Defects in Lamin are associated with advanced aging,[138] presumably due to genetic instability and accumulated DNA damage from external stressors,[139] and dilated cardiomyopathy.[15,140,141] The existence of mechanical couplings between the nucleus and the cytoskeletal network implies that some form of communication is required during development.[142]

Note that there exist structural overlaps in the actin and IF networks; ICD-localizing vinculin is also thought to stabilize gap junctions,[134] and perturbations of actin-binding nesprins, found at the nucleus, are also known to mute biomechanical signaling and induce cardiovascular dysfunction.[143,144] These findings suggest functional overlap in both cytoskeletal networks.

On the other hand, microtubules are crucial in guiding cytokinesis[145] and directing vesicular transport. The microtubule network also serves several load bearing and signaling functions which impact cardiac function.[146–148] Much focus has been placed on how signal transduction is altered following application of colchicine, a pharmacological agent that induces deconstruction of microtubule filaments, as well as how microtubule dysregulation occurs during cardiac hypertrophy.[149–152] While it is unclear whether mechanical perturbation, impaired vesicular transport, or both are responsible for alterations in function, what is known is that microtubules contribute to passive stiffness[20] and are remodeled during age[38]; both of these observations point to microtubules as being important regulators of cardiomyocyte function.

5.3. The cardiomyocyte membrane

In addition to intracellular and transmembrane mechanosensitive compartments, the cardiac myocyte membrane is also enriched with stretch-sensitive structures.[153–157] Stretch-activated channels have been implicated in modulation of calcium handling and rhythmicity, although the strains required to observed stretch activation are often superphysiological.[158] The membrane is partially buffered from stretch via caveolins, cytoskeletally regulated invaginations which can add additional material to the membrane through rapid disassembly under osmotic loads[159] or stretch.[160–163] Caveolin is known to alter its expression with age.[164,165] Deletion of Cav3[166,167] is associated with progressive dilated cardiomyopathy, while Cav1 appears to play a greater role in endothelial function.[168,169]

6. ECM AND MECHANOSENSING

The cardiac interstitium provides cardiomyocytes with avenues of mechanical and biochemical communication with their environment.[22,115,170–176] In particular, the insoluble ECM is secreted by cardiac fibroblasts,[170,171] which are the most abundant cells in the heart by number.[177] This ECM plays a crucial role in providing architectural support for cardiomyocytes and allowing for efficient transmission of forces during contraction. ECM organization is known to direct integrin assembly and signaling and vice versa[178] which can influence cell morphology and sarcomere alignment.[1,179,180] This interstitium, like the cardiomyocytes, remodels during development[12] and disease.[170,171,181] Deposition of fibrillar ECM proteins, such as collagens, laminins, and fibronectin,[182,183] is vital for the initiation of cardiogenesis and wound healing but is also associated with maladaptive hypertrophic growth. In addition to ECM deposition and paracrine signaling, cardiac fibroblasts are known to directly communicate with cardiomyocytes via connexins[184,185] and N-cadherin,[186,187] allowing for electromechanical coupling and modulation of action potential propagation.[187] Cardiac fibroblasts also respond to external forces by altering their internal cytoskeletal expression,[188] extracellular biochemical signaling,[187,189–191] and increasing matrix production, which can lead to the induction of a smooth muscle phenotype.[192] These "myofibroblasts" (Fig. 9.1, top) experience calcium-dependent contractility,[193] elevated ECM deposition,[171,194] and influence cardiac conduction.[186,195] Increased substrate stiffness can also increase differentiation of fibroblasts into myofibroblasts,[196–198] suggesting a positive feedback loop following ECM deposition.

7. THE INFLUENCE OF MECHANOTRANSDUCTION ON APPLICATIONS OF CARDIAC REGENERATION

In this chapter, we have broadly discussed how maintenance of cardiac structure and function is underlined by spatiotemporal patterning of biochemical and mechanical signaling. We have seen cases in which mutation, alternative splicing, maladaptive posttranslational modifications, and altered cytoskeletal and ECM assembly can dramatically alter morphology and function from the cell to the organ. One final consideration to note is how these changes, induced by development or disease, guide or can be used to guide regeneration.

The numerous heart shapes and sizes in the animal kingdom suggest that biology knows of many ways to build a heart. However, the sensitivity of the developing and aging human heart to genetic perturbations and external stresses further suggests that there is a narrow window to form a human tissue that is competently functional. Minor alterations in calcium handling or cytoskeletal ultrastructure and organization appear to have broad impacts on both basal state and response to mechanical and biochemical or pharmacological stress. For example, the use of time-dependent soluble cues appears to be sufficient for initial differentiation of cardiac lineage but insufficient for maturation.[7,199,200] The latter can be assisted by the application of external loads which in turn promote intracellular remodeling.[14,201–203] These mechanical cues may need to be dynamic; the limitations of static mechanical cues have become apparent in recent years. Attempts to revive cardiac function postinfarct via stem cell injection into the stiffened, infarcted niche, for example, result in differentiation into an osteogenic lineage instead of cardiac.[204] Bulk hydrogels do little to provide mechanical support and may introduce arrhythmogenic defects by disrupting electrical coupling.[205] Adult-like phenotype can be induced in developing cells through micropatterning of ECM in 2D.[206,207] In this way, a symmetry-breaking event can be used to guide integrin clustering and sarcomerogenesis downstream. However, this kind of ECM-mediated "boundary condition" guidance in cardiomyocytes is also dynamic *in vivo*, suggesting that the same will be required in 3D cultures of mature cardiac tissue, which cannot be created through current micropatterning technologies. All of these concerns are ameliorated if it is shown that, for a given concern, the response and function of engineered tissues are similar to those observed *in vivo*. Such criteria have held for the use of simpler animal models with otherwise limited homology to human structure and function.[78,208–211]

8. CONCLUSION

What is most evident from a review of current literature is that our understanding of precisely how the cardiomyocyte closes its mechanical feedback loop remains unknown; what are the basic signals that induce hypertrophy and self-assembly into a mature organ and what pathways signal the heart to cease or undergo aberrant growth? In the coming years, additional mechanosensitive molecules in cardiomyocytes will likely highlight cross talk between subcompartments, such as cell–cell and cell–matrix cooperation or myocyte–fibroblast communication. What is likely more important, however, is improving our understanding of the precise timing of mechanotransduction and its downstream pathways during the lifespan of the heart. Improved understanding of cardiac differentiation from stem cells and the mechanotransductive signaling that enables this may reveal indirect therapeutic targets and/or enable better direct engineering of cells and tissue for repair and regeneration.

REFERENCES

1. McCain ML, Parker KK. Mechanotransduction: the role of mechanical stress, myocyte shape, and cytoskeletal architecture on cardiac function. *Pflugers Archiv*. 2011; 462:89–104.
2. Mammoto T, Mammoto A, Ingber DE. Mechanobiology and developmental control. *Annu Rev Cell Dev Biol*. 2013;29:27–61.
3. Heisenberg CP, Bellaiche Y. Forces in tissue morphogenesis and patterning. *Cell*. 2013;153:948–962.
4. Patwari P, Lee RT. Mechanical control of tissue morphogenesis. *Circ Res*. 2008; 103:234–243.
5. Lopez-Sanchez C, Garcia-Martinez V. Molecular determinants of cardiac specification. *Cardiovasc Res*. 2011;91:185–195.
6. Lopez-Sanchez C, Garcia-Masa N, Ganan CM, Garcia-Martinez V. Movement and commitment of primitive streak precardiac cells during cardiogenesis. *Int J Dev Biol*. 2009;53:1445–1455.
7. Yuasa S, Itabashi Y, Koshimizu U, et al. Transient inhibition of BMP signaling by Noggin induces cardiomyocyte differentiation of mouse embryonic stem cells. *Nat Biotechnol*. 2005;23:607–611.
8. Marvin MJ, Di Rocco G, Gardiner A, Bush SM, Lassar AB. Inhibition of Wnt activity induces heart formation from posterior mesoderm. *Genes Dev*. 2001;15:316–327.
9. Tzahor E, Lassar AB. Wnt signals from the neural tube block ectopic cardiogenesis. *Genes Dev*. 2001;15:255–260.
10. Dyer LA, Patterson C. A novel ex vivo culture method for the embryonic mouse heart. *J Vis Exp*. 2013;e50359.
11. Granados-Riveron JT, Brook JD. The impact of mechanical forces in heart morphogenesis. *Circ Cardiovasc Genet*. 2012;5:132–142.
12. Hirschy A, Schatzmann F, Ehler E, Perriard JC. Establishment of cardiac cytoarchitecture in the developing mouse heart. *Dev Biol*. 2006;289:430–441.
13. Mollova M, Bersell K, Walsh S, et al. Cardiomyocyte proliferation contributes to heart growth in young humans. *Proc Natl Acad Sci USA*. 2013;110:1446–1451.
14. McCain ML, Sheehy SP, Grosberg A, Goss JA, Parker KK. Recapitulating maladaptive, multiscale remodeling of failing myocardium on a chip. *Proc Natl Acad Sci USA*. 2013;110:9770–9775.
15. Schaper J, Kostin S, Hein S, Elsasser A, Arnon E, Zimmermann R. Structural remodelling in heart failure. *Exp Clin Cardiol*. 2002;7:64–68.
16. Torella D, Ellison GM, Nadal-Ginard B. Adult c-kit(pos) cardiac stem cells fulfill Koch's postulates as causal agents for cardiac regeneration. *Circ Res*. 2014;114:e24–e26.
17. Ferreira-Martins J, Ogorek B, Cappetta D, et al. Cardiomyogenesis in the developing heart is regulated by c-kit-positive cardiac stem cells. *Circ Res*. 2012;110:701–715.
18. Hosoda T. C-kit-positive cardiac stem cells and myocardial regeneration. *Am J Cardiovasc Dis*. 2012;2:58–67.
19. Sonnenblick EH, Spotnitz HM, Spiro D. Role of the sarcomere in ventricular function and the mechanism of heart failure. *Circ Res*. 1964;15(suppl 2):70–81.
20. Granzier HL, Irving TC. Passive tension in cardiac muscle: contribution of collagen, titin, microtubules, and intermediate filaments. *Biophys J*. 1995;68:1027–1044.
21. Nicol RL, Frey N, Olson EN. From the sarcomere to the nucleus: role of genetics and signaling in structural heart disease. *Annu Rev Genomics Hum Genet*. 2000;1:179–223.
22. Kresh JY, Chopra A. Intercellular and extracellular mechanotransduction in cardiac myocytes. *Pflugers Archiv*. 2011;462:75–87.
23. Ehler E, Gautel M. The sarcomere and sarcomerogenesis. *Adv Exp Med Biol*. 2008; 642:1–14.

24. Swank DM, Wells L, Kronert WA, Morrill GE, Bernstein SI. Determining structure/function relationships for sarcomeric myosin heavy chain by genetic and transgenic manipulation of Drosophila. *Microsc Res Tech.* 2000;50:430–442.
25. Turczynska KM, Hellstrand P, Sward K, Albinsson S. Regulation of vascular smooth muscle mechanotransduction by microRNAs and L-type calcium channels. *Commun Integr Biol.* 2013;6:e22278.
26. Rovetti R, Cui X, Garfinkel A, Weiss JN, Qu Z. Spark-induced sparks as a mechanism of intracellular calcium alternans in cardiac myocytes. *Circ Res.* 2010;106:1582–1591.
27. Stern MD, Rios E, Maltsev VA. Life and death of a cardiac calcium spark. *J Gen Physiol.* 2013;142:257–274.
28. Chalovich JM, Chock PB, Eisenberg E. Mechanism of action of troponin·tropomyosin. Inhibition of actomyosin ATPase activity without inhibition of myosin binding to actin. *J Biol Chem.* 1981;256:575–578.
29. Bogeholz N, Muszynski A, Pott C. The physiology of cardiac calcium handling. *Wien Med Wochenschr.* 2012;162:278–282.
30. Judice CC, Marin TM, Franchini KG. Calcium and the mechanotransduction in cardiac myocytes. *Front Biosci.* 2009;1:189–199.
31. O'Neill L, Holbrook NJ, Fargnoli J, Lakatta EG. Progressive changes from young adult age to senescence in mRNA for rat cardiac myosin heavy chain genes. *Cardioscience.* 1991;2:1–5.
32. Wiesner RJ, Ehmke H, Faulhaber J, Zak R, Ruegg JC. Dissociation of left ventricular hypertrophy, beta-myosin heavy chain gene expression, and myosin isoform switch in rats after ascending aortic stenosis. *Circulation.* 1997;95:1253–1259.
33. Kiriazis H, Kranias EG. Genetically engineered models with alterations in cardiac membrane calcium-handling proteins. *Annu Rev Physiol.* 2000;62:321–351.
34. Douglas PM, Dillin A. Protein homeostasis and aging in neurodegeneration. *J Cell Biol.* 2010;190:719–729.
35. Schips TG, Wietelmann A, Hohn K, et al. FoxO3 induces reversible cardiac atrophy and autophagy in a transgenic mouse model. *Cardiovasc Res.* 2011;91:587–597.
36. Bos JM, Poley RN, Ny M, et al. Genotype-phenotype relationships involving hypertrophic cardiomyopathy-associated mutations in titin, muscle LIM protein, and telethonin. *Mol Genet Metab.* 2006;88:78–85.
37. Le Guennec JY, Cazorla O, Lacampagne A, Vassort G. Is titin the length sensor in cardiac muscle? Physiological and physiopathological perspectives. *Adv Exp Med Biol.* 2000;481:337–348, discussion 348–351.
38. Hein S, Kostin S, Heling A, Maeno Y, Schaper J. The role of the cytoskeleton in heart failure. *Cardiovasc Res.* 2000;45:273–278.
39. Sorimachi H, Freiburg A, Kolmerer B, et al. Tissue-specific expression and alpha-actinin binding properties of the Z-disc titin: implications for the nature of vertebrate Z-discs. *J Mol Biol.* 1997;270:688–695.
40. Mohapatra B, Jimenez S, Lin JH, et al. Mutations in the muscle LIM protein and alpha-actinin-2 genes in dilated cardiomyopathy and endocardial fibroelastosis. *Mol Genet Metab.* 2003;80:207–215.
41. LeWinter MM, Wu Y, Labeit S, Granzier H. Cardiac titin: structure, functions and role in disease. *Clin Chim Acta.* 2007;375:1–9.
42. Granzier H, Labeit S. Cardiac titin: an adjustable multi-functional spring. *J Physiol.* 2002;541:335–342.
43. Granzier H, Labeit D, Wu Y, Labeit S. Titin as a modular spring: emerging mechanisms for elasticity control by titin in cardiac physiology and pathophysiology. *J Muscle Res Cell Motil.* 2002;23:457–471.

44. Kruger M, Babicz K, von Frieling-Salewsky M, Linke WA. Insulin signaling regulates cardiac titin properties in heart development and diabetic cardiomyopathy. *J Mol Cell Cardiol.* 2010;48:910–916.
45. Lewinter MM, Granzier HL. Cardiac titin and heart disease. *J Cardiovasc Pharmacol.* 2014;63:207–212.
46. Miller MK, Bang ML, Witt CC, et al. The muscle ankyrin repeat proteins: CARP, ankrd2/Arpp and DARP as a family of titin filament-based stress response molecules. *J Mol Biol.* 2003;333:951–964.
47. Clark KA, Bland JM, Beckerle MC. The Drosophila muscle LIM protein, Mlp84B, cooperates with D-titin to maintain muscle structural integrity. *J Cell Sci.* 2007;120:2066–2077.
48. Buyandelger B, Ng KE, Miocic S, et al. MLP (muscle LIM protein) as a stress sensor in the heart. *Pflugers Archiv.* 2011;462:135–142.
49. Gunkel S, Heineke J, Hilfiker-Kleiner D, Knoll R. MLP: a stress sensor goes nuclear. *J Mol Cell Cardiol.* 2009;47:423–425.
50. Clark KA, Lesage-Horton H, Zhao C, Beckerle MC, Swank DM. Deletion of Drosophila muscle LIM protein decreases flight muscle stiffness and power generation. *Am J Physiol Cell Physiol.* 2011;301:C373–C382.
51. Lorenzen-Schmidt I, Stuyvers BD, ter Keurs HE, et al. Young MLP deficient mice show diastolic dysfunction before the onset of dilated cardiomyopathy. *J Mol Cell Cardiol.* 2005;39:241–250.
52. Faul C, Dhume A, Schecter AD, Mundel P. Protein kinase A, Ca2+/calmodulin-dependent kinase II, and calcineurin regulate the intracellular trafficking of myopodin between the Z-disc and the nucleus of cardiac myocytes. *Mol Cell Biol.* 2007;27:8215–8227.
53. Linnemann A, van der Ven PF, Vakeel P, et al. The sarcomeric Z-disc component myopodin is a multiadapter protein that interacts with filamin and alpha-actinin. *Eur J Cell Biol.* 2010;89:681–692.
54. Weins A, Schwarz K, Faul C, Barisoni L, Linke WA, Mundel P. Differentiation- and stress-dependent nuclear cytoplasmic redistribution of myopodin, a novel actin-bundling protein. *J Cell Biol.* 2001;155:393–404.
55. Kontrogianni-Konstantopoulos A, Ackermann MA, Bowman AL, Yap SV, Bloch RJ. Muscle giants: molecular scaffolds in sarcomerogenesis. *Physiol Rev.* 2009;89: 1217–1267.
56. Littlefield RS, Fowler VM. Thin filament length regulation in striated muscle sarcomeres: pointed-end dynamics go beyond a nebulin ruler. *Semin Cell Dev Biol.* 2008;19:511–519.
57. Fowler VM, McKeown CR, Fischer RS. Nebulin: does it measure up as a ruler? *Curr Biol.* 2006;16:R18–R20.
58. Ehler E, Horowits R, Zuppinger C, et al. Alterations at the intercalated disk associated with the absence of muscle LIM protein. *J Cell Biol.* 2001;153:763–772.
59. Panaviene Z, Moncman CL. Linker region of nebulin family members plays an important role in targeting these molecules to cellular structures. *Cell Tissue Res.* 2007;327:353–369.
60. Agarkova I, Perriard JC. The M-band: an elastic web that crosslinks thick filaments in the center of the sarcomere. *Trends Cell Biol.* 2005;15:477–485.
61. Schoenauer R, Bertoncini P, Machaidze G, et al. Myomesin is a molecular spring with adaptable elasticity. *J Mol Biol.* 2005;349:367–379.
62. Lange S, Agarkova I, Perriard JC, Ehler E. The sarcomeric M-band during development and in disease. *J Muscle Res Cell Motil.* 2005;26:375–379.
63. Geske JB, Bos JM, Gersh BJ, Ommen SR, Eidem BW, Ackerman MJ. Deformation patterns in genotyped patients with hypertrophic cardiomyopathy. *Eur Heart J Cardiovasc Imaging.* 2014;15:456–465.

64. Witjas-Paalberends ER, Piroddi N, Stam K, et al. Mutations in MYH7 reduce the force generating capacity of sarcomeres in human familial hypertrophic cardiomyopathy. *Cardiovasc Res*. 2013;99:432–441.
65. Cheng Y, Wan X, McElfresh TA, et al. Impaired contractile function due to decreased cardiac myosin binding protein C content in the sarcomere. *Am J Physiol Heart and Circ Physiol*. 2013;305:H52–H65.
66. Tian T, Liu Y, Zhou X, Song L. Progress in the molecular genetics of hypertrophic cardiomyopathy: a mini-review. *Gerontology*. 2013;59:199–205.
67. Landstrom AP, Ackerman MJ. Beyond the cardiac myofilament: hypertrophic cardiomyopathy- associated mutations in genes that encode calcium-handling proteins. *Curr Mol Med*. 2012;12:507–518.
68. Davis J, Yasuda S, Palpant NJ, et al. Diastolic dysfunction and thin filament dysregulation resulting from excitation-contraction uncoupling in a mouse model of restrictive cardiomyopathy. *J Mol Cell Cardiol*. 2012;53:446–457.
69. Parvatiyar MS, Pinto JR, Dweck D, Potter JD. Cardiac troponin mutations and restrictive cardiomyopathy. *J Biomed Biotechnol*. 2010;2010:350706.
70. Song W, Dyer E, Stuckey D, et al. Investigation of a transgenic mouse model of familial dilated cardiomyopathy. *J Mol Cell Cardiol*. 2010;49:380–389.
71. Chang AN, Parvatiyar MS, Potter JD. Troponin and cardiomyopathy. *Biochem Biophys Res Commun*. 2008;369:74–81.
72. Bos JM, Ommen SR, Ackerman MJ. Genetics of hypertrophic cardiomyopathy: one, two, or more diseases? *Curr Opin Cardiol*. 2007;22:193–199.
73. Palmiter KA, Solaro RJ. Molecular mechanisms regulating the myofilament response to Ca2+: implications of mutations causal for familial hypertrophic cardiomyopathy. *Basic Res Cardiol*. 1997;92(Suppl 1):63–74.
74. Kim SJ, Iizuka K, Kelly RA, et al. An alpha-cardiac myosin heavy chain gene mutation impairs contraction and relaxation function of cardiac myocytes. *Am J Physiol*. 1999;276:H1780–H1787.
75. Blanchard E, Seidman C, Seidman JG, LeWinter M, Maughan D. Altered crossbridge kinetics in the alphaMHC403/+ mouse model of familial hypertrophic cardiomyopathy. *Circ Res*. 1999;84:475–483.
76. Semsarian C, Ahmad I, Giewat M, et al. The L-type calcium channel inhibitor diltiazem prevents cardiomyopathy in a mouse model. *J Clin Invest*. 2002;109:1013–1020.
77. Akhter S, Bueltmann Jr K, Huang X, Jin JP. Restrictive cardiomyopathy mutations demonstrate functions of the C-terminal end-segment of troponin I. *Arch Biochem Biophys*. 2013;552–553:3–10.
78. Viswanathan MC, Kaushik G, Engler AJ, Lehman W, Cammarato A. A Drosophila melanogaster model of diastolic dysfunction and cardiomyopathy based on impaired troponin-T function. *Circ Res*. 2014;114:e6–e17.
79. Liu B, Lee RS, Biesiadecki BJ, Tikunova SB, Davis JP. Engineered troponin C constructs correct disease-related cardiac myofilament calcium sensitivity. *J Biol Chem*. 2012;287:20027–20036.
80. Frank D, Kuhn C, Brors B, et al. Gene expression pattern in biomechanically stretched cardiomyocytes: evidence for a stretch-specific gene program. *Hypertension*. 2008;51:309–318.
81. De Deyne PG. Formation of sarcomeres in developing myotubes: role of mechanical stretch and contractile activation. *Am J Physiol Cell Physiol*. 2000;279:C1801–C1811.
82. Dorn 2nd GW, Robbins J, Sugden PH. Phenotyping hypertrophy: eschew obfuscation. *Circ Res*. 2003;92:1171–1175.
83. Wilson AJ, Schoenauer R, Ehler E, Agarkova I, Bennett PM. Cardiomyocyte growth and sarcomerogenesis at the intercalated disc. *Cell Mol Life Sci*. 2014;71:165–181.
84. Young JL, Tuler J, Braden R, et al. In vivo response to dynamic hyaluronic acid hydrogels. *Acta Biomater*. 2013;9:7151–7157.

85. Young JL, Engler AJ. Hydrogels with time-dependent material properties enhance cardiomyocyte differentiation in vitro. *Biomaterials*. 2011;32:1002–1009.
86. Simpson DG, Majeski M, Borg TK, Terracio L. Regulation of cardiac myocyte protein turnover and myofibrillar structure in vitro by specific directions of stretch. *Circ Res*. 1999;85:e59–e69.
87. Simpson DG, Sharp WW, Borg TK, Price RL, Terracio L, Samarel AM. Mechanical regulation of cardiac myocyte protein turnover and myofibrillar structure. *Am J Physiol*. 1996;270:C1075–C1087.
88. Holle AW, Tang X, Vijayraghavan D, et al. In situ mechanotransduction via vinculin regulates stem cell differentiation. *Stem Cells*. 2013;31:2467–2477.
89. Spanjaard E, de Rooij J. Mechanotransduction: vinculin provides stability when tension rises. *Curr Biol*. 2013;23:R159–R161.
90. Leerberg JM, Yap AS. Vinculin, cadherin mechanotransduction and homeostasis of cell-cell junctions. *Protoplasma*. 2013;250:817–829.
91. Margadant F, Chew LL, Hu X, et al. Mechanotransduction in vivo by repeated talin stretch-relaxation events depends upon vinculin. *PLoS Biol*. 2011;9:e1001223.
92. Borghi N, Sorokina M, Shcherbakova OG, et al. E-cadherin is under constitutive actomyosin-generated tension that is increased at cell-cell contacts upon externally applied stretch. *Proc Natl Acad Sci USA*. 2012;109:12568–12573.
93. Le Duc Q, Shi Q, Blonk I, et al. Vinculin potentiates E-cadherin mechanosensing and is recruited to actin-anchored sites within adherens junctions in a myosin II-dependent manner. *J Cell Biol*. 2010;189:1107–1115.
94. Choquet D, Felsenfeld DP, Sheetz MP. Extracellular matrix rigidity causes strengthening of integrin-cytoskeleton linkages. *Cell*. 1997;88:39–48.
95. Zhuang J, Yamada KA, Saffitz JE, Kleber AG. Pulsatile stretch remodels cell-to-cell communication in cultured myocytes. *Circ Res*. 2000;87:316–322.
96. Perriard JC, Hirschy A, Ehler E. Dilated cardiomyopathy: a disease of the intercalated disc? *Trends Cardiovasc Med*. 2003;13:30–38.
97. Schaper J, Froede R, Hein S, et al. Impairment of the myocardial ultrastructure and changes of the cytoskeleton in dilated cardiomyopathy. *Circulation*. 1991;83:504–514.
98. Tangney JR, Chuang JS, Janssen MS, et al. Novel role for vinculin in ventricular myocyte mechanics and dysfunction. *Biophys J*. 2013;104:1623–1633.
99. Konhilas JP, Irving TC, de Tombe PP. Myofilament calcium sensitivity in skinned rat cardiac trabeculae: role of interfilament spacing. *Circ Res*. 2002;90:59–65.
100. Wang Y, Fuchs F. Interfilament spacing, Ca2+ sensitivity, and Ca2+ binding in skinned bovine cardiac muscle. *J Muscle Res Cell Motil*. 2001;22:251–257.
101. McDonald KS, Moss RL. Osmotic compression of single cardiac myocytes eliminates the reduction in Ca2+ sensitivity of tension at short sarcomere length. *Circ Res*. 1995;77:199–205.
102. Farman GP, Walker JS, de Tombe PP, Irving TC. Impact of osmotic compression on sarcomere structure and myofilament calcium sensitivity of isolated rat myocardium. *Am J Physiol Heart Circ Physiol*. 2006;291:H1847–H1855.
103. Hannigan GE, Coles JG, Dedhar S. Integrin-linked kinase at the heart of cardiac contractility, repair, and disease. *Circ Res*. 2007;100:1408–1414.
104. Bendig G, Grimmler M, Huttner IG, et al. Integrin-linked kinase, a novel component of the cardiac mechanical stretch sensor, controls contractility in the zebrafish heart. *Genes Dev*. 2006;20:2361–2372.
105. Franchini KG. Focal adhesion kinase—the basis of local hypertrophic signaling domains. *J Mol Cell Cardiol*. 2012;52:485–492.
106. Bettink SI, Werner C, Chen CH, et al. Integrin-linked kinase is a central mediator in angiotensin II type 1- and chemokine receptor CXCR4 signaling in myocardial hypertrophy. *Biochem Biophys Res Commun*. 2010;397:208–213.

107. DiMichele LA, Hakim ZS, Sayers RL, et al. Transient expression of FRNK reveals stage-specific requirement for focal adhesion kinase activity in cardiac growth. *Circ Res.* 2009;104:1201–1208.
108. Umar S, van der Valk EJ, Schalij MJ, van der Wall EE, Atsma DE, van der Laarse A. Integrin stimulation-induced hypertrophy in neonatal rat cardiomyocytes is NO-dependent. *Mol Cell Biochem.* 2009;320:75–84.
109. de Jonge HW, Dekkers DH, Houtsmuller AB, Sharma HS, Lamers JM. Differential signaling and hypertrophic responses in cyclically stretched vs endothelin-1 stimulated neonatal rat cardiomyocytes. *Cell Biochem Biophys.* 2007;47:21–32.
110. Brancaccio M, Hirsch E, Notte A, Selvetella G, Lembo G, Tarone G. Integrin signalling: the tug-of-war in heart hypertrophy. *Cardiovasc Res.* 2006;70:422–433.
111. Lu H, Fedak PW, Dai X, et al. Integrin-linked kinase expression is elevated in human cardiac hypertrophy and induces hypertrophy in transgenic mice. *Circulation.* 2006;114:2271–2279.
112. McCain ML, Lee H, Aratyn-Schaus Y, Kleber AG, Parker KK. Cooperative coupling of cell-matrix and cell-cell adhesions in cardiac muscle. *Proc Natl Acad Sci USA.* 2012;109:9881–9886.
113. Dabiri BE, Lee H, Parker KK. A potential role for integrin signaling in mechanoelectrical feedback. *Prog Biophys Mol Biol.* 2012;110:196–203.
114. Chopra A, Lin V, McCollough A, et al. Reprogramming cardiomyocyte mechanosensing by crosstalk between integrins and hyaluronic acid receptors. *J Biomech.* 2012;45:824–831.
115. Bendall JK, Heymes C, Ratajczak P, Samuel JL. Extracellular matrix and cardiac remodelling. *Arch Mal Coeur Vaiss.* 2002;95:1226–1229.
116. Kabaeva Z, Meekhof KE, Michele DE. Sarcolemma instability during mechanical activity in Largemyd cardiac myocytes with loss of dystroglycan extracellular matrix receptor function. *Hum Mol Genet.* 2011;20:3346–3355.
117. Ervasti JM, Sonnemann KJ. Biology of the striated muscle dystrophin-glycoprotein complex. *Int Rev Cytol.* 2008;265:191–225.
118. Kaprielian RR, Severs NJ. Dystrophin and the cardiomyocyte membrane cytoskeleton in the healthy and failing heart. *Heart Fail Rev.* 2000;5:221–238.
119. Carlson CG. Spontaneous changes in acetylcholine receptor and calcium leakage activity in cell-attached patches from cultured dystrophic myotubes. *Pflugers Archiv.* 1999;437:371–380.
120. Allen DG, Gervasio OL, Yeung EW, Whitehead NP. Calcium and the damage pathways in muscular dystrophy. *Can J Physiol Pharmacol.* 2010;88:83–91.
121. Williams IA, Allen DG. Intracellular calcium handling in ventricular myocytes from mdx mice. *Am J Physiol Heart Circ Physiol.* 2007;292:H846–H855.
122. Kreplak L, Herrmann H, Aebi U. Tensile properties of single desmin intermediate filaments. *Biophys J.* 2008;94:2790–2799.
123. Wagner OI, Rammensee S, Korde N, Wen Q, Leterrier JF, Janmey PA. Softness, strength and self-repair in intermediate filament networks. *Exp Cell Res.* 2007;313:2228–2235.
124. Steinert PM, North AC, Parry DA. Structural features of keratin intermediate filaments. *J Invest Dermatol.* 1994;103:19S–24S.
125. Pawlak A, Gil RJ, Walczak E, Seweryniak P. Desmin expression in human cardiomyocytes and selected clinical and echocardiographic parameters in patients with chronic heart failure. *Kardiol Pol.* 2009;67:955–961.
126. Auda-Boucher G, Bernard B, Fontaine-Perus J, Rouaud T, Mericksay M, Gardahaut MF. Staging of the commitment of murine cardiac cell progenitors. *Dev Biol.* 2000;225:214–225.
127. van der Loop FT, Schaart G, Langmann H, Ramaekers FC, Viebahn C. Rearrangement of intercellular junctions and cytoskeletal proteins during rabbit myocardium development. *Eur J Cell Biol.* 1995;68:62–69.

128. Brodehl A, Dieding M, Klauke B, et al. The novel desmin mutant p.A120D impairs filament formation, prevents intercalated disk localization, and causes sudden cardiac death. *Circ Cardiovasc Genet*. 2013;6:615–623.
129. Chang AN, Potter JD. Sarcomeric protein mutations in dilated cardiomyopathy. *Heart Fail Rev*. 2005;10:225–235.
130. Paulin D, Huet A, Khanamyrian L, Xue Z. Desminopathies in muscle disease. *J Pathol*. 2004;204:418–427.
131. Lorenzon A, Beffagna G, Bauce B, et al. Desmin mutations and arrhythmogenic right ventricular cardiomyopathy. *Am J Cardiol*. 2013;111:400–405.
132. Otten E, Asimaki A, Maass A, et al. Desmin mutations as a cause of right ventricular heart failure affect the intercalated disks. *Heart Rhythm*. 2010;7:1058–1064.
133. Tsatsopoulou AA, Protonotarios NI, McKenna WJ. Arrhythmogenic right ventricular dysplasia, a cell adhesion cardiomyopathy: insights into disease pathogenesis from preliminary genotype–phenotype assessment. *Heart*. 2006;92:1720–1723.
134. Swift J, Ivanovska IL, Buxboim A, et al. Nuclear lamin-A scales with tissue stiffness and enhances matrix-directed differentiation. *Science*. 2013;341:1240104.
135. Capo-chichi CD, Cai KQ, Smedberg J, Ganjei-Azar P, Godwin AK, Xu XX. Loss of A-type lamin expression compromises nuclear envelope integrity in breast cancer. *Chin J Cancer*. 2011;30:415–425.
136. De Vos WH, Houben F, Hoebe RA, et al. Increased plasticity of the nuclear envelope and hypermobility of telomeres due to the loss of A-type lamins. *Biochim Biophys Acta*. 1800;2010:448–458.
137. Talwar S, Jain N, Shivashankar GV. The regulation of gene expression during onset of differentiation by nuclear mechanical heterogeneity. *Biomaterials*. 2014;35:2411–2419.
138. Smith ED, Kudlow BA, Frock RL, Kennedy BK. A-type nuclear lamins, progerias and other degenerative disorders. *Mech Ageing Dev*. 2005;126:447–460.
139. Gonzalez-Suarez I, Gonzalo S. Nurturing the genome: A-type lamins preserve genomic stability. *Nucleus*. 2010;1:129–135.
140. Broers JL, Peeters EA, Kuijpers HJ, et al. Decreased mechanical stiffness in LMNA −/− cells is caused by defective nucleo-cytoskeletal integrity: implications for the development of laminopathies. *Hum Mol Genet*. 2004;13:2567–2580.
141. Takeda N. Cardiomyopathy: molecular and immunological aspects (review). *Int J Mol Med*. 2003;11:13–16.
142. Li F, Wang X, Gerdes AM. Formation of binucleated cardiac myocytes in rat heart: II. Cytoskeletal organisation. *J Mol Cell Cardiol*. 1997;29:1553–1565.
143. Banerjee I, Zhang J, Moore-Morris T, et al. Targeted ablation of nesprin 1 and nesprin 2 from murine myocardium results in cardiomyopathy, altered nuclear morphology and inhibition of the biomechanical gene response. *PLoS Genet*. 2014;10:e1004114.
144. Stroud MJ, Banerjee I, Veevers J, Chen J. Linker of nucleoskeleton and cytoskeleton complex proteins in cardiac structure, function, and disease. *Circ Res*. 2014;114:538–548.
145. Lee KY, Davies T, Mishima M. Cytokinesis microtubule organisers at a glance. *J Cell Sci*. 2012;125:3495–3500.
146. Garcia L, Verdejo HE, Kuzmicic J, et al. Impaired cardiac autophagy in patients developing postoperative atrial fibrillation. *J Thorac Cardiovasc Surg*. 2012;143:451–459.
147. Zhao Y, Xue T, Yang X, et al. Autophagy plays an important role in sunitinib-mediated cell death in H9c2 cardiac muscle cells. *Toxicol Appl Pharmacol*. 2010;248:20–27.
148. Webster DR. Microtubules in cardiac toxicity and disease. *Cardiovasc Toxicol*. 2002;2:75–89.
149. Gomez AM, Kerfant BG, Vassort G, Pappano AJ. Autonomic regulation of calcium and potassium channels is oppositely modulated by microtubules in cardiac myocytes. *Am J Physiol Heart Circ Physiol*. 2004;286:H2065–H2071.

150. Liu H, Ma A, Wang C, Liu Y, Tian H, Bai L. Variation and significance of microtubules in rat volume overload cardiac hypertrophy. *Chin Med J.* 2003;116:337–340.
151. Malan D, Gallo MP, Bedendi I, Biasin C, Levi RC, Alloatti G. Microtubules mobility affects the modulation of L-type I(Ca) by muscarinic and beta-adrenergic agonists in guinea-pig cardiac myocytes. *J Mol Cell Cardiol.* 2003;35:195–206.
152. Palmer BM, Valent S, Holder EL, Weinberger HD, Bies RD. Microtubules modulate cardiomyocyte beta-adrenergic response in cardiac hypertrophy. *Am J Physiol.* 1998;275:H1707–H1716.
153. Iribe G, Kohl P. Axial stretch enhances sarcoplasmic reticulum Ca2+ leak and cellular Ca2+ reuptake in guinea pig ventricular myocytes: experiments and models. *Prog Biophys Mol Biol.* 2008;97:298–311.
154. Lammerding J, Kamm RD, Lee RT. Mechanotransduction in cardiac myocytes. *Ann N Y Acad Sci.* 2004;1015:53–70.
155. Bett GC, Sachs F. Cardiac mechanosensitivity and stretch-activated ion channels. *Trends Cardiovasc Med.* 1997;7:4–8.
156. Akay M, Craelius W. Mechanoelectrical feedback in cardiac myocytes from stretch-activated ion channels. *IEEE Trans Biomed Eng.* 1993;40:811–816.
157. Bustamante JO, Ruknudin A, Sachs F. Stretch-activated channels in heart cells: relevance to cardiac hypertrophy. *J Cardiovasc Pharmacol.* 1991;17(Suppl 2):S110–S113.
158. Senatore S, Ramieddy V, Semeriva M, Perrin L, Lalevee N. Response to mechanical stress is mediated by the TRPA channel painless in the Drosophila heart. *PLoS Genet.* 2010;6:e1001088.
159. Sinha B, Koster D, Ruez R, et al. Cells respond to mechanical stress by rapid disassembly of caveolae. *Cell.* 2011;144:402–413.
160. Albinsson S, Nordstrom I, Sward K, Hellstrand P. Differential dependence of stretch and shear stress signaling on caveolin-1 in the vascular wall. *Am J Physiol Cell Physiol.* 2008;294:C271–C279.
161. Zhang B, Peng F, Wu D, Ingram AJ, Gao B, Krepinsky JC. Caveolin-1 phosphorylation is required for stretch-induced EGFR and Akt activation in mesangial cells. *Cell Signal.* 2007;19:1690–1700.
162. Bellott AC, Patel KC, Burkholder TJ. Reduction of caveolin-3 expression does not inhibit stretch-induced phosphorylation of ERK2 in skeletal muscle myotubes. *J Appl Physiol.* 2005;98:1554–1561.
163. Kawabe J, Okumura S, Lee MC, Sadoshima J, Ishikawa Y. Translocation of caveolin regulates stretch-induced ERK activity in vascular smooth muscle cells. *Am J Physiol Heart Circ Physiol.* 2004;286:H1845–H1852.
164. Fridolfsson HN, Patel HH. Caveolin and caveolae in age associated cardiovascular disease. *J Geriatr Cardiol.* 2013;10:66–74.
165. Zou H, Stoppani E, Volonte D, Galbiati F. Caveolin-1, cellular senescence and age-related diseases. *Mech Ageing Dev.* 2011;132:533–542.
166. Hnasko R, Lisanti MP. The biology of caveolae: lessons from caveolin knockout mice and implications for human disease. *Mol Interv.* 2003;3:445–464.
167. Woodman SE, Park DS, Cohen AW, et al. Caveolin-3 knock-out mice develop a progressive cardiomyopathy and show hyperactivation of the p42/44 MAPK cascade. *J Biol Chem.* 2002;277:38988–38997.
168. Chow AK, Daniel EE, Schulz R. Cardiac function is not significantly diminished in hearts isolated from young caveolin-1 knockout mice. *Am J Physiol Heart Circ Physiol.* 2010;299:H1183–H1189.
169. Murata T, Lin MI, Huang Y, et al. Reexpression of caveolin-1 in endothelium rescues the vascular, cardiac, and pulmonary defects in global caveolin-1 knockout mice. *J Exp Med.* 2007;204:2373–2382.

170. Ma Y, de Castro Bras LE, Toba H, et al. Myofibroblasts and the extracellular matrix network in post-myocardial infarction cardiac remodeling. *Pflugers Arch.* 2014;466:1113–1127.
171. Fan D, Takawale A, Lee J, Kassiri Z. Cardiac fibroblasts, fibrosis and extracellular matrix remodeling in heart disease. *Fibrogenesis Tissue Repair.* 2012;5:15.
172. Jourdan-Lesaux C, Zhang J, Lindsey ML. Extracellular matrix roles during cardiac repair. *Life Sci.* 2010;87:391–400.
173. McCurdy S, Baicu CF, Heymans S, Bradshaw AD. Cardiac extracellular matrix remodeling: fibrillar collagens and Secreted Protein Acidic and Rich in Cysteine (SPARC). *J Mol Cell Cardiol.* 2010;48:544–549.
174. Zamilpa R, Lindsey ML. Extracellular matrix turnover and signaling during cardiac remodeling following MI: causes and consequences. *J Mol Cell Cardiol.* 2010;48:558–563.
175. Goldsmith EC, Borg TK. The dynamic interaction of the extracellular matrix in cardiac remodeling. *J Card Fail.* 2002;8:S314–S318.
176. Borg TK, Rubin K, Carver W, Samarel A, Terracio L. The cell biology of the cardiac interstitium. *Trends Cardiovasc Med.* 1996;6:65–70.
177. Nag AC. Study of non-muscle cells of the adult mammalian heart: a fine structural analysis and distribution. *Cytobios.* 1980;28:41–61.
178. Coyer SR, Singh A, Dumbauld DW, et al. Nanopatterning reveals an ECM area threshold for focal adhesion assembly and force transmission that is regulated by integrin activation and cytoskeleton tension. *J Cell Sci.* 2012;125:5110–5123.
179. Parker KK, Tan J, Chen CS, Tung L. Myofibrillar architecture in engineered cardiac myocytes. *Circ Res.* 2008;103:340–342.
180. Grosberg A, Alford PW, McCain ML, Parker KK. Ensembles of engineered cardiac tissues for physiological and pharmacological study: heart on a chip. *Lab Chip.* 2011;11:4165–4173.
181. Patterson NL, Iyer RP, de Castro Bras LE, et al. Using proteomics to uncover extracellular matrix interactions during cardiac remodeling. *Proteomics Clin Appl.* 2013;7:516–527.
182. Konstandin MH, Volkers M, Collins B, et al. Fibronectin contributes to pathological cardiac hypertrophy but not physiological growth. *Basic Res Cardiol.* 2013;108:375.
183. Konstandin MH, Toko H, Gastelum GM, et al. Fibronectin is essential for reparative cardiac progenitor cell response after myocardial infarction. *Circ Res.* 2013; 113:115–125.
184. Lu D, Soleymani S, Madakshire R, Insel PA. ATP released from cardiac fibroblasts via connexin hemichannels activates profibrotic P2Y2 receptors. *FASEB J.* 2012;26: 2580–2591.
185. Doble BW, Kardami E. Basic fibroblast growth factor stimulates connexin-43 expression and intercellular communication of cardiac fibroblasts. *Mol Cell Biochem.* 1995;143:81–87.
186. Thompson SA, Blazeski A, Copeland CR, et al. Acute slowing of cardiac conduction in response to myofibroblast coupling to cardiomyocytes through N-cadherin. *J Mol Cell Cardiol.* 2014;68:29–37.
187. Thompson SA, Copeland CR, Reich DH, Tung L. Mechanical coupling between myofibroblasts and cardiomyocytes slows electric conduction in fibrotic cell monolayers. *Circulation.* 2011;123:2083–2093.
188. Leslie KO, Taatjes DJ, Schwarz J, vonTurkovich M, Low RB. Cardiac myofibroblasts express alpha smooth muscle actin during right ventricular pressure overload in the rabbit. *Am J Pathol.* 1991;139:207–216.
189. Dalla Costa AP, Clemente CF, Carvalho HF, Carvalheira JB, Nadruz Jr W, Franchini KG. FAK mediates the activation of cardiac fibroblasts induced by mechanical stress through regulation of the mTOR complex. *Cardiovasc Res.* 2010;86:421–431.

190. Wang J, Chen H, Seth A, McCulloch CA. Mechanical force regulation of myofibroblast differentiation in cardiac fibroblasts. *Am J Physiol Heart Circ Physiol.* 2003;285:H1871–H1881.
191. Herum KM, Lunde IG, Skrbic B, et al. Syndecan-4 signaling via NFAT regulates extracellular matrix production and cardiac myofibroblast differentiation in response to mechanical stress. *J Mol Cell Cardiol.* 2013;54:73–81.
192. Lefebvre P, Nusgens BV, Lapiere CM. Cultured myofibroblasts display a specific phenotype that differentiates them from fibroblasts and smooth muscle cells. *Dermatology.* 1994;189(Suppl. 2):65–67.
193. Pipelzadeh MH, Naylor IL. The in vitro enhancement of rat myofibroblast contractility by alterations to the pH of the physiological solution. *Eur J Pharmacol.* 1998;357:257–259.
194. de Haas HJ, Arbustini E, Fuster V, Kramer CM, Narula J. Molecular imaging of the cardiac extracellular matrix. *Circ Res.* 2014;114:903–915.
195. Kizana E, Chang CY, Cingolani E, et al. Gene transfer of connexin43 mutants attenuates coupling in cardiomyocytes: novel basis for modulation of cardiac conduction by gene therapy. *Circ Res.* 2007;100:1597–1604.
196. Guvendiren M, Perepelyuk M, Wells RG, Burdick JA. Hydrogels with differential and patterned mechanics to study stiffness-mediated myofibroblastic differentiation of hepatic stellate cells. *J Mech Behav Biomed Mater.* 2013. http://dx.doi.org/10.1016/j.jmbbm.2013.11.008.
197. Solon J, Levental I, Sengupta K, Georges PC, Janmey PA. Fibroblast adaptation and stiffness matching to soft elastic substrates. *Biophys J.* 2007;93:4453–4461.
198. Atance J, Yost MJ, Carver W. Influence of the extracellular matrix on the regulation of cardiac fibroblast behavior by mechanical stretch. *J Cell Physiol.* 2004;200:377–386.
199. Ren Y, Lee MY, Schliffke S, et al. Small molecule Wnt inhibitors enhance the efficiency of BMP-4-directed cardiac differentiation of human pluripotent stem cells. *J Mol Cell Cardiol.* 2011;51:280–287.
200. Klaus A, Saga Y, Taketo MM, Tzahor E, Birchmeier W. Distinct roles of Wnt/beta-catenin and Bmp signaling during early cardiogenesis. *Proc Natl Acad Sci USA.* 2007;104:18531–18536.
201. Feinberg AW, Ripplinger CM, van der Meer P, et al. Functional differences in engineered myocardium from embryonic stem cell-derived versus neonatal cardiomyocytes. *Stem Cell Rep.* 2013;1:387–396.
202. Feinberg AW, Alford PW, Jin H, et al. Controlling the contractile strength of engineered cardiac muscle by hierarchal tissue architecture. *Biomaterials.* 2012;33:5732–5741.
203. Sheehy SP, Grosberg A, Parker KK. The contribution of cellular mechanotransduction to cardiomyocyte form and function. *Biomech Model Mechanobiol.* 2012;11:1227–1239.
204. Yoon YS, Park JS, Tkebuchava T, Luedeman C, Losordo DW. Unexpected severe calcification after transplantation of bone marrow cells in acute myocardial infarction. *Circulation.* 2004;109:3154–3157.
205. Rane AA, Chuang JS, Shah A, et al. Increased infarct wall thickness by a bio-inert material is insufficient to prevent negative left ventricular remodeling after myocardial infarction. *PLoS One.* 2011;6:e21571.
206. Bray MA, Sheehy SP, Parker KK. Sarcomere alignment is regulated by myocyte shape. *Cell Motil Cytoskeleton.* 2008;65:641–651.
207. Grosberg A, Kuo PL, Guo CL, et al. Self-organization of muscle cell structure and function. *PLoS Comput Biol.* 2011;7:e1001088.
208. Bier E, Bodmer R. Drosophila, an emerging model for cardiac disease. *Gene.* 2004;342:1–11.

209. Hinits Y, Pan L, Walker C, Dowd J, Moens CB, Hughes SM. Zebrafish Mef2ca and Mef2cb are essential for both first and second heart field cardiomyocyte differentiation. *Dev Biol.* 2012;369:199–210.
210. Fink M, Callol-Massot C, Chu A, et al. A new method for detection and quantification of heartbeat parameters in Drosophila, zebrafish, and embryonic mouse hearts. *Biotechniques.* 2009;46:101–113.
211. Kaushik G, Zambon AC, Fuhrmann A, et al. Measuring passive myocardial stiffness in Drosophila melanogaster to investigate diastolic dysfunction. *J Cell Mol Med.* 2012;16:1656–1662.

CHAPTER TEN

Matrix Regulation of Tumor-Initiating Cells

Sophie Y. Wong, Sanjay Kumar
Department of Bioengineering, University of California, Berkeley, Berkeley, California, USA

Contents

Abstract

The recognition that the progression of many tumors may be driven by specific subpopulations of cells with stem/progenitor-like properties (tumor-initiating cells or TICs, a.k.a. cancer stem cells) represents an important recent paradigm shift in cancer biology and therapeutics. TICs in solid tissues are expected to interface with the extracellular matrix (ECM), which can strongly influence cell behavior through a variety of biochemical and biophysical mechanisms. Understanding ECM regulation of TIC behavior is important for developing strategies to isolate, expand, and characterize TICs in a laboratory setting and for understanding the roles ECM-based inputs may play in disease progression and therapy. In this chapter, we discuss how the ECM regulates TICs, starting with a brief overview of TIC biology, isolation, and characterization, molecular mechanisms through which TICs may be regulated by ECM-based signals, and the potential importance of these signals to TIC-driven tumor progression and metastasis.

Progress in Molecular Biology and Translational Science, Volume 126
ISSN 1877-1173
http://dx.doi.org/10.1016/B978-0-12-394624-9.00010-5

1. INTRODUCTION

1.1. What are tumor-initiating cells?

It is an unfortunate reality in cancer that a single treatment strategy is rarely if ever effective for all patients at all times. For example, a chemotherapeutic regimen that initially produces tumor regression may fail later as a resistant population of tumor cells emerges. Moreover, two patients with clinically and histologically similar tumors may exhibit dramatically different responses to a given chemotherapeutic regimen depending on the genomic and proteomic profile of the patient and tumor. This immense heterogeneity between and within tumors frustrates efforts to identify reliable molecular and cellular targets and thus represents a key therapeutic barrier to treatment.[1–3] While the nature and implications of this heterogeneity remain incompletely understood, one important and recently appreciated manifestation of this heterogeneity is the variable ability of cells within a given tumor to propagate the tumor and seed new tumors. In particular, it is becoming clear that for many tumors, a privileged and comparatively rare subpopulation of cells is uniquely able to seed new tumors, whereas the vast majority of tumor cells, presumably this rare subpopulation's more differentiated progeny, contribute to the tumor "bulk" (Fig. 10.1). In this sense, this tumor-initiating subpopulation shares conceptual similarities with stem or

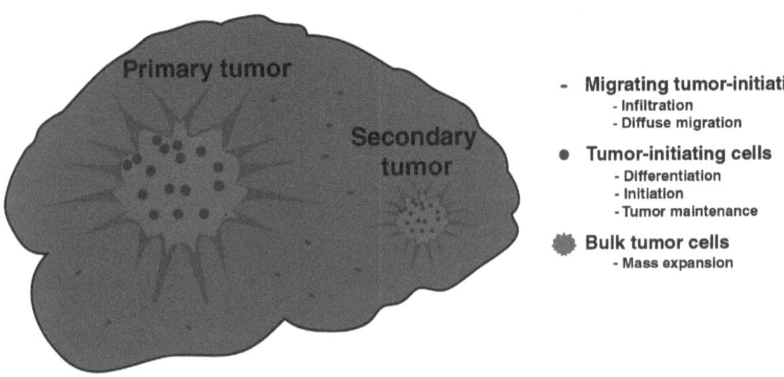

Figure 10.1 Tumor-initiating cells drive secondary tumor formation. Tumors are recognized to consist of a highly heterogeneous population of cells, only some of which can propagate and seed new tumors. This population of tumor-initiating cells (TICs) can diffusely infiltrate tissue (in this example, brain parenchyma), leading to secondary tumor formation. TICs also give rise to more differentiated progeny, which can both add to the tumor "bulk" as well as contribute to more specialized stromal functions such as angiogenesis. (See the color plate.)

progenitor cells, in that these cells can both self-renew and differentiate to yield more specialized progeny.

Some of the first evidence that a population of cancer cells may have stem-like capabilities emerged in 1994 when Lapidot and colleagues found that only the $CD34^+/CD38^-$ subpopulation of leukemia cells generated new tumors in mice.[4] The authors also found that the frequency of immature tumor cells was 1000 times less than mature tumor cells, suggesting that this tumor-initiating population could be rare. These findings led to what the field now recognizes as the cancer stem cell (CSC) hypothesis,[5–7] which states that a subpopulation of tumor-initiating cells (TICs) is more tumorigenic than the rest of the tumor cell population and is capable of recapitulating all components of the tumor (e.g. those involved in growth, invasion, and metastasis) and may be highly resistant to therapeutics. TICs have been identified in several cancers including breast cancer,[8,9] prostate cancer,[10,11] lung cancer,[12] and brain cancer.[13,14] These cells now go by many names, including TICs, CSCs,[6,15,16] and tumor-propagating cells. Whatever the terminology, the common concept is that these cells share key functional properties of stem cells and can initiate new tumors when introduced in small numbers to tumor-free tissue. Most importantly, TICs can initiate and propagate tumors that are histologically equivalent to their tumors of origin when orthotopically implanted into immunocompromised animals.

1.2. Significance of TICs

In addition to lending fundamental new insight into the pathophysiology of tumor progression, a deeper understanding of TIC biology may accelerate the optimization and discovery of treatment regimens. There are at least two ways in which deeper engagement of TICs could aid therapy: first, the development of modalities that directly and specifically target TICs would theoretically represent the most effective way to contain or eradicate a tumor. For example, even after surgical resection of a primary tumor and aggressive follow-up chemo- and/or radiation therapy, the tumor would be expected to recur if a small number of TICs are left behind that could seed new tumors, perpetuate angiogenesis, and invade surrounding tissue. In principle, precise neutralization of TICs would effectively stop tumor initiation[17]. Second, TICs offer a route to personalized medicine, in that they may be specifically isolated from a given patient's tumor and used for patient-specific molecular profiling, drug screening, and disease modeling. Molecular sequencing technologies can be combined with tumor sampling techniques to quantify tumor heterogeneity, trace cell population ancestry, and measure tumor-specific characteristics. For example, Sottoriva and colleagues[18] developed a framework that could generate patient-specific profiles days after tissue collection and did not require xenotransplantation. TIC characteristics considered in this model included a variety of factors such as: fraction of TICs in the tumor, TIC symmetric division rate, methylation/demethylation rate per cell division, relative tumor age from malignant transformation, and rate of apoptosis. The modeling results matched those reported from xenotransplantation assays, supporting the clinical relevance of this model and its potential for designing patient-specific treatments.

The role of TICs in driving glioblastoma (GBM) has been an especially active area of study. GBM is the most aggressive primary brain tumor and has a median survival time of about 15 months, even with surgery and aggressive chemo- and radiotherapy.[19] A variety of laboratories have isolated subpopulations of TICs from GBM tumors that can recapitulate characteristics of the original tumor when transplanted into immunocompromised mice, such as migratory and infiltrative capabilities, nest-like formations, vascular proliferation, nuclear pleomorphism with mitotic figures, and areas of pseudo-palisading necrosis.[5,6,20–24] The continuous cell lines that have been extensively used as culture models of GBM (e.g. U87-MG) typically grow *in vivo* by direct expansion and do not recapitulate the infiltrative character and other key histologic features of the original tumor when transplanted into mice.[22,25] Thus, while these lines may adequately capture more differentiated elements of the tumor that primarily participate in tissue infiltration, TICs may serve as a more clinically relevant model for investigating cellular aspects of the initiation and maintenance of GBM.

A number of studies have supported a role for TICs in tumor propagation[26] and correlated TIC presence with clinical outcome.[27,28] One study used matched TIC and nonstem tumor cells and followed the single cells from injection to tumor growth to show that TICs are more tumorigenic than more differentiated tumor cells.[26] The TICs proliferated faster than the nonstem tumor cells and more fully recapitulated tumor heterogeneity. Furthermore, analysis of secondary tumors contained a high population of TICs and their progeny. A clinical study that compared expression of the TIC marker CD133 (see below) and patient outcome using a panel of 95 gliomas found that high-grade glioma is strongly associated with high CD133 expression. This study also found that high frequency and clusters of CD133 positive cells—independent of tumor grade, extent of resection, and patient age—could be prognostic factors for gliomas.[27] Finally, another study found that low-grade gliomas have low expression of the neural stem cell marker nestin, whereas more aggressive, high-grade gliomas have higher nestin expression and produce shorter survival times. Xenotransplantation of tumor-derived spheroids in mice gave rise to tumors in which nestin-positive cells localized to the invasive front.[28]

2. IDENTIFICATION AND ISOLATION OF TICs

Manipulation of TICs in culture presupposes an ability to reliably identify and isolate these cells. As a result, much effort has been devoted to the search for sensitive and specific TIC markers that may be exploited in flow cytometry, fluorescence-activated cell sorting, immunofluorescence, and other applications. For identification and isolation of GBM TICs, many studies have used neural stem cell surface markers such as CD133,[29] CD15,[30] and A2B5.[31] Similarly, the integrin subunit α6 was shown to be expressed at high levels in GBM TICs and to play a functional role in GBM TIC maintenance and tumor formation capacity.[32] Although several markers have been identified, use of any one marker alone has proven to be somewhat unreliable. For example, both $CD133^+$ and $CD133^-$ glioma cells can display stem-like properties and can generate secondary tumors in orthotopic mouse models.[33,34] As a result, while the field continues to search for sensitive and specific molecular markers, the gold standard for verification of GBM TICs remains a functional one—i.e., GBM TICs are defined by their ability to recapitulate the tumor of origin when orthotopically implanted into immunocompromised mice. A variety of *in vitro* functional screens have been developed to streamline and augment *in vivo* implantation studies. For example, cell survival in neural stem cell medium over several passages has been successfully used to select GBM TICs from bulk tumor tissue.[20] Another study exploited the inverse correlation between proliferation rate (cell cycling speed) and tumorigenicity to select for GBM TICs.[5,35] Improved characterization of TIC properties and the development of new screening/isolation methodologies are critical for further studies that seek to better understand tumor pathogenesis.

3. ROLE OF EXTRACELLULAR MATRIX AND MECHANICAL SIGNALS IN REGULATING TIC FUNCTION

3.1. Extracellular matrix

Having described the identification and isolation of GBM TICs, we now turn to a more detailed discussion of how the extracellular matrix (ECM) may regulate TIC behavior, with a special focus on GBM. In GBM, it is now evident that TICs invade along perivascular spaces, which have a high concentration of ECM proteins such as collagen, fibronectin, and laminin.[36] TICs sense and process these matrix-bound factors through adhesion receptors such as integrins[32,37] and CD44.[38] For example, Lathia and colleagues discovered that integrin α6, a laminin receptor, is necessary for TIC survival and proliferation and directly correlates with TIC stem cell marker expression.[32] Another recent study showed that integrin α3, which adheres to laminin and fibronectin, is overexpressed in CD133-positive TICs. Suppression of α3 slowed random migration and reduced transwell invasion in glioma cell lines, which in turn depended on ERK1/2 phosphorylation.[37] In addition to integrins, the adhesion receptor CD44 has been widely studied and characterized in multiple cancers.[39] High expression of CD44 in GBM TICs correlates with poor clinical prognosis and has been shown to regulate TIC growth through Akt and other signals.[38,40] These adhesion receptor studies support an important role of ECM in TIC function and tumorigenesis.

While these and other studies clearly demonstrate that ECM ligation can trigger signals that modulate TIC behavior, it has also become clear over the past two decades that mechanical cues encoded within the ECM can also direct tumor invasion and growth. Features within the ECM such as matrix geometry, density, and rigidity have been shown to regulate fundamental cellular functions such as motility, proliferation, and gene expression.[41–45] For example, endothelial cells and fibroblasts have higher cell spreading area and motility on stiff matrices when compared to soft matrices.[44,46] It has also been shown that continuous GBM culture models have increased motility, spreading area, and proliferation on stiff matrices.[41] Interestingly, differences in ECM rigidity can also direct the differentiation of adult stem cells, including mesenchymal and neural stem cells.[42,47] In the first and perhaps best-known such study, Engler and colleagues showed that mesenchymal stem cells preferentially undergo neurogenesis on soft ECMs ranging in stiffness from 0.1 to 1 kPa, myogenesis on ECMs ranging from 8 to 17 kPa, and osteogenesis on stiff ECMs ranging from 25 to 40 kPa.[42] They also found that inhibition of nonmuscle myosin II blocked differentiation, thus implicating myosin-based contractile signaling in stiffness-dependent differentiation. Later, Keung and colleagues showed that soft matrices (0.1–0.7 kPa) directed neural differentiation of adult neural stem cells, whereas stiff matrices (1.5–75 kPa) produced relative enrichment of astrocytic differentiation.[48] Mechanistic studies then revealed that the GTPases RhoA and

Cdc42 were key to these effects, with suppression of these proteins rescuing neuronal differentiation on stiff ECMs.

3.2. Propagation of TICs in ECM-adherent cultures

Since the ECM can instruct or select for specific cellular behaviors, it is important to consider the role the ECM may play in culturing TICs in the laboratory setting.[49] For example, GBM TICs can be grown in adherent cultures[21] or as neurospheres in suspension[50,51] (Fig. 10.2). While TICs were long thought to retain their tumor-initiating capacity only when propagated long term as neurospheres, more recent studies reveal that TICs may be propagated as adherent cultures without loss of marker expression or tumor-initiating capacity.[21,22] Specifically, the authors of this study verified the tumorigenicity of each adherent TIC line by injecting 100,000 TICs intracranially into immunocompromised mice. After the mice were sacrificed, the resulting tumors had infiltrated brain tissue and expressed characteristic molecular markers (e.g. nestin) and displayed histopathological hallmarks of GBM. Remarkably, limiting dilution studies revealed that some TIC lines could form aggressive tumors upon transplantation of as few as 100 cells. In addition, adherent cells could be differentiated in culture into marker-positive neuronal, oligodendrocytic, and astrocytic lineages. Some important practical advantages of adherent culture over neurosphere culture include more straightforward quantification of cell proliferation, improved cellular homogeneity, and fewer gradients in oxygen, nutrients, and other

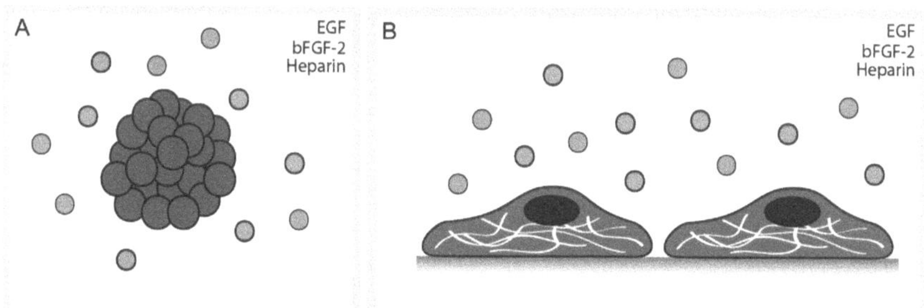

Figure 10.2 Comparison of adherent and neurosphere-based TIC cultures. Both neurosphere and ECM-adherent cultures are widely used to propagate TICs. (A) An acknowledged limitation of neurosphere culture is the possibility of gradients across the sphere in oxygen, nutrients in the culture, and cell-secreted factors, as well as uneven access to extracellular matrix. (B) Adherent culture has recently emerged as a complementary paradigm to neurosphere culture, with the prospect of minimizing these gradients while also offering greater scalability.

soluble factors. Perhaps most importantly, the adherent culture paradigm facilitates high-throughput screening; to illustrate this, the authors screened their TIC lines with 450 drugs from the NIH Clinical Collection and found that 23 of these drugs killed all TIC lines tested, including, unexpectedly, seven agents that target monoamine signaling (e.g. serotonin-specific reuptake inhibitors).

This is not to say that TICs in a small neurosphere (e.g. 150–200 μm) cannot maintain stem-like properties; however as the neurosphere grows larger, the percentage of stem-like cells rapidly decreases,[49] which has been ascribed to the increasingly uneven access to growth factors and oxygen as the neurosphere grows and may be further complicated by increases in juxtacrine and paracrine signaling. In adherent culture, all cells have effectively equal access to soluble factors in the medium, and cells may be plated at sufficiently low density as to minimize cell–cell contacts. In addition, the increased exposure to laminin in the matrix can promote maintenance of stem-like properties for adherent cells, which has been found to be an important factor in identifying TICs.[32] Although debate continues about which culture method is best for a given application, both are used to successfully propagate TICs *in vitro*.

3.3. Mechanisms of mechanotransduction

The finding that TIC behavior is regulated by ECM engagement and biophysical properties raises the question of whether the molecules that mediate these effects may bear value as drug targets. For example, Cilengitide, an αv integrin antagonist, inhibits GBM growth in preclinical models and is currently being evaluated in clinical trials.[52] Recent studies with breast and prostate cancer have used integrins to select for a tumor-initiating subpopulation from the bulk tumor.[8,11] As described earlier, the laminin receptor integrin α6 is highly expressed in GBM TICs and is necessary for BTIC self-renewal, proliferation, and tumor formation capacity.[32] Since laminin is abundant in the BTIC perivascular niche, this result is significant because it suggests a mechanism through which this ECM protein can contribute to maintenance of stemness.

Several actin binding proteins[53,54] and transcription factors relevant to integrin signaling and mechanotransduction[55,56,57,58] have been identified to have the capability of regulating GBM initiation, invasion, and chemosensitivity. For example, the transcription factor ZEB1 is highly expressed in GBM TICs and is known to be correlated with shorter survival and poor response to Temozolomide[58(p1),59]. ZEB1 is regulated by tyrosine receptor type A, and increased expression leads to increased binding to E-box regions of E-cadherin, resulting in highly motile cells and increased tumor invasion. Downstream targets of ZEB1 have subsequently been shown to include ROBO1, OLIG2, CD133, and MGMT. Knockdown of ZEB1 sensitizes cells to temozolomide and decreases expression of stem cell markers SOX2, OLIG2, and CD133.[58(p1)]

Another example is the influence of the actin binding protein, Girdin, which is activated by the PI-3-Kinase/Akt pathway. Activation of the Akt pathway can induce conversion from low-grade to high-grade glioma and regulates angiogenesis, apoptosis, and invasion.[60,61] Girdin is known to regulate cell migration, cell polarity, and epithelial–mesenchymal transition, and in GBM TICs contributes to self-renewal and tumorigenicity.[54] Tumor grade is positively correlated with Girdin expression, and knockdown of Girdin decreases motility and invasion, neurosphere formation, tumorigenicity, and expression of nestin and CD133, and induces differentiation. These studies collectively show that tumor initiation, invasion, and chemoresistance are linked by pathways that are activated by biochemical and potentially mechanical factors in the ECM.

Liu and colleagues recently tied together these concepts by showing that the composition and mechanics of the ECM used to culture TICs can exert powerful instructive and/or selective effects that can profoundly influence subsequent tumorigenicity.[62] The authors examined the formation of melanoma TICs in 3D fibrin gels of varying stiffnesses and found that the softest gel (0.09 kPa) generated the most and largest spheroids over a 5-day period when compared to the stiff gel (1.05 kPa). Subcutaneous transplantation of TICs propagated in the soft gel resulted in greater primary tumor formation and lung metastasis compared to TICs propagated on hard plastic. Furthermore, spheroids grown in the soft gel exhibited increased expression of stem cell markers CD133, nestin, and Bmi-1.

4. CONCLUSION

While it has long been recognized that tumors are highly heterogeneous, only recently has it been appreciated that this heterogeneity may reflect a hierarchy of cellular entities in which a comparatively rare subpopulation of TICs are capable of initiating and propagating the tumor. Over the past decade, significant effort has been devoted to identifying and clarifying the function of these TICs, which has allowed investigators to dissect specific contributions of individual factors to tumor progression. As described in this review, the field is only beginning to understand the importance of the ECM and other solid-state components of the microenvironment in regulating TIC behavior, and there is every reason to expect that mechanical inputs will prove to be an important dimension of this regulation. As our understanding of the role of ECM and mechanical signals to TIC biology advances, there are several open questions to address, each of which presents important opportunities to innovate. First, what are the defining characteristics of the TIC physical microenvironment *in vivo*, how is this different from the normal tissue microenvironment, and which characteristics are most important to tumor initiation and propagation? Second, if mechanical inputs are important to TIC function, how do the signaling systems that process these inputs interface with more canonical oncogenic signaling systems? Specifically, can aberrant mechanotransductive signaling "tip the balance" between TIC quiescence and tumorigenesis, and could this be leveraged in some way to identify new druggable targets? Third, combining these concepts, is it possible to develop advanced *in vitro* culture systems that enable one to investigate ECM and mechanobiological regulation of TICs in a systematic, high-throughput, and physiologically mimetic fashion? One envisions that advances in this last area could dramatically accelerate both fundamental discovery and therapeutic design, with standardized culture platforms serving as key enabling technologies for personalized molecular and chemotherapeutic screening. Realizing this vision will require a highly multidisciplinary effort, including input from biomaterials scientists, micro- and nanotechnologists, cell and ECM biologists, and of course cancer biologists. The coming years and decades are likely to be extremely exciting ones for this field, with advances in basic science directly informing technology and therapeutics and therapeutic advances opening new avenues for scientific inquiry.

REFERENCES

1. Rouzier R, Perou CM, Symmans WF, et al. Breast cancer molecular subtypes respond differently to preoperative chemotherapy. *Clin Cancer Res*. 2005;11(16):5678–5685. http://dx.doi.org/10.1158/1078-0432.CCR-04-2421.
2. Magee JA, Piskounova E, Morrison SJ. Cancer stem cells: impact, heterogeneity, and uncertainty. *Cancer Cell*. 2012;21(3):283–296. http://dx.doi.org/10.1016/j.ccr.2012.03.003.
3. Banerjee S, Reis-Filho JS, Ashley S, et al. Basal-like breast carcinomas: clinical outcome and response to chemotherapy. *J Clin Pathol*. 2006;59(7):729–735. http://dx.doi.org/10.1136/jcp.2005.033043.
4. Lapidot T, Sirard C, Vormoor J, et al. A cell initiating human acute myeloid leukaemia after transplantation into SCID mice. *Nature*. 1994;367(6464):645–648. http://dx.doi.org/10.1038/367645a0.

5. Deleyrolle LP, Harding A, Cato K, et al. Evidence for label-retaining tumour-initiating cells in human glioblastoma. *Brain.* 2011;134(5):1331–1343. http://dx.doi.org/10.1093/brain/awr081.
6. Dirks PB. Brain tumor stem cells: the cancer stem cell hypothesis writ large. *Mol Oncol.* 2010;4(5):420–430. http://dx.doi.org/10.1016/j.molonc.2010.08.001.
7. Vermeulen L, de Sousa e Melo F, Richel DJ, Medema JP. The developing cancer stem-cell model: clinical challenges and opportunities. *Lancet Oncol.* 2012;13(2): e83–e89.
8. Vaillant F, Asselin-Labat M-L, Shackleton M, Forrest NC, Lindeman GJ, Visvader JE. The mammary progenitor marker CD61/β3 integrin identifies cancer stem cells in mouse models of mammary tumorigenesis. *Cancer Res.* 2008;68(19):7711–7717. http://dx.doi.org/10.1158/0008-5472.CAN-08-1949.
9. Ponti D, Costa A, Zaffaroni N, et al. Isolation and in vitro propagation of tumorigenic breast cancer cells with stem/progenitor cell properties. *Cancer Res.* 2005;65(13): 5506–5511. http://dx.doi.org/10.1158/0008-5472.CAN-05-0626.
10. Collins AT, Berry PA, Hyde C, Stower MJ, Maitland NJ. Prospective identification of tumorigenic prostate cancer stem cells. *Cancer Res.* 2005;65(23):10946–10951. http://dx.doi.org/10.1158/0008-5472.CAN-05-2018.
11. Patrawala L, Calhoun-Davis T, Schneider-Broussard R, Tang DG. Hierarchical organization of prostate cancer cells in xenograft tumors: the CD44+α2β1+ cell population is enriched in tumor-initiating cells. *Cancer Res.* 2007;67(14):6796–6805. http://dx.doi.org/10.1158/0008-5472.CAN-07-0490.
12. Eramo A, Lotti F, Sette G, et al. Identification and expansion of the tumorigenic lung cancer stem cell population. *Cell Death Differ.* 2007;15(3):504–514. http://dx.doi.org/10.1038/sj.cdd.4402283.
13. Beier F, Beier CP, Aschenbrenner I, Hildebrandt GC, Brümmendorf TH, Beier D. Identification of CD133−/telomerase low progenitor cells in glioblastoma-derived cancer stem cell lines. *Cell Mol Neurobiol.* 2010;31(3):337–343. http://dx.doi.org/10.1007/s10571-010-9627-4.
14. Huang Q, Zhang Q-B, Dong J, et al. Glioma stem cells are more aggressive in recurrent tumors with malignant progression than in the primary tumor, and both can be maintained long-term in vitro. *BMC Cancer.* 2008;8:304. http://dx.doi.org/10.1186/1471-2407-8-304.
15. Beck B, Blanpain C. Unravelling cancer stem cell potential. *Nat Rev Cancer.* 2013;13(10):727–738. http://dx.doi.org/10.1038/nrc3597.
16. Reya T, Morrison SJ, Clarke MF, Weissman IL. Stem cells, cancer, and cancer stem cells. *Nature.* 2001;414(6859):105–111. http://dx.doi.org/10.1038/35102167.
17. Jordan CT, Guzman ML, Noble M. Cancer stem cells. *N Engl J Med.* 2006;355(12):1253–1261. http://dx.doi.org/10.1056/NEJMra061808.
18. Sottoriva A, Spiteri I, Shibata D, Curtis C, Tavaré S. Single-molecule genomic data delineate patient-specific tumor profiles and cancer stem cell organization. *Cancer Res.* 2013;73(1):41–49. http://dx.doi.org/10.1158/0008-5472.CAN-12-2273.
19. Nakada M, Nakada S, Demuth T, Tran NL, Hoelzinger DB, Berens ME. Molecular targets of glioma invasion. *Cell Mol Life Sci.* 2007;64(4):458–478. http://dx.doi.org/10.1007/s00018-007-6342-5.
20. Galli R, Binda E, Orfanelli U, et al. Isolation and characterization of tumorigenic, stem-like neural precursors from human glioblastoma. *Cancer Res.* 2004;64(19):7011–7021. http://dx.doi.org/10.1158/0008-5472.CAN-04-1364.
21. Pollard SM, Yoshikawa K, Clarke ID, et al. Glioma stem cell lines expanded in adherent culture have tumor-specific phenotypes and are suitable for chemical and genetic screens. *Cell Stem Cell.* 2009;4(6):568–580. http://dx.doi.org/10.1016/j.stem.2009.03.014.
22. Lee J, Kotliarova S, Kotliarov Y, et al. Tumor stem cells derived from glioblastomas cultured in bFGF and EGF more closely mirror the phenotype and genotype of primary

tumors than do serum-cultured cell lines. *Cancer Cell.* 2006;9(5):391–403. http://dx.doi.org/10.1016/j.ccr.2006.03.030.
23. Kleihues P, Sobin LH. World Health Organization classification of tumors. *Cancer.* 2000;88(12):2887. http://dx.doi.org/10.1002/1097-0142(20000615)88:12<2887::AID-CNCR32>3.0.CO;2-F.
24. Lathia JD, Heddleston JM, Venere M, Rich JN. Deadly teamwork: neural cancer stem cells and the tumor microenvironment. *Cell Stem Cell.* 2011;8(5):482–485. http://dx.doi.org/10.1016/j.stem.2011.04.013.
25. Niclou SP, Danzeisen C, Eikesdal HP, et al. A novel eGFP-expressing immunodeficient mouse model to study tumor-host interactions. *FASEB J.* 2008;22(9):3120–3128. http://dx.doi.org/10.1096/fj.08-109611.
26. Lathia JD, Gallagher J, Myers JT, et al. Direct in vivo evidence for tumor propagation by glioblastoma cancer stem cells. *PLoS One.* 2011;6(9):e24807. http://dx.doi.org/10.1371/journal.pone.0024807.
27. Zeppernick F, Ahmadi R, Campos B, et al. Stem cell marker CD133 affects clinical outcome in glioma patients. *Clin Cancer Res.* 2008;14(1):123–129. http://dx.doi.org/10.1158/1078-0432.CCR-07-0932.
28. Strojnik T, Røsland GV, Sakariassen PO, Kavalar R, Lah T. Neural stem cell markers, nestin and musashi proteins, in the progression of human glioma: correlation of nestin with prognosis of patient survival. *Surg Neurol.* 2007;68(2):133–143. http://dx.doi.org/10.1016/j.surneu.2006.10.050.
29. Singh SK, Hawkins C, Clarke ID, et al. Identification of human brain tumour initiating cells. *Nature.* 2004;432(7015):396–401. http://dx.doi.org/10.1038/nature03128.
30. Son MJ, Woolard K, Nam D-H, Lee J, Fine HA. SSEA-1 is an enrichment marker for tumor-initiating cells in human glioblastoma. *Cell Stem Cell.* 2009;4(5):440–452. http://dx.doi.org/10.1016/j.stem.2009.03.003.
31. Ogden ATMD, Waziri AEMD, Lochhead RA, et al. Identification of A2B5+CD133– tumor-initiating cells in adult human gliomas. *Neurosurgery.* 2008;62(2):505–515. http://dx.doi.org/10.1227/01.neu.0000316019.28421.95 [Miscellaneous Article].
32. Lathia JD, Gallagher J, Heddleston JM, et al. Integrin alpha 6 regulates glioblastoma stem cells. *Cell Stem Cell.* 2010;6(5):421–432. http://dx.doi.org/10.1016/j.stem.2010.02.018.
33. Clément V, Dutoit V, Marino D, Dietrich P-Y, Radovanovic I. Limits of CD133 as a marker of glioma self-renewing cells. *Int J Cancer.* 2009;125(1):244–248. http://dx.doi.org/10.1002/ijc.24352.
34. Hermansen SK, Christensen KG, Jensen SS, Kristensen BW. Inconsistent immunohistochemical expression patterns of four different CD133 antibody clones in glioblastoma. *J Histochem Cytochem.* 2011;59(4):391–407. http://dx.doi.org/10.1369/0022155411400867.
35. Lyons AB. Analysing cell division in vivo and in vitro using flow cytometric measurement of CFSE dye dilution. *J Immunol Methods.* 2000;243(1–2):147–154. http://dx.doi.org/10.1016/S0022-1759(00)00231-3.
36. Venere M, Fine HA, Dirks PB, Rich JN. Cancer stem cells in gliomas: identifying and understanding the apex cell in cancer's hierarchy. *Glia.* 2011;59(8):1148–1154. http://dx.doi.org/10.1002/glia.21185.
37. Nakada M, Nambu E, Furuyama N, et al. Integrin α3 is overexpressed in glioma stem-like cells and promotes invasion. *Br J Cancer.* 2013;108(12):2516–2524. http://dx.doi.org/10.1038/bjc.2013.218.
38. Zhu X, Morales FC, Agarwal NK, Dogruluk T, Gagea M, Georgescu M-M. Moesin is a glioma progression marker that induces proliferation and Wnt/β-catenin pathway activation via interaction with CD44. *Cancer Res.* 2013;73(3):1142–1155. http://dx.doi.org/10.1158/0008-5472.CAN-12-1040.
39. Marhaba R, Zöller M. CD44 in cancer progression: adhesion, migration and growth regulation. *J Mol Histol.* 2004;35(3):211–231. http://dx.doi.org/10.1023/B:HIJO.0000032354.94213.69.

40. Jijiwa M, Demir H, Gupta S, et al. CD44v6 regulates growth of brain tumor stem cells partially through the AKT-mediated pathway. *PLoS One*. 2011;6(9):e24217. http://dx.doi.org/10.1371/journal.pone.0024217.
41. Ulrich TA, de Juan Pardo EM, Kumar S. The mechanical rigidity of the extracellular matrix regulates the structure, motility, and proliferation of glioma cells. *Cancer Res*. 2009;69(10):4167–4174. http://dx.doi.org/10.1158/0008-5472.CAN-08-4859.
42. Engler AJ, Sen S, Sweeney HL, Discher DE. Matrix elasticity directs stem cell lineage specification. *Cell*. 2006;126(4):677–689. http://dx.doi.org/10.1016/j.cell.2006.06.044.
43. Tilghman RW, Cowan CR, Mih JD, et al. Matrix rigidity regulates cancer cell growth and cellular phenotype. *PLoS One*. 2010;5(9):e12905. http://dx.doi.org/10.1371/journal.pone.0012905.
44. Georges PC, Janmey PA. Cell type-specific response to growth on soft materials. *J Appl Physiol*. 2005;98(4):1547–1553. http://dx.doi.org/10.1152/japplphysiol.01121.2004.
45. Cheng C-M, Steward Jr RL, LeDuc PR. Probing cell structure by controlling the mechanical environment with cell–substrate interactions. *J Biomech*. 2009;42(2): 187–192. http://dx.doi.org/10.1016/j.jbiomech.2008.10.014.
46. Yeung T, Georges PC, Flanagan LA, et al. Effects of substrate stiffness on cell morphology, cytoskeletal structure, and adhesion. *Cell Motil Cytoskeleton*. 2005;60(1):24–34. http://dx.doi.org/10.1002/cm.20041.
47. Keung AJ, Kumar S, Schaffer DV. Presentation counts: microenvironmental regulation of stem cells by biophysical and material cues. *Annu Rev Cell Dev Biol*. 2010;26(1):533–556. http://dx.doi.org/10.1146/annurev-cellbio-100109-104042.
48. Keung AJ, de Juan-Pardo EM, Schaffer DV, Kumar S. Rho GTPases mediate the mechanosensitive lineage commitment of neural stem cells. *Stem Cells*. 2011;29(11): 1886–1897. http://dx.doi.org/10.1002/stem.746.
49. Woolard K, Fine HA. Glioma stem cells: better flat than round. *Cell Stem Cell*. 2009;4(6):466–467. http://dx.doi.org/10.1016/j.stem.2009.05.013.
50. Deleyrolle LP, Reynolds BA. Identifying and enumerating neural stem cells: application to aging and cancer. In: Verhaagen J, Hol EM, Huitenga I, Wijnholds J, Bergen AB, Boer GJ, Swaab DF, eds. *Neurotherapy: progress in restorative neuroscience and neurology*. Vol. 175. New York, NY: Elsevier; 2009:43–51 Available at: http://www.sciencedirect.com/science/article/B7CV6-4WXGYYC-9/2/e6a7879250e9bf6126f9504ae6c027ce. Accessed April 5, 2011.
51. Azari H, Millette S, Ansari S, Rahman M, Deleyrolle LP, Reynolds BA. Isolation and expansion of human glioblastoma multiforme tumor cells using the neurosphere assay. *J Vis Exp*. 2011;56:e3633. http://dx.doi.org/10.3791/3633.
52. Reardon DA, Nabors LB, Stupp R, Mikkelsen T. Cilengitide: an integrin-targeting arginine–glycine–aspartic acid peptide with promising activity for glioblastoma multiforme. *Expert Opin Investig Drugs*. 2008;17(8):1225–1235. http://dx.doi.org/10.1517/13543784.17.8.1225.
53. Terakawa Y, Agnihotri S, Golbourn B, et al. The role of drebrin in glioma migration and invasion. *Exp Cell Res*. 2013;319(4):517–528. http://dx.doi.org/10.1016/j.yexcr.2012.11.008.
54. Natsume A, Kato T, Kinjo S, et al. Girdin maintains the stemness of glioblastoma stem cells. *Oncogene*. 2012;31(22):2715–2724. http://dx.doi.org/10.1038/onc.2011.466.
55. Kobayashi K, Takahashi H, Inoue A, et al. Oct-3/4 promotes migration and invasion of glioblastoma cells. *J Cell Biochem*. 2012;113(2):508–517. http://dx.doi.org/10.1002/jcb.23374.
56. Alonso MM, Diez-Valle R, Manterola L, et al. Genetic and epigenetic modifications of Sox2 contribute to the invasive phenotype of malignant gliomas. *PLoS One*. 2011;6(11): e26740. http://dx.doi.org/10.1371/journal.pone.0026740.
57. Oppel F, Müller N, Schackert G, et al. SOX2-RNAi attenuates S-phase entry and induces RhoA-dependent switch to protease-independent amoeboid migration in

human glioma cells. *Mol Cancer.* 2011;10:137. http://dx.doi.org/10.1186/1476-4598-10-137.
58. Siebzehnrubl FA, Silver DJ, Tugertimur B, et al. The ZEB1 pathway links glioblastoma initiation, invasion and chemoresistance. *EMBO Mol Med.* 2013;5(8):1196–1212. http://dx.doi.org/10.1002/emmm.201302827.
59. Edwards LA, Woolard K, Son MJ, et al. Effect of brain- and tumor-derived connective tissue growth factor on glioma invasion. *J Natl Cancer Inst.* 2011;103(15):1162–1178. http://dx.doi.org/10.1093/jnci/djr224.
60. Enomoto A, Ping J, Takahashi M. Girdin, a novel actin-binding protein, and its family of proteins possess versatile functions in the Akt and Wnt signaling pathways. *Ann NY Acad Sci.* 2006;1086(1):169–184. http://dx.doi.org/10.1196/annals.1377.016.
61. Jiang P, Enomoto A, Jijiwa M, et al. An actin-binding protein girdin regulates the motility of breast cancer cells. *Cancer Res.* 2008;68(5):1310–1318. http://dx.doi.org/10.1158/0008-5472.CAN-07-5111.
62. Liu J, Tan Y, Zhang H, et al. Soft fibrin gels promote selection and growth of tumorigenic cells. *Nat Mater.* 2012;11(8):734–741. http://dx.doi.org/10.1038/nmat3361.

CHAPTER ELEVEN

Biomaterials Approaches in Stem Cell Mechanobiology

Nikolche Gjorevski, Matthias Lutolf
Institute of Bioengineering, Ecole Polytechnique Fédérale de Lausanne, Lausanne, Switzerland

Contents

Abstract

Multiple molecular and physical cues engage in complex interplay to afford the fine control over stem cell behavior that is required during tissue development, homeostasis, and repair. As a result of a close collaboration between biologists, engineers, and physicists over the past decade, the types of mechanical parameters which influence stem cells, along with the corresponding cellular outcomes, have started to emerge. Moreover, the field has begun to identify the molecular structures and mechanisms whereby stem cells sense physical signals and transduce them into biological response.

In this chapter, we survey the existing knowledge of stem cell mechanobiology and discuss bioengineering approaches which hold the potential to address outstanding questions and concepts. In particular, we cover approaches which may help dissect the context-dependent influence of the physical environment, and resolve its interactions with molecular signals. We also describe platforms which could recreate the spatial and temporal variations in mechanics which are present *in vivo* and may play a role in patterning stem cell behaviors. Comprehensive understanding of the complex mechanical regulation of stem cell fate may bring us closer to taking full advantage of their vast regenerative potential.

Progress in Molecular Biology and Translational Science, Volume 126
ISSN 1877-1173
http://dx.doi.org/10.1016/B978-0-12-394624-9.00011-7

1. INTRODUCTION

Stem cells, characterized by their ability to replicate and give rise to specialized progeny, are indispensable contributors to tissue morphogenesis, repair, and homeostatic cell turnover in the body. Owing to these properties, they have long been recognized as a prime cellular candidate for the regeneration of injured, aged, or diseased tissue. However, numerous obstacles have kept us from harnessing the full potential of stem cells and reaching this venerable goal.[1] Among these hurdles, particularly limiting is our incomplete understanding of the complex array of signals that cooperate to instruct stem cell fate decisions, that is, self-renewal and maintenance of their multilineage potential, or differentiation into the desired type of adult, functional cells. Hence, securing comprehensive knowledge of the mechanisms whereby our bodies exercise tight control over stem cell behavior is critical for unlocking their full clinical potential.

Aside from a range of soluble and tethered molecular cues, cells and tissues are exposed to and interpret various physical signals from the microenvironment. Accumulating evidence implicates the mechanical environment, in particular, as a potent and pervasive regulator of biological processes. Mechanical factors control a number of basic cellular processes, including cell death and viability,[2] proliferation,[3] and epithelial-to-mesenchymal transition.[4,5] Mechanical influences have also emerged as important regulators of multicellular processes, including most stages of embryogenesis.[6] The proper development of the *Xenopus laevis* embryo is critically dependent upon the mechanical properties of its mesoderm and notochord.[7] Cellular contractions are required for dorsal closure of the *Drosophila* embryo,[8] as are tensile force which accumulate within the epithelium upon extrusion of apoptotic cells.[9] As demonstrated recently, externally imposed compressive forces contribute to the establishment of the anterior–posterior axis in the early mouse embryo.[10] Further, mechanical factors have been shown to control the development of several organs, including the lung,[11] the kidney,[12] the salivary gland,[13] and the small intestine.[14] Importantly, deregulation of the normal mechanical environment can disrupt normal tissue architecture and function and result in pathologies. For example, perturbations in the mechanical properties of breast tissue can trigger invasiveness in otherwise noninvasive breast cancer cells, and drive tumor progression.[15,16]

Unsurprisingly, stem cells and stem cell-containing tissues are not immune to the influences of their mechanical environment, as evidenced by an ever-growing number of studies.[17,18] While the link between upstream mechanical triggers and resulting biological outcomes is clear, our understanding of the intracellular mechanisms which mediate these effects is nascent. More specifically, we do not have a full grasp of how cells can sense and interpret physical signals, ultimately converting them into a functional downstream response, processes termed mechanosensing and mechanotransduction. Understanding these processes may expand our arsenal of extrinsic tools for tightly controlling stem cell fate, which is an important step toward the ultimate goal of using them in the clinic safely and reliably. Here, we will begin by describing the factors, both intra- and extracellular, that influence the mechanical state of cells and tissues, and highlight important findings of the stem cell mechanobiology field. We will then cover the known modes of cellular mechanosensing and mechanotransduction and offer our perspective on how biomaterials approaches can be applied to address questions that evade current techniques.

2. MECHANICAL REGULATION OF STEM CELL FATE

2.1. Tensional homeostasis and the origins of cellular force

Cells can actively generate mechanical force, a phenomenon recognized for several decades, and first visualized by Harris and colleagues, who demonstrated that fibroblasts plated on deformable silicon membranes pull on the underlying substratum, generating visible wrinkles.[19] These cell-generated forces have since been extensively measured and characterized,[20–22] using techniques that required merging of several disciplines, including cell biology, biophysics, continuum mechanics, and materials science. In general, endogenous cellular forces are the result of cytoskeletal contractions, which are regulated by complex molecular machinery. The downstream-most events causing cytoskeletal contractions involve activated myosin II motors, which connect multiple actin filaments into contractile actomyosin bundles, termed stress fibers. A number of upstream molecular events can trigger stress fiber formation, but best-described are the Rho/ROCK pathway and myosin light-chain kinase (MLCK).[23,24] MLCK induces stress fiber formation by directly phosphorylating the myosin light chain (MLC), whereas ROCK has a dual positive effect: in addition to phosphorylating MLC directly, it inactivates MLC phosphatase.

Importantly, adherent cells are not physically isolated from their environment: the contractile cytoskeleton is physically linked to the extracellular matrix (ECM), at cell-ECM junctions termed focal adhesions, comprising the transmembrane heterodimer integrin receptors and a number of scaffolding proteins.[25,26] Forces generated at the cellular level are counterbalanced by a resisting ECM. This tendency of the ECM to resist deformation is quantified by its elastic modulus or stiffness. Thus, the ECM, focal adhesions (FAs), and actin cytoskeleton form a mechanical continuum which both establishes the mechanical homeostasis of the cell and enables bidirectional transmission of mechanical signals (Fig. 11.1). In fact, traction force microscopy, a technique routinely used to measure cell-generated forces, directly relies upon this continuum and the inside-out channeling of force. Here, continuum mechanics theory is used to back-calculate cell-generated forces from the extent to which they deform their two-dimensional (2D) substratum[20] or three-dimensional

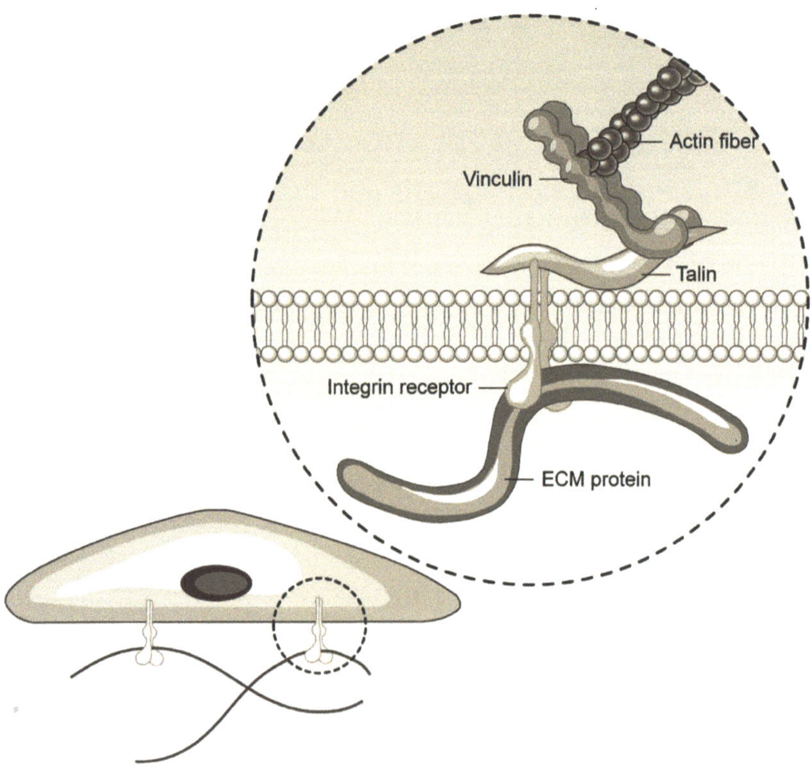

Figure 11.1 *The mechanical machinery of the cell.* The physical continuity that connects the cellular interior and the extracellular matrix via focal adhesion proteins and the actin cytoskeleton allows for bidirectional transmission of force. (See the color plate.)

(3D)[21] matrix. However, the response of the matrix is not merely a passive readout of the mechanical state of the cell; ECM mechanics and alterations therein can feedback to modulate the mechanical activity of the cell. ECM stiffening leads to further activation of the Rho pathway, integrin clustering and consequent strengthening of focal adhesions, and, ultimately, enhanced generation of force.[16,27]

Another key determinant of cellular mechanics and tensional homeostasis, often neglected by studies focusing on noncoherent cells, is interaction with neighboring cells. All cells during early development and epithelial cells in the adult do not functional individually, but are instead connected with their neighbors via cell–cell junctions, including the transmembrane cadherin receptors. The engagement between the extracellular cadherin domains of adjacent cells triggers the recruitment of structural and signaling proteins on the cytoplasmic face, which anchor the junction to actin and the force-generating structures of the cell.[28] Hence, mechanical continuity exists not only between the cells and ECM, but also between cells and their neighbors. Indeed, actin cables which circumscribe wounds in epithelial sheets appear to be continuous from cell to cell, collectively contracting to generate forces which ultimately seal the wound.[29] Keeping in mind the integral part that intercellular junctions play in the mechanical network of cells and tissues, their contribution in mechanobiological phenomena cannot be neglected, and mechanotransduction should be probed not only in single cells but also in multicellular structures.

2.2. Matrix stiffness, cell shape, and tension as regulators of stem cell fate

Most living organisms, including mammals, are mechanically heterogeneous. One does not need sophisticated instruments to assess that bone is stiffer than muscle, which is stiffer than fat. Likewise, various cells of the body assume a range of shapes, starkly different from one another: the elaborately branched neuron differs from both the spindle-like fibroblast and the hexagonal epithelial cell found in embryonic tissues. Determining whether these physical attributes were optimized to help cells and tissues perform their specialized functions, or they have a more profound formative role, required methods for extrinsic manipulation of cell shape and the mechanical properties of their surroundings. Such methods were indeed devised over the past two decades, and the role of bioengineers and biomaterials scientists in the process was indispensable. These platforms were used to demonstrate the far-reaching role that the stiffness–shape–tension axis plays in a myriad of cellular processes, including stem cell fate decisions. Before we discuss the latter in detail, we would like to underscore that matrix stiffness, cell shape, and cellular tension are intimately linked and co-dependent, which results in confounding effects on cellular behavior and difficulty in defining these processes mechanistically. High matrix stiffness increases cell spreading,[5] activates contractility pathways, and leads to enhanced adhesion and cellular tension.[16] Conversely, forces generated by cells residing within mechanically nonlinear matrices, including collagen and fibrin, result in compaction and subsequent stiffening of the matrix. Enhancing the strength of cell–matrix adhesions by overexpressing a constitutively active form of β_1 integrin dramatically increases the spread area of the cell.[16] In light of this interdependence, it is often difficult pinpoint the upstream- or downstream-most parameter responsible for the effect, especially *in vivo*. Nonetheless, the advent of sophisticated and highly controlled *in vitro* biomechanics approaches has greatly contributed toward unveiling and analyzing the mechanical control of stem cell fate.

To control cell spreading and shape, researchers have borrowed microfabrication technologies from the computer industry. In particular, using a combination of photolithography and microcontact printing, ECM islands of defined shape and size are generated. Cells plated on these micropatterned substrata selectively adhere to the ECM-coated regions, adopting their geometry.[30] In a pioneering effort, McBeath *et al.* showed that, when cultured in a medium containing both adipogenic and osteogenic differentiation inducers, single human mesenchymal stem cells (hMSCs) confined to small islands preferentially adopted an adipogenic fate, whereas those allowed to spread on large islands preferentially underwent osteogenesis.[31] At the molecular level, the mechanical control of the adipogenic–osteogenic switch was mediated by the Rho pathway and cellular contractility. Pharmacologic inhibition of contractility suppressed osteogenesis in spread cells, whereas overexpression of a constitutively active form of ROCK induced osteogenesis in poorly spread hMSCs. These results suggest that the cellular contractility pathways and cytoskeletal tension lie downstream in the mechanical regulation of hMSC fate. To further probe this phenomenon, Ruiz *et al.* used the same microfabrication approach to create multicellular structures of defined geometry, wherein cells at distinct geometric features, including corners and regions of high convex curvature, produced and experienced higher levels of mechanical tension. Consistently, cells experiencing higher tension preferentially committed to the osteogenic lineage, whereas those in low-tension regions became adipocytes.[32] Unsurprisingly, treatment with pharmacologic inhibitors of contractility abolished the spatial gradients of adipogenic versus osteogenic differentiation. In addition to demonstrating the role of mechanical tension in hMSC fate decisions, this study revealed a method for using mechanical cues to spatially pattern stem cell differentiation, thus opening exciting possibilities for the generation of complex multicellular tissues *ex vivo*.

In addition to the overall cell area and spreading, cell geometry appears to influence hMSC fate decisions. When cultured in shapes containing different geometric features (rectangles of varying aspect ratio, pentagons with varying subcellular symmetry), but occupying the same area, hMSCs display distinct adipogenic–osteogenic differentiation profiles in response to competing bi-potential differentiation medium.[33] It was determined that osteogenic differentiation occurs preferentially in shapes favoring the assembly of actomyosin stress fibers, confirming once again the overarching influence of cellular tension on hMSC fate decisions.

Matrix stiffness has received much attention as a regulator of stem cell fate. Probing the role of mechanical cues *per se* using natural ECMs, including collagen and fibrin, is complicated by the fact that variations in the stiffness of these materials are accompanied by variations in the concentration of adhesive ligands. To overcome this limitation, synthetic and bioinert polyacrylamide (PAA) gels were introduced.[34] The mechanical properties of PAA gels can be modulated in a wide range by controlling the extent of cross-linking, while maintaining a constant density of an adhesive protein or a peptide covalently bound to their surface. In a landmark study, Engler *et al.* reported that hMSCs undergo neurogenic differentiation when cultured on substrata with brain-like stiffness and undergo myogenic differentiation when cultured on substrata with muscle-like stiffness, whereas relatively stiff, bone-like substrata favor osteogenic differentiation.[35] The implication of this mechanomimicry is profound: stem cells in the body may be instructed to contribute to the local pool of specialized and functional cells based upon the mechanical properties of the environment in which they find themselves. Keeping in mind that most mesenchymal cells *in vivo* reside in a 3D environment, rather than on top of a flat surface, Huebsch *et al.* examined mechanically controlled mMSC fate decisions in a more physiological 3D context. In particular, cells were embedded within synthetic 3D alginate matrices with variable mechanical properties rendered adhesive by including an RGD peptide.[36] Consistently with results from 2D studies, osteogenic differentiation peaked within relatively stiff (11–30 kPa) matrices. Unlike the behavior observed on 2D substrata, and perhaps unsurprisingly, keeping in mind the confining nature of nonfibrillar 3D matrices, variations in stiffness were not accompanied by changes in cellular morphology and spreading. Nonetheless, elevated matrix stiffness lead to more extensive integrin–RGD bond formation and enhanced traction force generation, which was required for osteogenic differentiation.

Aside from MSCs, the influence of mechanical stimuli on the fate choices of several other types of stem cells has been examined. Mouse embryonic stem cells (mESCs) have been shown to preserve their self-renewal potential when cultured on soft PAA gels, by downregulating traction forces.[37] The latter correlated directly with increasing matrix stiffness. Adult neural stem cells (NSCs) are likewise profoundly affected by their mechanical environment. Soft (<1 kPa) matrices favor neuronal differentiation of NSCs in both 2D[38] and 3D.[39] Keung *et al.* reported that mechanical signals originating from the ECM act though the Rho/ROCK pathway to exert their influence on NSC lineage commitment: stiffer matrices result in Rho activation, cellular contraction, and stiffening and ultimately suppress neurogenesis.[40] Keung *et al.* went a step further and dissected the long-term behavior of NSCs plated on substrata of variable stiffness.[41] They found that, under unipotent neuronal differentiation conditions, the percentage of NSCs differentiating to neurons is unaffected by substratum stiffness, whereas neuronal maturation (i.e., neuronal gene expression) of already committed cells is optimal on substrata of intermediate stiffness. These findings illustrate the complexities of mechanically regulated stem cell fate, suggesting that the influence of microenvironmental mechanics is highly dependent upon the biochemical context.

Gilbert *et al.* implicated matrix stiffness as a potent regulator of muscle stem cell (MuSC) fate in culture.[42] MuSCs *in vivo* boast a robust regenerative potential, which is rapidly lost when cultured on plastic. However, culturing MuSCs on compliant hydrogels with stiffness approximating that of native muscle tissue rescues their self-renewal capacity. Gilbert *et al.* showed that MuSCs cultured on soft hydrogels contribute extensively to muscle regeneration when transplanted in mice. This effort demonstrates the feasibility of applying cell-based therapies to treat muscle-wasting diseases and shows how taking advantage of the mechanical regulation of stem cell behavior can help achieve this goal.

Importantly, while the upstream mechanical cues which direct stem cell fate vary in nature (e.g., cell spreading, geometry, and matrix stiffness), these signals converge downstream at cellular tension and intracellular contractility pathways, which appear to be pervasive regulators of the process.

However, the question remains: what is the intermediary between upstream mechanical events and downstream molecular changes, that is, how is mechanical information from the microenvironment sensed and relayed to the intracellular domain, and where it triggers signaling and gene expression responses. Whereas researchers have not yet pieced together a full picture of how mechanosensing and mechanotransduction works, a number of possible mechanisms are starting to emerge.

3. MECHANOSENSING AND MECHANOTRANSDUCTION

3.1. Focal adhesions as mechanical sensors

The multiprotein FAs are situated at the very interface between the cell and the extracellular environment and serve as the primer mechanical sensor and first responders to any changes in the mechanical environment. The overall assembly and maturation of FAs are tension dependent.[43,44] The FA-associated $\alpha_5\beta_1$ switches to its high-affinity state in response to mechanical force. Conversely, abolishing myosin II-mediated tension leads to rapid disassembly of FAs.[45]

Distinct components within the FA complex respond dramatically to mechanical cues. The mechanosensitive properties of focal adhesion kinase (FAK), Src, and p130Cas, in particular, have long been recognized. Application of exogenous mechanical force or enhanced tension in response to ECM stiffening activates FAK through phosphorylation at tyrosine 125[46] and 397.[16,47] Mechanosensing through FAK is required for the durotactic movements of fibroblasts.[48] Importantly, activated FAK phosphorylates Src and subsequently p130Cas, which are themselves mechanoresponsive. Förster resonance energy transfer has revealed that Src is activated at adhesion sites in response to mechanical tension.[49] Likewise, externally applied mechanical force leads to unfolding of the p130Cas proteins, potentially exposing cryptic binding and/or phosphorylation sites.[50]

The interacting talin and vinculin proteins are also mechanoresponsive. Talin exposes up to 11 otherwise cryptic vinculin-binding sites in response to force.[51] In turn, endogenous or external forces trigger the recruitment of vinculin at FAs.[52] In an elegant effort, Schwartz and colleagues took advantage of the mechanoresponsive properties of vinculin to develop a calibrated force biosensor which can be used to measure forces across specific proteins within the cell with piconewton sensitivity.[53] Importantly, vinculin was recently implicated as a key molecular mediator of the mechanical regulation of hMSC fate.[54] Holle *et al.* showed that knocking-down of vinculin significantly attenuates stiffness-mediated myogenesis and embarked on an effort to define the downstream pathways whereby the mechanical signals imposed externally and sensed via vinculin are converted into gene expression changes.

3.2. Integrin-mediated mechanical signaling

FAs and their components are not only structural proteins and mechanosensors but also important signaling molecules, capable of relaying signals to the intracellular space via a range of signaling cascades (Fig. 11.2A). Activation of the FAK–Src–p130Cas complex via phosphorylation (which, as discussed above, can be triggered mechanically) leads to induction of several signaling pathways, including MAPK and PI3K,[18,55] both of which are crucial regulators of stem cell behavior. For example, MAPK-mediated integrin signaling is required for stemness maintenance of both NSCs[56] and epidermal stem cells.[57] Intriguingly, MAPK appears to have a stage-dependent influence of stem cell fate. Adipogenic lineage commitment from mESCs requires an early window of MAPK activation, whereas MAPK has to be inactivated in order for terminal differentiation to occur.[58] A recent study suggests that the mechanical regulation of hMSC myogenesis may proceed through the stiffness-dependent unfolding of vinculin and its subsequent interaction with and phosphorylation of MAPK.[54] A kinase binding accessibility screen predicted a cryptic MAPK1 binding site on vinculin, which was postulated to be exposed in response to force. Indeed, knocking down vinculin both abolished myogenesis and attenuated MAPK1 phosphorylation on substrata of intermediate (11 kPa) stiffness. As expected, pharmacologic inhibition of global MAPK1 activity phenocopied the effect of vinculin knockdown on myogenic differentiation.

The PI3K pathway can be similarly activated through integrin signaling[59] and is required for maintaining the self-renewal capacity of mESCs.[60] Further, PI3K mediates the exit from quiescent state and subsequent activation of adult stem cells, including hematopoietic[61] and intestinal[62] stem cells.

3.3. Mechanically responsive ion channels, chromatin remodeling, and transcription factors

In addition to mechanical signaling through integrins, several other modes of mechanotransduction have been described. Physical signal transduction through mechanosensitive ion channels is a well-documented mechanism.[63] Importantly, a particular class of mechanically sensitive ion channels in the

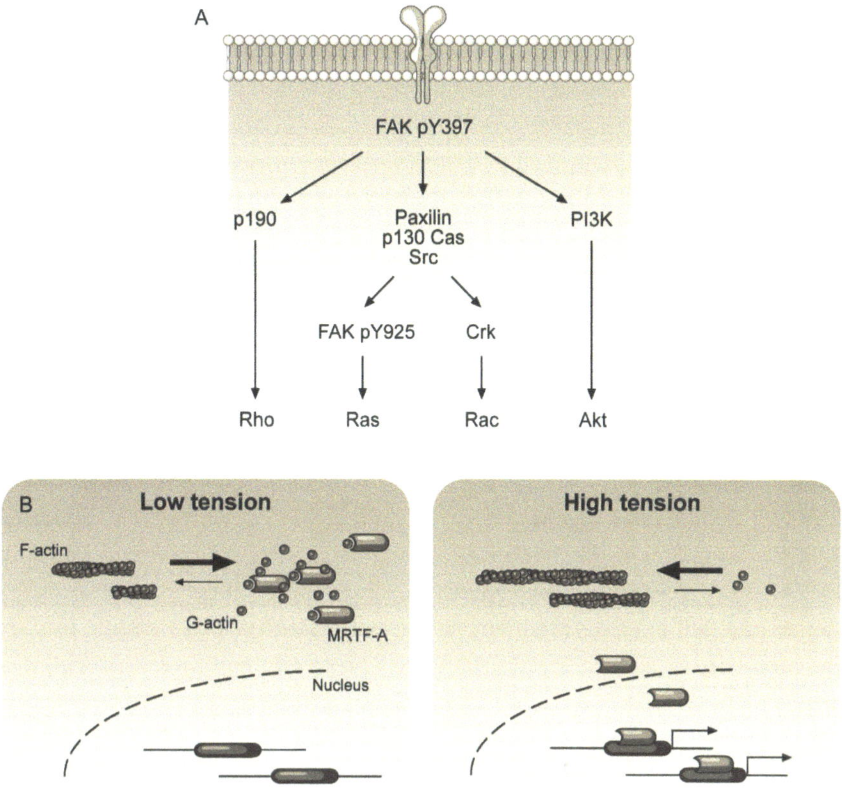

Figure 11.2 *Cellular modes of mechanotransduction.* (A) Integrin stimulation by external cues triggers kinase signaling cascades, activating major pathways which ultimately direct cell survival, proliferation, and stem cell fate. (B) Cytoskeletal regulation of transcription. The mechanical environment can regulate MRTF-A-mediated gene transcription by altering the balance of G- and F-actin. Under conditions of low tension, high levels of monomeric actin bind MRTF-A, preventing it from entering the nucleus. High mechanical tension induces actin polymerization and reduces the pool of monomeric actin in the cytoplasm, thus releasing bound MRTF-A, which is then free to enter the nucleus and transcribe target genes. (See the color plate.)

transient receptor potential family has been implicated in the regulation of stem cell fate.[64,65] This class of channels is activated in response to elevated cytoskeletal tension[66] or matrix stiffness,[67] leading primarily to changes and oscillations in the intracellular levels of Ca^{2+}. Oscillations in Ca^{2+} have been reported in a number of stem cell systems, including mESCs,[68] human preadipocytes,[69] and human cardiac progenitor cells.[70] Notably, physical manipulation of calcium signaling has been used to control MSC differentiation.[71] Intriguingly, growing evidence reveals a direct link between Rho/ROCK and TRP-mediated signaling,[72] suggesting that cross talk may exist between mechanical signaling through integrins and through mechanosensitive ion channels.

The cellular nucleus itself is a site of mechanotransduction, and several mechanisms whereby physical signals directly influence chromatin structure or modulate transcription have been detailed. For example, mechanical force has been shown to regulate the expression and localization of a class of chromatin-remodeling enzymes termed TIPs (tension-induced proteins). In particular, the mechanical regulation of embryonic lung MSC myogenesis is mediated by TIP1, whose expression and nuclear translocation can be triggered by externally imposed mechanical stress.[73]

Cytoskeletal tension also triggers the nuclear translocation and target gene expression by transcription factors MRTF-A[4,74,75] and YAP/TAZ.[76] A detailed picture of the molecular events which underlie MRTF-A-mediated mechanotransduction has emerged. Under conditions of low Rho activity and cytoskeletal tension, the pool of actin monomers within the cell is relatively high. Monomeric actin displays a high affinity for members of the myocardin family of proteins, including MRTF-A.[77] Under conditions of high cytoskeletal tension, activation of the Rho/ROCK pathway prompts actin polymerization, thus sequestering cytoplasmic actin monomers and preventing their association with MRTF-A, which is then free to enter the nucleus and regulate transcription[4,74,78] (Fig. 11.2B). Notably, using the microfabrication approaches described earlier to control cell spreading and tension, Connelly *et al.* showed that MRTF-A transduces mechanical signals to regulate epidermal stem cell fate.[79]

The transcriptional coactivators YAP and TAZ are potent regulators of cell proliferation and survival and exhibit striking responsiveness to mechanical signals from the microenvironment, including mechanical force and substratum stiffness.[76] As shown recently, YAP/TAZ activity is spatially confined to regions within multicellular sheets experiencing high mechanical stress, whether it is caused by substratum stiffness, endogenous tension, or externally applied force.[80] The molecular mechanisms which underlie YAP/TAZ-mediated mechanotransduction have not been fully unraveled. Nevertheless, as in the case of MRTF-A, the Rho pathway, cytoskeletal dynamics, and actin-modifying proteins are thought to play a critical role in the process: myosin-driven contractility and stress fiber formation are required for YAP/TAZ activation. Loss of the actin capping and severing proteins CapZ and Cofilin leads to the reappearance of actin stress fibers and rescues YAP/TAZ nuclear translocation and transcriptional activity within low-stress regions. Aside from controlling basic cellular processes and organ growth during development,[81] YAP and TAZ have recently been implicated as regulators of stem cell fate. YAP is inactivated during ESC differentiation and activated during induced pluripotent stem cell reprogramming.[82] Consistently, ectopic expression of YAP prevents the loss of stemness *in vitro*, even under differentiation conditions. These findings establish an important role for YAP in maintaining stem cell pluripotency. A recent study also demonstrated the involvement of the β_1 integrin–Rho–YAP/TAZ axis in directing skeletal stem cell fate decisions.[83]

4. PUSHING AHEAD: BIOMATERIALS APPROACHES TO PROBE STEM CELL MECHANOBIOLOGY

4.1. Resolving interactions between mechanical and molecular signals

In contrast with the highly simplified microenvironment encountered *in vitro*, stem cells *in vivo* reside in complex, multicomponent niches, wherein, in addition to several types of physical signals, they receive and interpret a wide array of biochemical cues, including soluble factors, cell–cell, and cell–ECM interactions. To complicate matters further, mechanical signals rarely exert their influence independently, but rather interact with molecular factors from the niche in a context-dependent manner. The mechanical environment often modulates the effect of molecular factors. For example, stiff substrata permit the TGFβ-driven transdifferentiation of fibroblasts to myofibroblasts, whereas compliant substrata block it.[84] Mechanical forces remodel the ECM, unraveling hidden protein motifs and releasing bound soluble factors, thus potentiating both cell–ECM and growth factor signaling.[85,86] Remarkably, the mechanotransduction machinery intersects with classic signaling pathways, including TGFβ and Wnt, both critical regulators of stem cell biology. TGFβ signaling, through its well-characterized antimitotic effects, is required for keeping NSCs[87] and HSCs[88] in their quiescent state as well as for maintaining the pluripotency of hESCs.[89] Importantly, TGFβ, secreted into the extracellular space in a latent form, can be activated by cellular forces, in an integrin-dependent manner.[90] Wnt plays a pervasive role in stem cell biology, regulating the fate decisions of ESCs, HSCs, MSCs, NSCs, small intestinal, and skin stem cells.[18,91] Signaling cross talk between Wnt and the mechanosensitive integrin and

YAP/TAZ pathways has been proposed: integrin engagement with the ECM and the subsequent activation of integrin-linked kinase leads to the nuclear accumulation of β-catenin,[92] as does activation of the Grb2–FAK–JNK axis,[93] a separate branch of integrin signaling. Curiously, Wnt has been recently placed upstream of the highly mechanically responsive YAP/TAZ transcriptional activators.[94] Moreover, Wnt signaling in osteoblasts can be directly regulated by physical cues, in a time-dependent manner—externally imposed cyclic stretch initially enhances, but ultimately attenuates Wnt signaling.[95] These complex interactions between physical and molecular signals are responsible for the exquisite control of stem cell behavior *in vivo*, but present a significant hurdle in securing comprehensive understanding of such control, and recapitulating it outside the body for ultimate use in the clinic. To fully untangle the intricacies of cellular mechanotransducive networks, we need methods whereby biophysical and salient molecular parameters can be varied in a wide range and independently of each other, allowing for the systematic interrogation of the individual and potential context-dependent (i.e., interactive) influences of mechanical signals on stem cell fate.

Motivated by an interest in breaking down the complexity of stem cell niches, and inspired by high-throughput approaches from the DNA microarray field, researchers over the past decade have developed "artificial niche microarrays."[96] Here, relevant molecular factors encountered *in vivo*, including ECM proteins, cell–cell interaction proteins, and growth factors, are immobilized at discrete spots, individually or in defined combinations. Sophisticated imaging and multivariate analysis approaches are then used to monitor and dissect the influence of individual cues or groups thereof on stem cell fate. In an early effort, a combinatorial library of acrylate-based polymers was generated and screened for effects on hESC attachment, spreading, proliferation, and differentiation.[97] LaBarge *et al.* used a similar approach to examine how protein constituents of the mammary gland microenvironment influence the proliferation and fate decisions of bipotent mammary progenitors.[98] Whereas these methods showcase the power of high-throughput combinatorial approaches in parsing the effects of multiple molecular inputs on stem cell behavior, they fall short in addressing the contribution of mechanical signals, as they do not afford the possibility to control and perturb the physical microenvironment. Using an indirect protein spotting technique, we generated a protein microwell array on the surface of poly(ethylene glycol) (PEG) hydrogels, thus moving the cells away from stiff glass and plastic to a more physiological substratum with modular mechanical properties.[99] In addition to exploring cell density effects in MSC adipogenesis on soft substrata, we used this platform to examine potential cooperative effects between substratum stiffness and adhesive ligand concentration. Tethering various inhibitors or activators of the signaling pathways discussed above to rigorously define their dialogue with mechanical signaling is readily imaginable. Furthermore, these approaches can be modified to offer control over a wider catalog of mechanical parameters. For example, exogenous mechanical stimulation can be applied either globally via mechanical actuators,[100] or with single-cell (or single-library member) resolution using laser-based ablation of cellular tension.[101] Finally, the microfabrication and automated liquid handling approaches introduced above can be merged with existing 3D synthetic, yet chemically and physically versatile matrices[102,103] to generate arrays of 3D combinatorial microenvironments.

4.2. Controlling mechanics in space and time

Cells and tissues *in vivo* reside in an environment which is mechanically dynamic and heterogeneous. Atomic force microscopy measurements have revealed that the mechanical properties do not only vary from organ to organ, but variations of stiffness are detectable even between the different tissue components of the same organ.[104,105] Tissue microfabrication approaches have shown that cellular contractility and nonspherical multicellular geometry give rise to spatial patterns of mechanical tension, which in turn generate mechanical heterogeneities in the surrounding strain-stiffening matrix.[106] Furthermore, the mechanical environment of cells and tissues is not static, but evolves in time according to the developmental or physiological state of the organism.[29,104] Keeping in mind the profound influence of mechanical signals on basic cellular processes, it has been postulated that spatiotemporal variations in mechanics can serve to pattern cell behaviors within multicellular tissues.[107] Spatial and temporal control of stem cell fate decisions is of critical importance during tissue development, regeneration, and homeostasis. The intestinal crypt is a particularly striking example of spatially patterned stem cell behavior. Here, the intestinal stem cell resides at the crypt bottom, where it undergoes continuous self-renewal. Exiting the crypt bottom location also means exiting the multipotent state: differentiated and functional intestinal cells are distributed in defined patterns along the crypt–villus axis. A potential causative link between spatial and temporal variations in mechanics and stem cell fate decisions is an intriguing notion. To rigorously explore the existence of such a link, platforms that offer tight spatiotemporal control of mechanical signals are required.

Recent advances in light-mediated patterning of hydrogels have furnished micrometer-scale control over the distribution of both biochemical[108,109] and biophysical signals in 3D, thus recapitulating the molecular and mechanical gradients seen in native tissues. In a strategy described by Anseth and colleagues, PEG hydrogels are rendered sensitive to light by incorporating photolabile chemical groups in the hydrogel backbone.[110] Regions exposed to light undergo degradation and subsequent softening, the extent of which can be controlled by the modulating the laser intensity. Conversely, defined regions of high stiffness can be introduced by light exposure in the presence of a photoinitiator, which locally triggers additional cross-linking of hydrogel.[111] Importantly, aside from furnishing spatial control over the mechanical properties of the ECM these approaches provide the possibility to tune matrix mechanics dynamically. As proof of concept, Kloxin *et al.* controlled the 3D spreading of hMSCs in real time by manipulating the local stiffness of the matrix.[110] These sophisticated biomaterial approaches will surely expand our toolkit for studying stem cell mechanobiology in a physiologically relevant 3D and multicellular context. Moreover, recapitulating the mechanical complexity of the native environment may bring us closer to recreating its cellular complexity and produce a new generation of transplantable tissues with histological fidelity superior to the current frontier.

5. SUMMARY AND OUTLOOK

Stem cells receive, interpret, and integrate dynamic information from their environment to ultimately make fate decisions and play out their key roles in tissue development, homeostasis, and repair. In addition to the multitude of soluble or tethered molecular cues, the microenvironment directs stem cell fate by communicating mechanical signals. Intense investigation of stem cell mechanobiology over the past decade began elucidating the physical rules which guide stem cell behavior, but also the cellular machinery responsible for deciphering extrinsic mechanical signals and relaying them to the nucleus. The remarkable progress in the field was critically facilitated by close collaboration between biologists, physicists, and engineers, and the development of sophisticated methods, including force microscopy, synthetic hydrogels, and microfabrication. These tools have been applied to study stem cell mechanobiology at the subcellular and single-cell scales, whereas mechanobiological phenomena in the context of tissues and organs remain underinvestigated. Indeed, the majority of the paradigms discussed here remain yet to be confirmed *in vivo*, a task which is hindered by the lack of tools which allow the interrogation and manipulation of mechanical parameters in the body. The recently introduced stem cell-derived organoids,[112] which mimic key developmental and histological features of real tissues while affording optical and physical accessibility, could serve as a bridge between the reductionist single-cell studies and the intractable *in vivo* environment. These models can be used as proxies for native tissues and organs to both validate existing knowledge and further explore stem cell mechanotransduction, in a more physiologically relevant 3D and multicellular context. Incidentally, self-organizing stem cells are considered prime candidates for cell-based therapies and disease modeling.[112,113] Understanding how these stem cell-containing tissues are influenced by their physical environment, and using this knowledge to modulate and direct their self-organization at will, may expedite their translation from the laboratory to the clinic.

REFERENCES

1. Daley GQ, Scadden DT. Prospects for stem cell-based therapy. *Cell*. 2008;132: 544–548.
2. Chen CS, Mrksich M, Huang S, Whitesides GM, Ingber DE. Geometric control of cell life and death. *Science*. 1997;276:1425–1428.
3. Nelson CM, Jean RP, Tan JL, et al. Emergent patterns of growth controlled by multicellular form and mechanics. *Proc Natl Acad Sci USA*. 2005;102:11594–11599.
4. Gomez EW, Chen QK, Gjorevski N, Nelson CM. Tissue geometry patterns epithelial-mesenchymal transition via intercellular mechanotransduction. *J Cell Biochem*. 2010;110:44–51.
5. Lee K, Chen QK, Lui C, Cichon MA, Radisky DC, Nelson CM. Matrix compliance regulates Rac1b localization, NADPH oxidase assembly, and epithelial-mesenchymal transition. *Mol Biol Cell*. 2012;23:4097–4108.
6. Gleghorn JP, Manivannan S, Nelson CM. Quantitative approaches to uncover physical mechanisms of tissue morphogenesis. *Curr Opin Biotechnol*. 2013;24:954–961.
7. Adams DS, Keller R, Koehl MA. The mechanics of notochord elongation, straightening and stiffening in the embryo of Xenopus laevis. *Development*. 1990;110:115–130.
8. Hutson MS, Tokutake Y, Chang MS, et al. Forces for morphogenesis investigated with laser microsurgery and quantitative modeling. *Science*. 2003;300:145–149.
9. Toyama Y, Peralta XG, Wells AR, Kiehart DP, Edwards GS. Apoptotic force and tissue dynamics during Drosophila embryogenesis. *Science*. 2008;321:1683–1686.
10. Hiramatsu R, Matsuoka T, Kimura-Yoshida C, et al. External mechanical cues trigger the establishment of the anterior-posterior axis in early mouse embryos. *Dev Cell*. 2013;27:131–144.
11. Moore KA, Polte T, Huang S, et al. Control of basement membrane remodeling and epithelial branching morphogenesis in embryonic lung by Rho and cytoskeletal tension. *Dev Dyn*. 2005;232:268–281.

12. Michael L, Sweeney DE, Davies JA. A role for microfilament-based contraction in branching morphogenesis of the ureteric bud. *Kidney Int.* 2005;68:2010–2018.
13. Daley WP, Gervais EM, Centanni SW, Gulfo KM, Nelson DA, Larsen M. ROCK1-directed basement membrane positioning coordinates epithelial tissue polarity. *Development.* 2012;139:411–422.
14. Shyer AE, Tallinen T, Nerurkar NL, et al. Villification: how the gut gets its villi. *Science.* 2013;342:212–218.
15. Levental KR, Yu H, Kass L, et al. Matrix crosslinking forces tumor progression by enhancing integrin signaling. *Cell.* 2009;139:891–906.
16. Paszek MJ, Zahir N, Johnson KR, et al. Tensional homeostasis and the malignant phenotype. *Cancer Cell.* 2005;8:241–254.
17. Guilak F, Cohen DM, Estes BT, Gimble JM, Liedtke W, Chen CS. Control of stem cell fate by physical interactions with the extracellular matrix. *Cell Stem Cell.* 2009;5:17–26.
18. Sun Y, Chen CS, Fu J. Forcing stem cells to behave: a biophysical perspective of the cellular microenvironment. *Annu Rev Biophys.* 2012;41:519–542.
19. Harris AK, Wild P, Stopak D. Silicone rubber substrata: a new wrinkle in the study of cell locomotion. *Science.* 1980;208:177–179.
20. Dembo M, Wang YL. Stresses at the cell-to-substrate interface during locomotion of fibroblasts. *Biophys J.* 1999;76:2307–2316.
21. Legant WR, Miller JS, Blakely BL, Cohen DM, Genin GM, Chen CS. Measurement of mechanical tractions exerted by cells in three-dimensional matrices. *Nat Methods.* 2010;7:969–971.
22. Maskarinec SA, Franck C, Tirrell DA, Ravichandran G. Quantifying cellular traction forces in three dimensions. *Proc Natl Acad Sci USA.* 2009;106:22108–22113.
23. Amano M, Ito M, Kimura K, et al. Phosphorylation and activation of myosin by Rho-associated kinase (Rho-kinase). *J Biol Chem.* 1996;271:20246–20249.
24. Kimura K, Ito M, Amano M, et al. Regulation of myosin phosphatase by Rho and Rho-associated kinase (Rho-kinase). *Science.* 1996;273:245–248.
25. Burridge K, Fath K, Kelly T, Nuckolls G, Turner C. Focal adhesions: transmembrane junctions between the extracellular matrix and the cytoskeleton. *Annu Rev Cell Biol.* 1988;4:487–525.
26. Miyamoto S, Akiyama SK, Yamada KM. Synergistic roles for receptor occupancy and aggregation in integrin transmembrane function. *Science.* 1995;267:883–885.
27. Wozniak MA, Desai R, Solski PA, Der CJ, Keely PJ. ROCK-generated contractility regulates breast epithelial cell differentiation in response to the physical properties of a three-dimensional collagen matrix. *J Cell Biol.* 2003;163:583–595.
28. McNeill H, Ryan TA, Smith SJ, Nelson WJ. Spatial and temporal dissection of immediate and early events following cadherin-mediated epithelial cell adhesion. *J Cell Biol.* 1993;120:1217–1226.
29. Adams CL, Nelson WJ. Cytomechanics of cadherin-mediated cell-cell adhesion. *Curr Opin Cell Biol.* 1998;10:572–577.
30. Chen CS, Mrksich M, Huang S, Whitesides GM, Ingber DE. Micropatterned surfaces for control of cell shape, position, and function. *Biotechnol Prog.* 1998;14:356–363.
31. McBeath R, Pirone DM, Nelson CM, Bhadriraju K, Chen CS. Cell shape, cytoskeletal tension, and RhoA regulate stem cell lineage commitment. *Dev Cell.* 2004;6:483–495.
32. Ruiz SA, Chen CS. Emergence of patterned stem cell differentiation within multicellular structures. *Stem Cells.* 2008;26:2921–2927.
33. Kilian KA, Bugarija B, Lahn BT, Mrksich M. Geometric cues for directing the differentiation of mesenchymal stem cells. *Proc Natl Acad Sci USA.* 2010;107:4872–4877.
34. Pelham Jr RJ, Wang Y. High resolution detection of mechanical forces exerted by locomoting fibroblasts on the substrate. *Mol Biol Cell.* 1999;10:935–945.

35. Engler AJ, Sen S, Sweeney HL, Discher DE. Matrix elasticity directs stem cell lineage specification. *Cell.* 2006;126:677–689.
36. Huebsch N, Arany PR, Mao AS, et al. Harnessing traction-mediated manipulation of the cell/matrix interface to control stem-cell fate. *Nat Mater.* 2010;9:518–526.
37. Chowdhury F, Li Y, Poh YC, Yokohama-Tamaki T, Wang N, Tanaka TS. Soft substrates promote homogeneous self-renewal of embryonic stem cells via downregulating cell-matrix tractions. *PLoS One.* 2010;5:e15655.
38. Saha K, Keung AJ, Irwin EF, et al. Substrate modulus directs neural stem cell behavior. *Biophys J.* 2008;95:4426–4438.
39. Banerjee A, Arha M, Choudhary S, et al. The influence of hydrogel modulus on the proliferation and differentiation of encapsulated neural stem cells. *Biomaterials.* 2009;30:4695–4699.
40. Keung AJ, de Juan-Pardo EM, Schaffer DV, Kumar S. Rho GTPases mediate the mechanosensitive lineage commitment of neural stem cells. *Stem Cells.* 2011;29: 1886–1897.
41. Keung AJ, Dong M, Schaffer DV, Kumar S. Pan-neuronal maturation but not neuronal subtype differentiation of adult neural stem cells is mechanosensitive. *Sci Rep.* 2013;3:1817.
42. Gilbert PM, Havenstrite KL, Magnusson KE, et al. Substrate elasticity regulates skeletal muscle stem cell self-renewal in culture. *Science.* 2010;329:1078–1081.
43. Eyckmans J, Boudou T, Yu X, Chen CS. A hitchhiker's guide to mechanobiology. *Dev Cell.* 2011;21:35–47.
44. Parsons JT, Horwitz AR, Schwartz MA. Cell adhesion: integrating cytoskeletal dynamics and cellular tension. *Nat Rev Mol Cell Biol.* 2010;11:633–643.
45. Kirfel G, Rigort A, Borm B, Herzog V. Cell migration: mechanisms of rear detachment and the formation of migration tracks. *Eur J Cell Biol.* 2004;83:717–724.
46. Yano Y, Geibel J, Sumpio BE. Tyrosine phosphorylation of pp125FAK and paxillin in aortic endothelial cells induced by mechanical strain. *Am J Physiol.* 1996;271: C635–C649.
47. Tang D, Mehta D, Gunst SJ. Mechanosensitive tyrosine phosphorylation of paxillin and focal adhesion kinase in tracheal smooth muscle. *Am J Physiol.* 1999;276:C250–C258.
48. Wang HB, Dembo M, Hanks SK, Wang Y. Focal adhesion kinase is involved in mechanosensing during fibroblast migration. *Proc Natl Acad Sci USA.* 2001;98: 11295–11300.
49. Wang Y, Botvinick EL, Zhao Y, et al. Visualizing the mechanical activation of Src. *Nature.* 2005;434:1040–1045.
50. Sawada Y, Tamada M, Dubin-Thaler BJ, et al. Force sensing by mechanical extension of the Src family kinase substrate p130Cas. *Cell.* 2006;127:1015–1026.
51. del Rio A, Perez-Jimenez R, Liu R, Roca-Cusachs P, Fernandez JM, Sheetz MP. Stretching single talin rod molecules activates vinculin binding. *Science.* 2009;323: 638–641.
52. Galbraith CG, Yamada KM, Sheetz MP. The relationship between force and focal complex development. *J Cell Biol.* 2002;159:695–705.
53. Grashoff C, Hoffman BD, Brenner MD, et al. Measuring mechanical tension across vinculin reveals regulation of focal adhesion dynamics. *Nature.* 2010;466:263–266.
54. Holle AW, Tang X, Vijayraghavan D, et al. In situ mechanotransduction via vinculin regulates stem cell differentiation. *Stem Cells.* 2013;31:2467–2477.
55. Cabodi S, del Pilar Camacho-Leal M, Di Stefano P, Defilippi P. Integrin signalling adaptors: not only figurants in the cancer story. *Nat Rev Cancer.* 2010;10:858–870.
56. Campos LS, Leone DP, Relvas JB, et al. Beta1 integrins activate a MAPK signalling pathway in neural stem cells that contributes to their maintenance. *Development.* 2004;131:3433–3444.

57. Zhu AJ, Haase I, Watt FM. Signaling via beta1 integrins and mitogen-activated protein kinase determines human epidermal stem cell fate in vitro. *Proc Natl Acad Sci U S A*. 1999;96:6728–6733.
58. Bost F, Aouadi M, Caron L, Binetruy B. The role of MAPKs in adipocyte differentiation and obesity. *Biochimie*. 2005;87:51–56.
59. Chen HC, Guan JL. Stimulation of phosphatidylinositol 3′-kinase association with foca adhesion kinase by platelet-derived growth factor. *J Biol Chem*. 1994;269: 31229–31233.
60. Paling NR, Wheadon H, Bone HK, Welham MJ. Regulation of embryonic stem cell self-renewal by phosphoinositide 3-kinase-dependent signaling. *J Biol Chem*. 2004;279:48063–48070.
61. Zhang J, Grindley JC, Yin T, et al. PTEN maintains haematopoietic stem cells and acts in lineage choice and leukaemia prevention. *Nature*. 2006;441:518–522.
62. He XC, Yin T, Grindley JC, et al. PTEN-deficient intestinal stem cells initiate intestinal polyposis. *Nat Genet*. 2007;39:189–198.
63. Martinac B. Mechanosensitive ion channels: molecules of mechanotransduction. *J Cell Sci*. 2004;117:2449–2460.
64. Christensen AP, Corey DP. TRP channels in mechanosensation: direct or indirect activation? *Nat Rev Neurosci*. 2007;8:510–521.
65. Liedtke W, Kim C. Functionality of the TRPV subfamily of TRP ion channels: add mechano-TRP and osmo-TRP to the lexicon! *Cell Mol Life Sci*. 2005;62:2985–3001.
66. Hayakawa K, Tatsumi H, Sokabe M. Actin stress fibers transmit and focus force to activate mechanosensitive channels. *J Cell Sci*. 2008;121:496–503.
67. Kim TJ, Seong J, Ouyang M, et al. Substrate rigidity regulates Ca2+ oscillation via RhoA pathway in stem cells. *J Cell Physiol*. 2009;218:285–293.
68. Kapur N, Mignery GA, Banach K. Cell cycle-dependent calcium oscillations in mouse embryonic stem cells. *Am J Physiol Cell Physiol*. 2007;292:C1510–C1518.
69. Hu H, He ML, Tao R, et al. Characterization of ion channels in human preadipocytes. *J Cell Physiol*. 2009;218:427–435.
70. Ferreira-Martins J, Rondon-Clavo C, Tugal D, et al. Spontaneous calcium oscillations regulate human cardiac progenitor cell growth. *Circ Res*. 2009;105:764–774.
71. Sun S, Liu Y, Lipsky S, Cho M. Physical manipulation of calcium oscillations facilitates osteodifferentiation of human mesenchymal stem cells. *FASEB J*. 2007;21:1472–1480.
72. Mehta D, Ahmmed GU, Paria BC, et al. RhoA interaction with inositol 1,4,5-trisphosphate receptor and transient receptor potential channel-1 regulates Ca2+ entry. Role in signaling increased endothelial permeability. *J Biol Chem*. 2003;278:33492–33500.
73. Jakkaraju S, Zhe X, Pan D, Choudhury R, Schuger L. TIPs are tension-responsive proteins involved in myogenic versus adipogenic differentiation. *Dev Cell*. 2005;9:39–49.
74. Miralles F, Posern G, Zaromytidou AI, Treisman R. Actin dynamics control SRF activity by regulation of its coactivator MAL. *Cell*. 2003;113:329–342.
75. Olson EN, Nordheim A. Linking actin dynamics and gene transcription to drive cellular motile functions. *Nat Rev Mol Cell Biol*. 2010;11:353–365.
76. Dupont S, Morsut L, Aragona M, et al. Role of YAP/TAZ in mechanotransduction. *Nature*. 2011;474:179–183.
77. Posern G, Treisman R. Actin' together: serum response factor, its cofactors and the link to signal transduction. *Trends Cell Biol*. 2006;16:588–596.
78. Somogyi K, Rorth P. Evidence for tension-based regulation of Drosophila MAL and SRF during invasive cell migration. *Dev Cell*. 2004;7:85–93.
79. Connelly JT, Gautrot JE, Trappmann B, et al. Actin and serum response factor transduce physical cues from the microenvironment to regulate epidermal stem cell fate decisions. *Nat Cell Biol*. 2010;12:711–718.

80. Aragona M, Panciera T, Manfrin A, et al. A mechanical checkpoint controls multicellular growth through YAP/TAZ regulation by actin-processing factors. *Cell.* 2013;154:1047–1059.
81. Pan D. The hippo signaling pathway in development and cancer. *Dev Cell.* 2010;19:491–505.
82. Lian I, Kim J, Okazawa H, et al. The role of YAP transcription coactivator in regulating stem cell self-renewal and differentiation. *Genes Dev.* 2010;24:1106–1118.
83. Tang Y, Rowe RG, Botvinick EL, et al. MT1-MMP-dependent control of skeletal stem cell commitment via a beta1-integrin/YAP/TAZ signaling axis. *Dev Cell.* 2013;25:402–416.
84. Wipff PJ, Rifkin DB, Meister JJ, Hinz B. Myofibroblast contraction activates latent TGF-beta1 from the extracellular matrix. *J Cell Biol.* 2007;179:1311–1323.
85. Sternlicht MD, Kouros-Mehr H, Lu P, Werb Z. Hormonal and local control of mammary branching morphogenesis. *Differentiation.* 2006;74:365–381.
86. Vogel V, Sheetz M. Local force and geometry sensing regulate cell functions. *Nat Rev Mol Cell Biol.* 2006;7:265–275.
87. Aigner L, Bogdahn U. TGF-beta in neural stem cells and in tumors of the central nervous system. *Cell Tissue Res.* 2008;331:225–241.
88. Blank U, Karlsson G, Karlsson S. Signaling pathways governing stem-cell fate. *Blood.* 2008;111:492–503.
89. James D, Levine AJ, Besser D, Hemmati-Brivanlou A. TGFbeta/activin/nodal signaling is necessary for the maintenance of pluripotency in human embryonic stem cells. *Development.* 2005;132:1273–1282.
90. Maeda T, Sakabe T, Sunaga A, et al. Conversion of mechanical force into TGF-beta-mediated biochemical signals. *Curr Biol.* 2011;21:933–941.
91. Sato T, van Es JH, Snippert HJ, et al. Paneth cells constitute the niche for Lgr5 stem cells in intestinal crypts. *Nature.* 2011;469:415–418.
92. Oloumi A, Syam S, Dedhar S. Modulation of Wnt3a-mediated nuclear beta-catenin accumulation and activation by integrin-linked kinase in mammalian cells. *Oncogene.* 2006;25:7747–7757.
93. Crampton SP, Wu B, Park EJ, et al. Integration of the beta-catenin-dependent Wnt pathway with integrin signaling through the adaptor molecule Grb2. *PLoS One.* 2009;4:e7841.
94. Azzolin L, Zanconato F, Bresolin S, et al. Role of TAZ as mediator of Wnt signaling. *Cell.* 2012;151:1443–1456.
95. Jansen JH, Eijken M, Jahr H, et al. Stretch-induced inhibition of Wnt/beta-catenin signaling in mineralizing osteoblasts. *J Orthop Res.* 2010;28:390–396.
96. Ranga A, Lutolf MP. High-throughput approaches for the analysis of extrinsic regulators of stem cell fate. *Curr Opin Cell Biol.* 2012;24:236–244.
97. Anderson DG, Levenberg S, Langer R. Nanoliter-scale synthesis of arrayed biomaterials and application to human embryonic stem cells. *Nat Biotechnol.* 2004;22:863–866.
98. LaBarge MA, Nelson CM, Villadsen R, et al. Human mammary progenitor cell fate decisions are products of interactions with combinatorial microenvironments. *Integr Biol (Camb).* 2009;1:70–79.
99. Gobaa S, Hoehnel S, Roccio M, Negro A, Kobel S, Lutolf MP. Artificial niche microarrays for probing single stem cell fate in high throughput. *Nat Methods.* 2011;8:949–955.
100. Gong Z, Niklason LE. Small-diameter human vessel wall engineered from bone marrow-derived mesenchymal stem cells (hMSCs). *FASEB J.* 2008;22:1635–1648.
101. Desprat N, Supatto W, Pouille PA, Beaurepaire E, Farge E. Tissue deformation modulates twist expression to determine anterior midgut differentiation in Drosophila embryos. *Dev Cell.* 2008;15:470–477.

102. Nguyen KT, West JL. Photopolymerizable hydrogels for tissue engineering applications. *Biomaterials*. 2002;23:4307–4314.
103. Tibbitt MW, Anseth KS. Dynamic microenvironments: the fourth dimension. *Sci Transl Med*. 2012;4:160ps24.
104. Plodinec M, Loparic M, Monnier CA, et al. The nanomechanical signature of breast cancer. *Nat Nanotechnol*. 2012;7:757–765.
105. Lopez JI, Kang I, You WK, McDonald DM, Weaver VM. In situ force mapping of mammary gland transformation. *Integr Biol (Camb)*. 2011;3:910–921.
106. Gjorevski N, Nelson CM. Mapping of mechanical strains and stresses around quiescent engineered three-dimensional epithelial tissues. *Biophys J*. 2012;103:152–162.
107. Gjorevski N, Nelson CM. Integrated morphodynamic signalling of the mammary gland. *Nat Rev Mol Cell Biol*. 2011;12:581–593.
108. Mosiewicz KA, Kolb L, van der Vlies AJ, et al. In situ cell manipulation through enzymatic hydrogel photopatterning. *Nat Mater*. 2013;12:1072–1078.
109. Wylie RG, Ahsan S, Aizawa Y, Maxwell KL, Morshead CM, Shoichet MS. Spatially controlled simultaneous patterning of multiple growth factors in three-dimensional hydrogels. *Nat Mater*. 2011;10:799–806.
110. Kloxin AM, Kasko AM, Salinas CN, Anseth KS. Photodegradable hydrogels for dynamic tuning of physical and chemical properties. *Science*. 2009;324:59–63.
111. Guvendiren M, Burdick JA. Stiffening hydrogels to probe short- and long-term cellular responses to dynamic mechanics. *Nat Commun*. 2012;3:792.
112. Sasai Y. Next-generation regenerative medicine: organogenesis from stem cells in 3D culture. *Cell Stem Cell*. 2013;12:520–530.
113. Sato T, Clevers H. Growing self-organizing mini-guts from a single intestinal stem cell: mechanism and applications. *Science*. 2013;340:1190–1194.

PART FIVE

Multi-Cellular Sensing

CHAPTER TWELVE

Mechanotransduction in *C. elegans* Morphogenesis and Tissue Function

Erin J. Cram
Department of Biology, Northeastern University, Boston, Massachusetts, USA

Contents

Abstract

Mechanobiology is an emerging field that investigates how living cells sense and respond to their physical surroundings. Recent interest in the field has been sparked by the finding that stem cells differentiate along different lineages based on the stiffness of the cell surroundings (Engler *et al.*, 2006), and that metastatic behavior of cancer cells is strongly influenced by the mechanical properties of the surrounding tissue (Kumar and Weaver, 2009). Many questions remain about how cells convert mechanical information, such as viscosity, stiffness of the substrate, or stretch state of the cells, into the

Progress in Molecular Biology and Translational Science, Volume 126
ISSN 1877-1173
http://dx.doi.org/10.1016/B978-0-12-394624-9.00012-9

biochemical signals that control tissue function. *Caenorhabditis elegans* researchers are making significant contributions to the understanding of mechanotransduction *in vivo*. This review summarizes recent insights into the role of mechanical forces in morphogenesis and tissue function. Examples of mechanical regulation across length scales, from the single-celled zygote, to the intercellular coordination that enables cohesive tissue function, to the mechanical influences between tissues, are considered. The power of the *C. elegans* system as a gene discovery and *in vivo* quantitative bioimaging platform is enabling an important discoveries in this exciting field.

1. INTRACELLULAR SENSATION AND RESPONSE TO MECHANICAL INPUT

1.1. Introduction

Mechanotransduction, the sensation of and response to mechanical forces such as stretch or matrix stiffness, is an essential regulator of cell biological processes including cell survival, cell migration, mitosis, and differentiation. Biophysical studies and cell biology on engineered substrates have demonstrated that force-induced changes in molecular conformation alter biochemical activity, and that mechanical properties of the cellular microenvironment influence cell behavior.[1–6] Cell signaling is induced by mechanical stimuli through the activation of stretch-gated ion channels,[7,8] and through conformational changes in proteins that reveal cryptic binding sites, stimulate catch bond formation, or result in changes in phosphorylation state or protein stability.[9–12] The initial signal often results in the production of a second messenger such as cAMP or calcium, which eventually results in actomyosin-based contraction[13] and/or changes in gene regulation[14] that alter cell phenotypes.

How mechanotransduction is coordinated within and between cells remains a central question in biology.[10] Analysis of developmental processes in model organisms, such as *Caenorhabditis elegans*, *Drosophila melanogaster*, *Danio rerio*, and *Xenopus laevis*,[15–22] has provided insight into the importance of mechanical forces for multicellular rearrangements and tissue morphogenesis. This chapter focuses on contributions made using *C. elegans* as a model system to our understanding of the role of mechanical forces in animal development, morphogenesis, and the function of differentiated, multicellular tissues. For neural aspects of mechanotransduction in *C. elegans*, such as the role of TRP channels in ciliated mechanosensory neurons, I refer the reader to several excellent reviews.[23–25]

The nematode *C. elegans*, an animal with a simple body plan and an invariant developmental program, has been used as a model system since the 1970s to elucidate the mechanisms underlying developmental morphogenesis.[26,27] The majority of the population consists of self-fertile hermaphrodites; however, males also exist and can be used to conduct genetic crosses. The rapid life cycle, small size, large brood size, transparent body, and assortment of powerful genetic tools, including simple and effective RNA interference (RNAi), make *C. elegans* amenable to genetic and genomic approaches such as classical mutagenesis and genome-wide screens using RNAi.[28]

C. elegans was the first animal genome to be fully sequenced and extensively annotated.[29,30] Most mammalian genes have identifiable orthologs in *C. elegans*. Due to its translucent body and convenient methods for producing transgenic animals by DNA microinjection, the use of green fluorescent protein (GFP) as a marker was pioneered in *C. elegans*.[31] The expanding assortment of genetically encoded fluorescent markers and biosensors enables quantitative live imaging techniques and makes *C. elegans* an excellent organism for the *in vivo* study of the cell biology of mechanotransduction.

1.2. Mechanical influences in the *C. elegans* zygote

The earliest mechanically influenced events in the life of the worm occur in the single-celled zygote. Polarization of the zygote is induced by multiple redundant mechanisms involving sperm entry, actomyosin contractility, cortical geometry and viscosity, and asymmetric distribution of the PAR proteins.[32–34] Before fertilization, the oocyte cortical actomyosin cytoskeleton is a dynamic structure consisting of large, regularly spaced myosin foci interconnected by actin fibers. Entry of the sperm centrosome breaks this symmetry by locally disrupting the actomyosin network and inhibiting myosin contraction.[35,36] The resulting nonuniform contractility and movement of myosin foci toward the anterior of the zygote produce a cortical flow dependent on the gradient of myosin contractility and the cortical viscosity.[37–39] The conserved polarity proteins PAR-3, PAR-6, and PKC-6 move with the cortical flow toward the anterior, allowing accumulation of PAR-1 and PAR-2 on the posterior cortex.[35,40–42] Once asymmetrically distributed, the anterior and posterior PAR proteins mutually antagonize each other, maintaining their asymmetric distribution.[43] The PAR proteins control downstream events such as the differential distribution of cell fate determinants and the posterior displacement of the mitotic spindle.[33]

Recently, multiple genes have been identified that affect myosin activity, cortex viscosity, and cortical distribution of myosin in the zygote, including the known actin or myosin regulators *cdc-42*/Cdc42, *csnk-1*/casein kinase, *mlc-5*/myosin light chain, *rga-3*/Rho GAP, *erm-1*/ezrin, *plst-1*/plastin (an actin bundling protein), and the myosin chaperone *unc-45*. Identification of novel regulators, including potential signaling proteins like *cnt-2*/Arf GAP, *sac*-1/PIP phosphatase, and *gsp-1*/Glc7-like phosphatase; regulators of gene expression such as *ntl-2*/CCR4/NOT; the nuclear pore protein *npp-2*/nucleoporin; the cytokinesis protein *unc-59*/septin; and Y54H5A.2, a novel conserved protein of unknown function, significantly expand the known repertoire of actomyosin and polarity regulators and open up many new questions about how polarity, contractility, and cortical viscosity are established.[44]

After polarity is established, an asymmetric cell division produces two differently sized and partially differentiated daughter cells, AB and P1.[45] Positioning of the mitotic spindle determines placement of the cytokinesis furrow and is essential for determining the relative size and shape of the resulting daughter cells.[46] In the *C. elegans* zygote, the spindle assembles in the cell center, but is displaced toward the posterior during metaphase and anaphase. A combination of pulling forces on the plus-ends of astral microtubules and the regulation of microtubule dynamics position the spindle toward the posterior.[33,47]

A membrane-anchored and well-conserved complex consisting of LIN-5/NuMA, GPR-1, 2/GoLoco, and Gα is required to recruit dynein to the correct cortical location and regulate its activity.[47] Spindle positioning defects in *lin-5* mutants are suppressed by loss of genes such as *zyg-9*/XMAP215, *tac-1*/TACC, γ-tubulin, the prefoldin tubulin chaperones, and *cls-2*/CLASP. RNAi depletion of these genes in the *lin-5* background results in shorter microtubules, suggesting that LIN-5 might normally increase spindle pulling forces through increased microtubule plus-end depolymerization, in addition to localization of dynein.[44]

Forces produced by astral microtubules undergoing catastrophe make an important contribution to the placement of the spindle. In *C. elegans*, astral microtubules are very unstable, residing at the cell cortex for only ~1 s before depolymerizing. An estimated force of ~50 pN is exerted on individual astral microtubules during spindle positioning, similar to the force generated by shrinking microtubules *in vitro*. Therefore, dynein may serve primarily to anchor depolymerizing microtubules to the cell cortex, with microtubule depolymerization generating the actual pulling force.

Cooperative action of cortical dynein molecules could also exert sufficient pulling forces to shift the mitotic spindle.[47]

These results suggest that establishment and maintenance of cell polarity, regulation of actomyosin contraction and cortical flow, and asymmetric placement of the mitotic spindle all have mechanical components. Further genetic, biophysical, and modeling studies will lead to a more complete understanding of how multiple inputs are integrated to produce robust and reproducible cell polarization outcomes.

1.3. Cell shape changes

Cell shape changes and rearrangements are critical for morphogenesis[48] and require highly polarized changes in cell shape and position. In *C. elegans*, these events drive the processes of gastrulation, ventral enclosure, dorsal intercalation, and embryonic elongation. Each of these processes involves cell shape changes driven by remodeling of the actomyosin cytoskeleton, dynamics of cortical actomyosin networks, and precise control of activation of myosin contractility. The microtubule cytoskeleton also contributes significantly to cell shape changes, perhaps by transporting key regulators of actin polymerization and myosin activation.[49] As the cells change shape, they exert mechanical influences on neighboring cells.[48]

In the following sections, cell shape changes will be discussed in the context of mechanical contributions to *C. elegans* embryonic morphogenesis and tissue function. For a more detailed treatment of the biomechanics of cell shape change, a number of excellent reviews are available including Refs. 50–52.

1.4. Nuclear response

In addition to changing the shape of cells and influencing the cell movements that lead to morphogenetic changes, mechanical inputs can also affect cell fates and lineage commitments through altered gene expression patterns. For example, altered stiffness of 2D or 3D substrates,[1,53] exposure to tension gradients,[54] compression,[55] and forces produced by cell contractility and cell migration[55,56] can result in significant changes in gene expression and cell specification.[19,57] Although many transcription factors are likely to be involved in response to mechanical stimuli, several have emerged as key regulators of mechanotransduction during development and may also play a role in *C. elegans* morphogenesis.

In *Drosophila*, the transcriptional activator Armadillo/β-catenin relocates from the cadherin complex to the nucleus in response to compressive stress, where it induces expression of the transcription factor Twist.[55] Twist expression is high in cells undergoing mechanical deformation during embryogenesis, and localized mechanical compression is sufficient for nuclear localization of Armadillo and upregulation of *twist* expression.[55] *C. elegans* expresses one Twist-like transcription factor, HLH-8.[58] The bHLH domain of HLH-8 is ~60% identical to Twist in other species.[59] HLH-8 is expressed in tissues that respond to and exert mechanical forces, such as undifferentiated cells of the M (mesoderm) lineage and differentiated body wall, vulval, and intestinal muscles,[58] and is required for proper specification in these lineages.[60] However, potential regulation of HLH-8/CeTwist by β-catenin or by mechanical forces has not been investigated in *C. elegans*.

Another important mechanotransduction pathway is the Hippo pathway, which controls the organ size in part by regulating the subcellular localization of the transcription factor YAP.[61,62] The Hippo kinase activates another kinase, Warts/LATS, which phosphorylates and inhibits YAP by inducing association with 14-3-3 and retaining YAP in the cytoplasm. Hippo signaling and YAP localization are also regulated by cadherin signaling,[63,64] possibly in response to forces exerted on cell–cell junctions.[65] In human cells, YAP expression and nuclear localization are enhanced when cells are grown on stiff substrates.[66] YAP (and its paralog, TAZ) associates in the nucleus with coactivators such as TEAD1-4 and β-catenin to activate transcriptional programs that drive cell proliferation and inhibit apoptosis.[67]

Although *C. elegans* seems to lack a canonical YAP,[67] a possible homolog has been identified. *C. elegans* YAP-1, which plays a role in thermal stress response, possesses a TEAD-binding and YAP domain.[68] WTS-1/LATS and FTT-2/14-3-3 are required for proper nuclear/cytoplasmic shuttling of YAP-1, suggesting that YAP-1 may be regulated similarly to YAP in other animals. YAP-1 is expressed in the muscle, hypodermis, gonadal sheath, and spermatheca.[68] Many of these tissues are mechanosensitive and/or contractile. Therefore, it would be interesting to determine what role, if any, YAP-1 plays in the function of these tissues and in response to mechanical stresses in *C. elegans*.

2. MECHANICAL INFLUENCES IN EMBRYONIC DEVELOPMENT

During animal development, cells undergo shape changes, move to new locations, and assemble into tissues and organs. The mechanosensory properties of the cells themselves and their cell–cell and cell–matrix adhesions have a central function in these processes.

2.1. *C. elegans* integrins and mechanotransduction

Much of the mechanotransduction work in mammalian cell culture has focused on the role of integrins, heterodimeric transmembrane receptors that connect the extracellular matrix to the cytoskeleton.[69,70] Integrins are the foundation of focal adhesions and hemidesmosomes, both of which are multiprotein structural and signaling complexes.[71,72] Integrins are bidirectional signaling molecules.[73] For example, in "outside-in" signaling, binding to stiffer matrix leads to force-dependent integrin clustering and activation of integrin signaling through FAK and Src, activation of Rho, ROCK, and actomyosin contractility, or to changes in gene expression.[14] Conversely, "inside-out" signaling through integrins leads to remodeling of matrix and altered cell environments.[74]

C. elegans expresses two α (PAT-2 and INA-1) and one β (PAT-3) integrin.[75,76] In body wall muscles, PAT-3 is localized to muscle attachment sites called dense bodies.[77] In addition to integrins, many cytoskeletal and signaling molecules, such as DEB-1/vinculin,[78] talin,[79,80] ZYX-1/zyxin,[81] and UNC-97/PINCH,[82] colocalize to dense bodies, suggesting similarity to vertebrate focal adhesions. Attachment of muscle cells to the nearby hypodermis is essential for muscle filament assembly, and function and attachment defects caused by loss-of-function mutations in *pat-2* or *pat-3*, or in matrix components such as *unc-52*/perlecan, give a *p*aralyzed and *a*rrested at *t*wo-fold (Pat) embryonic phenotype.[83,84] However, other than the requirement of integrins and matrix attachment for productive muscle contraction,[85] and the role of extracellular matrix (ECM) in the process by which muscle contractions stimulate the remodeling of hemidesmosome-like junctions in the *C. elegans* epidermis[86] (discussed below), very little concerning the role of integrins and ECM in mechanotransduction in *C. elegans* is known. The next section summarizes insights from the study of apical ECM on the role of matrix in cell morphogenesis and development.

2.2. Mechanical stability and attachment to apical ECM

C. elegans senses chemicals in its environment using a collection of sensory neurons, the amphid neurons, whose cell bodies are located near the nerve ring in the head of the animal. Amphid neurons extend unbranched, ciliated sensory dendrites anteriorly to the tip of the animal's nose.[87] The amphid neurons are encased by and form junctions with a glial cell called the sheath cell.[87,88] The neuron is born at the anterior nose tip, and the dendrite is formed by retrograde extension; as the cell body migrates toward the pharyngeal bulb, the dendrite remains attached at the nose tip, elongating as the cell body migrates toward the pharyngeal bulb (Fig. 12.1).[89] In other words, rather than extending to its target location, the dendrite starts at its target location and the cell body moves away. This process requires anchorage of the dendrite at the nose tip.

Two genes, *dex-1* and *dyf-7*, encode components of the apical extracellular matrix and are required for anchoring the dendrite at the tip of the nose. In the absence of DEX-1 or DYF-7, the unanchored dendrite is dragged along behind the migrating cell body, resulting in a short, stubby dendrite that does not extend to the nose tip.[89] DEX-1 is a homolog of zonadhesin

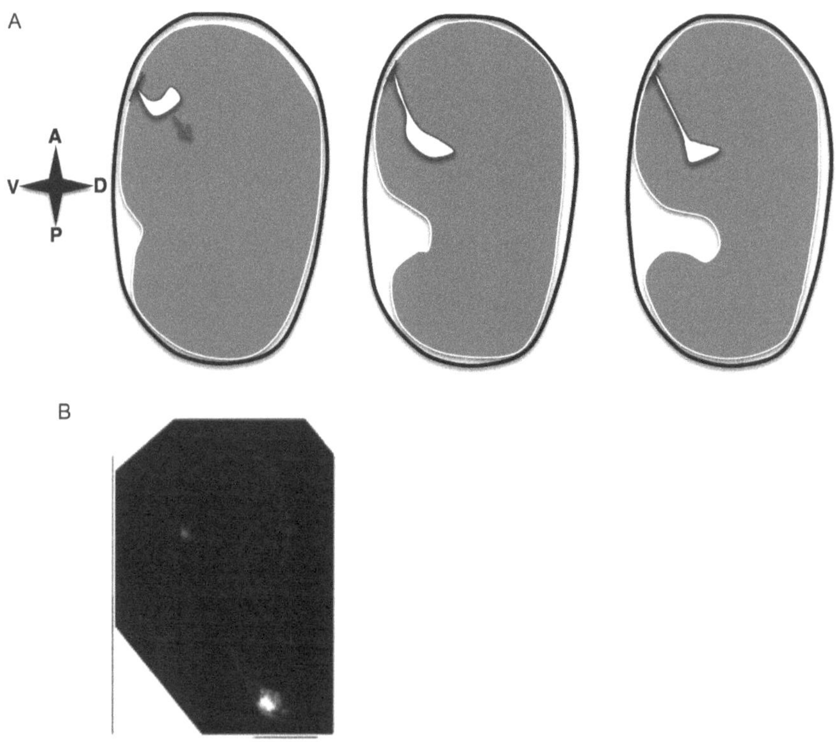

Figure 12.1 *The amphid neuron undergoes retrograde extension.* The amphid neuron is born at the anterior of the embryo, and the dendrite is formed as the cell body (white cell) migrates away. The diagrams in (A) depict a *C. elegans* embryo over time. The amphid neuron is in white and the apical matrix in black. The arrow indicates the direction of retrograde extension. (B) A fluorescently labeled amphid neuron undergoing retrograde extension. *Adapted with permission from Ref. 89.*

and DYF-7 is a zona pellucida (ZP) domain-containing protein. In vertebrates, zonadhesin and ZP domain proteins are best known for their role in mediating the interaction of sperm and egg during fertilization.[90] DEX-1 and DYF-7 also resemble the tectorins, which anchor stereocilia in the vertebrate inner ear,[91] suggesting that this anchoring system may have been incorporated into a mechanosensory structure in the ear. Although DEX-1 and DYF-7 have domains that may interact, DEX-1 is unlikely to be the cell receptor responsible for anchoring the dendrite. DEX-1 is expressed by nonneuronal cells in the head and acts cell-nonautonomously. In addition, a secreted form of DEX-1, in which the transmembrane segment is deleted, can rescue a *dex-1* mutant.[89] Therefore, the receptor responsible for anchoring, or for transmitting the information that the dendrite should begin to elongate, remains at large.

ZP domain proteins, which can drive polymerization and can self-assemble into a filamentous matrix,[92] have been found at the apical surfaces of nearly all animal epithelia that have been examined.[93] Apical ECM is generally thought of as serving as a protective barrier, and apical ECM molecules can also play a role in epithelial junction remodeling.[94,95] However, the role of apical ECM proteins in morphogenesis and tissue function is not well understood. In the amphid neuron system, DYF-7 may be required in the apical ECM not only for attachment to the dendrite but also to disperse loads and prevent rupture of the interactions between the glia and dendrite during morphogenesis. The apical ECM protein SYM-1/eLRRon may be serving a similar role in *C. elegans* epithelia. Mammalian homologs, such as the eLRRon proteins decorin or LRRTM1-3, play a role in ECM assembly[96] and as cell adhesion molecules at synaptic junctions.[97–99] In *C. elegans* embryos lacking SYM-1, epithelial junctions form correctly, but rupture during cuticle secretion and remodeling of cell-matrix interactions.[100] These results suggest that apical matrix proteins may be important for cell–matrix, cell–cell attachments, and stability of epithelial junctions.

2.3. Introduction to cell–cell junctions in *C. elegans*

Apical junction complexes are specialized adhesion structures that separate the apical and basolateral compartments of epithelial cells.[101] *C. elegans* apical junctions are a single electron-dense unit that contains spatially distinct macromolecular complexes. At the apex is a cortical complex including the apical and subapical cytocortex proteins, actin, the intermediate filament protein IFB-1, the actin binding protein ERM-1/ezrin, and the EGF and laminin domain-containing protein known as CRB-1/Crumbs. Slightly more basal is the cadherin complex and its associated proteins, and most basal is a complex that includes the *C. elegans* homologue of Discs large, DLG-1, and its binding partner, AJM-1.[102]

2.4. Role of adherens junctions in *C. elegans* embryonic elongation

Cadherin-based cell–cell junctions play a critical role in the morphogenesis of epithelial tissues, establishment of cell polarity,[99,103] and transduction of mechanical forces across tissues.[104] Cadherins are Ca^{2+}-dependent homodimeric cell–cell adhesion receptors integral to the apical adherens junctions in epithelial cells (Fig. 12.2).[105] The cadherin–catenin complex

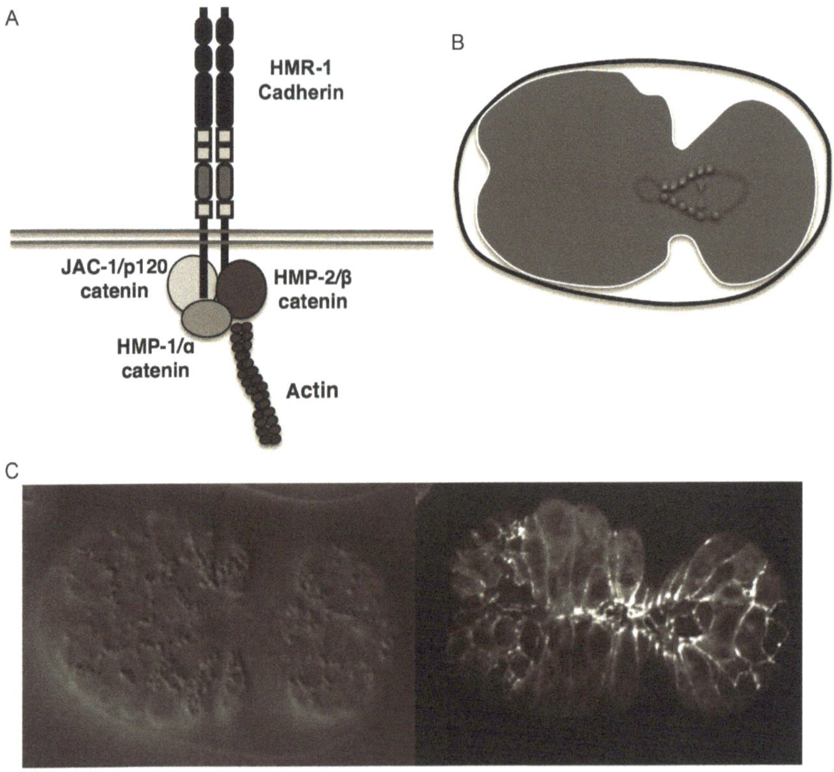

Figure 12.2 *The C. elegans cadherin-catenin complex localizes to cell-cell junctions.* (A) Diagram depicting the *C. elegans* cadherin-catenin complex. (B) Diagrammatic representation of the embryo in (C). Dots represent the cadherin-catenin complex, and arrows indicate direction of closure. (C) Left panel, DIC image and right panel, fluorescence image of an embryo expressing HMP-1::GFP and undergoing ventral enclosure. *Image courtesy of Dr. Jeff Hardin.*

(CCC) connects to the actin cytoskeleton and mediates the transmission of forces between adjacent cells, thereby integrating single-cell behaviors to produce the tissue-level responses critical for development and tissue function.[17,104,106–110] Once adherens junctions form, actin is linked to the cell surface via α-catenin-containing complexes.[111]

C. elegans expresses 13 cadherins including the classic cadherin HMR-1 and the less well-understood Fat-like, Flamingo, and Calsyntenin families of cadherins.[112,113] HMR-1 is functionally equivalent to E-cadherin, with an extracellular Ca^{2+} binding dimerization domain, and a classic cadherin cytoplasmic domain that serves as a site for macromolecular complex assembly. HMR-1/E-cadherin, HMP-1/alpha catenin, HMP-2/beta catenin, and JAC-1/p120 catenin comprise the core functional unit of the CCC.[108,112,114] Although null alleles of *hmr-1*, *hmp-1*, and *hmp-2* result in embryonic lethality with enclosure and elongation defects,[114,115] prior to lethality, the epithelium retains apical–basal polarity and remains adherent,[114] perhaps because of other cell–cell adhesion molecules such as SAX-7/L1CAM[116] and components of the DLG-1/discs large complex.[117] The CCC is nevertheless crucial for several key morphogenetic events in *C. elegans*, including epidermal sheet sealing, actomyosin force transmission, and maintenance of tissue integrity.[108,112]

C. elegans transitions from a round embryo to a worm shape in late embryogenesis through anterior–posterior (A–P) elongation. Unlike many organisms, in which convergent extension movements drive elongation, lengthening of the *C. elegans* embryo is accomplished primarily by concerted epidermal (hypodermal) cell shape changes.[118] Elongation is driven by coordinated contraction of circumferential actin filament bundles regulated by activation or suppression of nonmuscle myosin activity within specific epidermal cells. Elevated activity of the myosin regulatory light chain, MLC-4, presumably downstream of RHO-1 and LET-502/ROCK, results in contraction and elongation of the lateral seam epidermal cells,[119,120] whereas myosin activity is suppressed in other epidermal cells via the RhoGAP RGA-2[120] and the myosin phosphatase MEL-11.[119]

Forces generated within elongating epidermal cells are communicated to cell neighbors via the CCC. Loss of *hmp-1*/alpha catenin, or other core components, severely affects elongation, leading to dorsal folds in the epidermis (the Humpback, or Hmp, phenotype) and failure of morphogenesis early in elongation.[114,121] Coordination between the CCC and the actin cytoskeleton is required for elongation. Embryos expressing a mutant version of HMP-1 missing its C-terminal actin-binding domain (ABD) can form circumferential actin bundles, but fail to form mechanically robust connections between the junction and the actin network, leading to loss of junctional–proximal F-actin and detachment of circumferential actin from the cell periphery.[121]

Withstanding and responding to mechanical stress requires robust cell–cell adhesion. Studies from mammalian cell culture show that cadherin-based junctions are strengthened and expanded in a force-dependent manner,[122] primarily through activation of myosin, increased time of actin–myosin interaction under load, or strengthening of myosin II filaments.[105,123] Mechanical load also stimulates clustering and recruitment of cadherins[124] and associated proteins such as vinculin.[125,126]

In *C. elegans*, enhancer screens in hypomorphic *hmp-1*/alpha catenin backgrounds have identified modulators of cell response to stress including UNC-94/tropomodulin,[127] ZOO-1/ZO1,[128] JAC-1/p120 catenin,[129] and SRGP-1/Slit/Robo GAP.[130] UNC-94/tropomodulin, better known for its role in thin filament length regulation in muscle sarcomeres, is enriched at stressed cell junctions during epithelial morphogenesis, where it protects actin filaments recruited by alpha catenin from minus-end subunit loss, improving the stability of the actin cytoskeleton at cell–cell junctions.[127,131] ZOO-1/ZO1 is a membrane-associated guanylate kinase that in *C. elegans* localizes to the CCC. Knockdown of *zoo-1* enhances the severity of the actin defects in hypomorphic *hmp-1* and *hmp-2* mutants and leads to rupture of cell–cell junctions under stress.[128] It is not known if the actin defect is entirely responsible for the weakening of the junctions, or if some other function of ZOO-1 is also necessary to strengthen the junctions. JAC-1/p120 catenin similarly is not essential for embryonic morphogenesis, but does contribute to the stability of stressed epithelial junctions.[129] JAC-1 may help to strengthen adhesions by mediating lateral clustering,[132,133] improving anchorage of the CCC to the cytoskeleton, and, although not demonstrated in *C. elegans*, may also regulate the activity of Rho, as does its vertebrate homologue.[134]

SRGP-1 is a conserved GAP with an F-BAR (Bin1, Amphiphysin, and RVS167) domain.[130] The F-bar domain produces outward membrane bends and could contribute to adhesion by increasing the surface area of contact between adjacent cells.[135] Alternatively (and not mutually exclusively) as a Rho GAP, SRGP-1 may also modulate actin dynamics in some situations[136]; however, the GAP domain of SRGP-1 is dispensable for epidermal morphogenesis.[130] During embryonic closure in *C. elegans*, SRGP-1 promotes the rapid formation of new junctions by increasing membrane dynamics at cell–cell contacts.[130] SRGP-1 also plays an important role in membrane bending and GAP activation in cell corpse engulfment.[136]

During epithelial remodeling, the main roles of the adherens junctions are to provide adhesion between cells and to stabilize actomyosin-driven cell shape changes.[137] However, cadherins may also have an active signaling role. For example, in mammalian cell culture and in *Drosophila*, strain on the adherens junctions can activate Rac and Cdc42 signaling.[105,123,138] Clear evidence for signaling via cadherins is currently lacking in *C. elegans*. However, in the case of Ea/Ep ingression during gastrulation, genetic evidence suggests that cadherin-based adhesion and Rac signaling functionally interact. HMR-1/E-cadherin acts along with SAX-7/L1CAM during Ea/Ep ingression. In addition, knockdown of CCC components and components of the Rac pathway, such as CED-5/Dock180, CED-12/ELMO, and CED-10/Rac, synergistically affect Ea/Ep ingression. How this synergy occurs is unclear, but it may involve alterations in coupling of the actomyosin cytoskeleton to the cortex, or in the production of lamellipodia by cells adjacent to Ea and Ep that aids ingression.[108]

2.5. Mechanical coordination between tissues

Cells in the embryo must respond in a temporally and spatially coordinated manner to mechanical and biochemical inputs and coordinate cell and tissue movements across multiple germ layers that are often separated by layers of extracellular matrix.[22,139] Although significant insights into this process have been derived from the study of *Xenopus* elongation and *Drosophila* dorsal closure and germ band extension,[19,21,110] many questions remain. Recently, a mechanical feedback loop that coordinates muscle and epidermal development has been identified in *C. elegans*.[140]

Although actomyosin-mediated contractile forces, transmitted through the CCC, are crucial for the early phases of elongation, later stages depend on interactions between the hypodermis and the underlying muscle cells. Muscle cells are mechanically connected to the hypodermis, and ultimately to the cuticle, through tissue-spanning attachment structures known as fibrous organelles (FOs) (Fig. 12.3). FOs are hemidesmosome-like structures that connect the apical and basal surfaces of hypodermal cells via intermediate filaments.[101] Accumulating evidence suggests that this FO-mediated mechanical and functional connection between muscle and epidermis is required for embryonic elongation. Embryos that cannot contract their muscle cells, or that have compromised attachment of muscle cells to the extracellular matrix, fail to elongate and exhibit the paralyzed at twofold (Pat) phenotype.[75] Contraction of the muscle cells induces correct spatial organization of the FOs.[77] Mutations that disrupt the organization of the FOs, such as *let-805*/myotactin,[142] *vab-10*/spectraplakin,[143] *vab-19*/Kank, and *eps-8*,[144] or conditions that alter the abundance of FO proteins also cause later defects in elongation.[145] VAB-10, which can bind to actin, microtubules, and

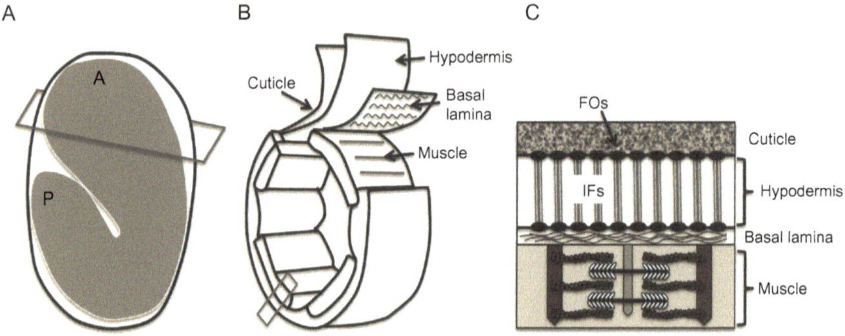

Figure 12.3 *The fibrous organelles convey mechanical information from the underlying muscle cells to the hypodermis.* (A) Diagram of a *C. elegans* embryo at the beginning of embryonic elongation. The blue polygon indicates the position of the cross section shown in (B). A indicates the anterior and P the posterior of the animal. (B) Cross section of the animal and (C) detailed diagram indicating the relative position of the muscle, hypodermis, and cuticle. The blue polygon in (B) indicates the position of the detail shown in (C). FOs = fibrous organelles; IFs = intermediate filaments. *Adapted with permission from Ref. 141.* (See the color plate.)

intermediate filaments,[143] is ideally placed to coordinate the response of the cell cytoskeleton in response to changing mechanical input.

Communication between the muscle and hypodermis is mediated in part by Rac activation of a cassette including the p21-activated kinase PAK-1, the adaptor protein GIT-1/G-protein-coupled receptor kinase-interacting protein, and the GEF PIX-1/βPIX. GIT-1 acts through PIX-1 and Rac to stimulate PAK-1 activity. PAK-1 strengthens hemidesmosomes by phosphorylating the intermediate filament protein IFA-3, thereby affecting the organization and integrity of the FOs.[140] Embryos with loss-of-function mutations in *pak-1* fail to elongate properly, and the loss of PIX-1 or GIT-1 in a sensitized genetic background results in a similar phenotype, suggesting that this cassette may be a key mediator of mechanical information between cells. Mechanical tension exerted by muscle cells may be required to maintain GIT-1 localization at FOs. In Pat mutants, GIT-1 does not accumulate at cell boundaries, intermediate filaments are not phosphorylated, and FOs become disorganized. Importantly, the GIT-1 localization and FO organization can be rescued if external forces are applied via direct micromanipulation of embryos. These results suggest that muscle contractions drive the maturation of the hemidesmosomes into mature structures that can resist mechanical stress. Regulated colocalization with signaling proteins such as PAK-1 suggests that the hemidesmosome is not simply an attachment structure but also acts as a mechanotransducer that responds to tension and induces signaling.

Mechanical interplay between tissues likely goes beyond the effects mediated by the FO. For example, earlier in embryogenesis, interactions between cell layers may also regulate ventral enclosure. During this process, epidermal cells migrate over the ventral surface, enclosing the embryo in a layer of epidermis. Recent evidence suggests that ANI-1/anillin, an actin and a myosin-binding protein that coordinates actomyosin contractility, is required for division of the neuroblast cells that underlie the ventral epidermal cells in the enclosing embryo. The defective ventral enclosure phenotypes in *ani-1* RNAi embryos, and similar phenotypes in embryos lacking RHO-1 and other regulators of myosin activity, suggest that the position and/or shape of the neuroblasts is important for directing migration of the epidermal cells.[146]

3. MECHANICAL INFLUENCES IN LARVAL DEVELOPMENT AND TISSUE FUNCTION

3.1. Excretory canal development

Many organs, such as the vascular, lymphatic, intestinal, reproductive, and excretory systems, include tube elements. Although tubes are often constructed of multiple cells linked by cell–cell junctions, single cells can also form tubes. Single-celled tubes can form when a cell forms a junction with itself (autocellular tubes), by fusion of autocellular or multicellular tubes, or by vesicle fusion that forms a hollowed-out lumen within a single cell. These latter types are known as seamless tubes. Seamless tubes are found in the zebrafish vascular system, the fly respiratory system, and the nematode excretory system.[147–149]

The *C. elegans* excretory canal consists of a single, mononucleate, H-shaped excretory cell that runs the entire length of the animal (Fig. 12.4). In addition to executing the amazing feat of forming a single-celled polarized epithelial tube, the cell maintains and adjusts lumen diameter throughout the life of the organism and is required for osmotic homeostasis.[149,150] During development, the four arms of the "H" extend outward, in a process similar to growth cone extension. The hollow lumens then form by vesicle fusion, although the mechanism driving luminal extension is not well understood.[151,152]

During the first larval stage (L1), the lumen extends from the center of the animal to the rectum at the posterior.[152] Canal outgrowth is osmotically regulated. For example, hyperosmotic shock induced by salt or sucrose treatment stimulates vesicles to dock with the lumen and accelerates excretory canal extension. The STE20 family kinase GCK-3, required for recovery

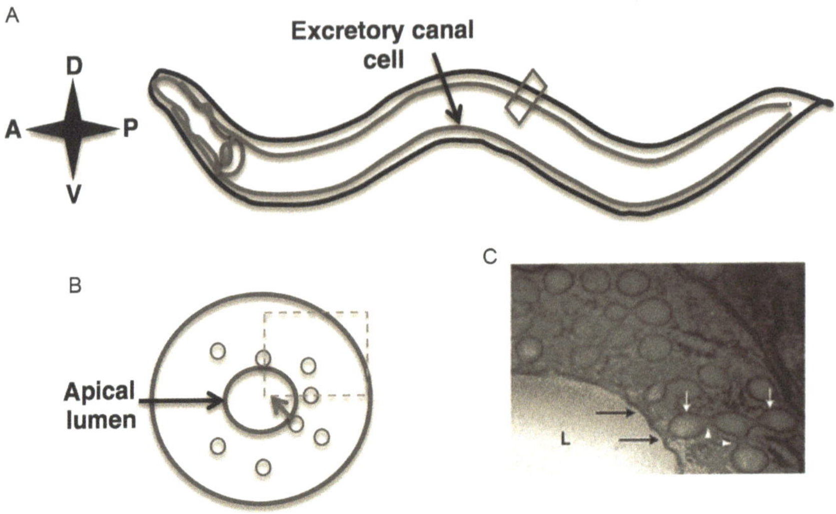

Figure 12.4 *The lumen of the excretory canal forms by vesicle fusion.* (A) The excretory canal in *C. elegans* consists of a single, H-shaped cell that extends the length of the animal. The detail in (B) is indicated by the polygon. B) Cross section of one branch of the excretory canal depicting lumen formation. The circles represent vesicles and the arrow indicates the flow of water into the lumen. The gray square indicates the area shown in (C). (C) Electron micrograph of vesicles (indicated by arrows) fusing with the excretory canal lumen (L). *Reprinted with permission from Macmillan Publishers Ltd.: Nature Cell Biology,153 2013.*

after hyperosmotic shock, is required for docking of vesicles to the lumen and for proper extension of the excretory canals.[152] These data, and data from a paper investigating the role of the actin-binding protein ERM-1 in canal formation, suggest that osmotically regulated vesicle fusion brings new membrane to the lumen, thereby enabling canal extension.[152,153]

The actin-binding protein ERM-1/ezrin–radixin–moesin localizes to lumenal membranes and is required for recruiting, incorporating, and coating lumenal membrane-forming vesicles.[153–155] The correct amount of ERM-1 is required for proper lumen formation.[154] Cells with insufficient ERM-1 have incomplete lumenal membrane expansion and accumulation of cytoplasmic vacuoles. Cells with too much ERM-1 have excess luminal actin coating and form large intraluminal cysts. This ERM-1-driven cyst formation requires the membrane water channel AQP-8/aquaporin. Water flux through AQP-8 is required not only for cyst formation but also for normal canal extension.[153] AQP-8 and ERM-1 interact at the interface between canalicular vesicles and the lumen, which may allow localized entry of water to the lumen. Consistent with this idea, excess AQP-8, and presumably excess water, results in distention of the canal lumen. These results indicate that aquaporins play a role in the formation and function of the excretory canal, and suggest that water flux, or hydrostatic pressure, might have a direct effect on lumenal morphogenesis.

Although aquaporin-regulated hydrostatic pressure regulates plant development,[156] not much is known about how hydrostatic pressure might act as a morphogenetic force in animals. Hydrostatic pressure is involved in mitotic cell rounding[157] and can help shape blood vessel formation in zebrafish,[148] and the AQP-8 results suggest that water influx can act as a direct morphogenetic force.[153] Many questions remain. Aquaporins can travel in secretory vesicles, with or without incorporation into the plasma membrane.[158,159] However, how aquaporins are excluded from membranes such as the excretory canal luminal membrane is not known. Aquaporins are passive water channels.[158] How is flux through AQP-8 regulated? Clearly, hyper- and hypo-osmolarity can impact excretory canal development, but the source and role of osmolarity differences during normal development are not well understood. Osmotic differences, perhaps generated by differences in activity or polarized localization of ion channels to vesicle or lumen,[151,152] may regulate the direction of water flow through the aquaporins and the development of seamless tubes.

3.2. Mechanical inputs into spermathecal function

The *C. elegans* reproductive system consists of a simple tube of germ line cells surrounded by a basement membrane and a single layer of contractile myoepithelial cells. The ovary, in which oocytes develop, is connected to the spermatheca, the site of oocyte fertilization. The spermatheca is an accordion-like multicellular tube that contains sperm (Fig. 12.5). It is composed of a monolayer of 24 myoepithelial cells surrounded by a basement membrane comprising 2 rows of 4 cells each that form a narrow entry corridor or neck and 16 cells that form a wider bag-like chamber. To exit the spermatheca, fertilized embryos pass through the spermatheca-uterine (sp-ut) valve, a contractile toroid generated by the fusion of four cells.[160,161] Ovulation is regulated by signaling between the oocytes, the sheath cells, and the sperm.[162,163] LET-23/EGF receptor signaling in the sheath cells is thought to activate PLC-3/phospholipase C-gamma, which generates IP_3 and triggers calcium release via the IP_3 receptor ITR-1.[164,165] Ca^{2+} release stimulates the sheath cell contractions that push the oocyte into

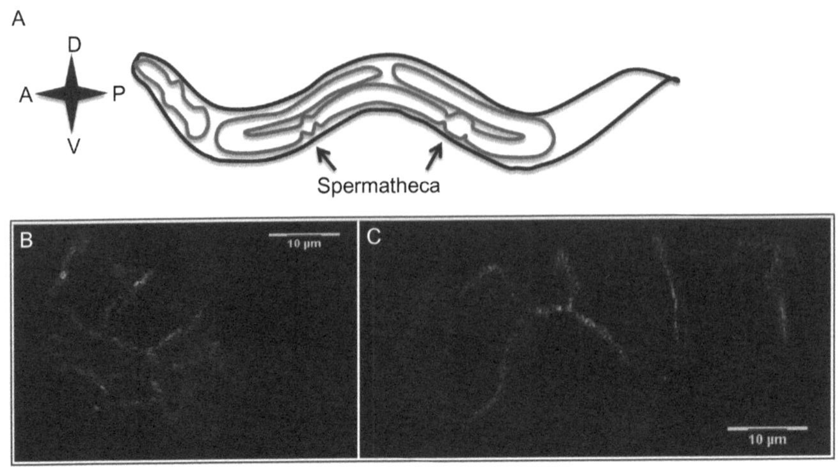

Figure 12.5 *The spermatheca is stretched by oocyte entry.* (A) The spermatheca is a tissue in the nematode reproductive system (indicated by the blue line) that stores sperm and is the site of oocyte fertilization. (B, C) *C. elegans* spermathecae expressing the gap junction protein INX-12::GFP, enabling visualization of cell-cell interfaces. (B) The unoccupied spermatheca is a compact structure. (C) Oocyte entry stretches the spermatheca. (See the color plate.)

the spermatheca. Oocyte entry dramatically stretches the spermathecal cells and initiates a process that culminates in spermathecal cell contraction and expulsion of the fertilized egg into the uterus. The next oocyte is then propelled into the spermatheca, reinitiating the "oocyte transit" cycle, which repeats every ~20 min.

In contrast to the oocyte maturation and ovulation process, the molecular details of spermathecal transit are poorly understood. What is known is that oocyte transit requires the phospholipase PLC-1/PLC-ε[166] and the actin-binding protein FLN-1/filamin. In *plc-1* or *fln-1* mutant animals, entry is normal, but the embryo does not exit, becoming trapped in the spermatheca.[167–174] Like PLC-3 in the sheath cells, PLC-1 generates IP_3 and triggers calcium release via ITR-1.[164,165] In contrast, however, PLC-1[166,175] possesses unique RasGEF and Ras association domains in addition to the catalytic domain.[176,177] In mammals, PLC-ε is regulated by small GTPases and may act as a GEF for Rap1 and Ras.[178,179] In *C. elegans*, PLC-1 has been shown to interact with LET-60/Ras in a GTP-dependent manner, but the functional consequence of this interaction is unknown.[180]

The protein FLN-1/filamin is also essential for spermathecal transit. Filamins are large, dimeric proteins composed of an N-terminal ABD and multiple C-terminal immunoglobulin-like filamin (IgFLN) repeats arranged into two rods separated by flexible hinges.[169–171] Filamins organize the actin cytoskeleton into an elastic three-dimensional network[181]; connect the cytoskeleton to transmembrane proteins, including stretch-gated ion channels and integrins[182,183]; and interact with many intracellular signaling proteins, such as migfilin, FilGAP, Trio, and small GTPases.[184–186] Filamins can act as mechanical stress sensors[170,187–190] through stretch-sensitive changes in conformation or arrangement of the IgFLN repeats that reveal cryptic binding sites for downstream effectors.[174,191,192] Because filamins can concurrently bind transmembrane proteins, filamentous actin, and intracellular signaling components, they are ideally placed to regulate the response of cells or tissues to changing mechanical forces.

Using transgenic nematodes expressing the genetically encoded Ca^{2+} sensor GCaMP,[193,194] we have discovered that oocyte entry triggers a distinctive series of Ca^{2+} oscillations that propagate across the tissue, culminating in coordinated contraction and expulsion of the fertilized egg.[173] Both FLN-1 and PLC-1 are necessary for proper induction and coordination of Ca^{2+} signaling in this tissue.[173] Our favored working model is that stretching FLN-1 reveals cryptic binding sites, scaffolding effectors such as Rho, and leading to the activation of PLC-1. Active PLC-1 then hydrolyzes membrane lipids to generate the second IP_3,[195] stimulating Ca^{2+} release from the endoplasmic reticulum and triggering actomyosin contraction.[196] Also consistent with current data is a more indirect model in which the effects of FLN-1 are mediated through the actin cytoskeleton, with PLC-1 acting in parallel to stimulate the release of calcium from internal stores.

In the spermatheca, actomyosin contractility is generated through the concerted action of the small GTPase Rho and the calmodulin target myosin light-chain kinase. RHO-1/Rho activation of the kinase LET-502/Rho-kinase promotes contraction through the phosphorylation of the nonmuscle myosins NMY-1 and NMY-2. MEL-11, a subunit of the myosin light-chain phosphatase, MEL-11 inhibits myosin contraction.[197,198] MEL-11 is initially cytoplasmic, but in response to LET-502/ROCK activity, it becomes sequestered away from the contractile apparatus to the plasma membrane.[198] Depletion of the *nmy-1*/nonmuscle myosin produces a relaxed, noncontractile spermatheca, and *mel-11*/myosin phosphatase RNAi causes the spermatheca to pull itself apart due to excessive constriction.[173] NMY-1, MEL-11, and LET-502 have been extensively studied in the context of embryonic elongation where they are required to fine-tune contractility of the hypodermal cells[119,120,197–200] and act similarly in the spermatheca to contract the spermathecal cells and expel the fertilized egg.[173]

How is the directional contraction of the spermathecal tissue coordinated? In wild-type spermathecae, oocyte entry leads to the initiation of Ca^{2+} signaling in the distal "neck" of the spermatheca. Signal then spreads proximally in waves across the tissue in the direction of the sp-ut valve. After a stereotypical set of pulses, the tissue contracts in this distal-to-proximal direction. This coordination of Ca^{2+} signaling and directional contraction of the spermatheca requires the gap junction protein, INX-12/innexin.[173] Gap junctions are channels composed of connexins (innexins in invertebrates) that directly connect the cytoplasm of neighboring cells and allow the passage of small (<1 kDa) molecules such as Ca^{2+} and IP_3.[201] Gap junctions are critical for coordinated activity of organs such as the heart[202] in higher animals, and similarly, in the function of the spermatheca in *C. elegans*. Although loss of INX-12 disrupts the coordination of Ca^{2+} signaling, individual spermathecal cells do produce stochastic calcium pulses, suggesting that the gap junctions are not needed for Ca^{2+} release per se, but instead are needed for tissue-level coordination.

In addition to communication via gap junctions, physical adhesions between cells play an important role in coordinating tissue responses.[109] Given the strong perturbation of Ca^{2+} signaling in *inx-12* RNAi animals, we hypothesize that the residual, localized coordination of Ca^{2+} signaling in *inx-12* RNAi may be mediated by cell–cell adhesions. For example, the Ca^{2+} signal in one cell may cause that cell to contract, with the strain of cell contraction propagated mechanically to neighbors through cell–cell junctions. In the spermathecal cells, actin fibers are arranged circumferentially, aligning around the cylinder of the organ. Fibers under tension can rapidly transmit mechanical information between cells.[203] Therefore, circumferential communication of mechanical information may be more rapid than lateral communication between cells, which may rely more on diffusion of molecules through gap junctions.

As the egg is expelled, the cells of the spermatheca relax. One possibility is that a contractile force-mediated negative feedback provides a "reset" mechanism to relax the tissue. Contractile forces are unable to alter cell size when physically obstructed by the oocyte. However, as the embryo moves forward, contractile and elastic relaxation forces should productively diminish cell size and reduce the extent of cell stretching. As cells productively contract to expel the oocyte, mechanical negative feedback may promote a return to the relaxed state, perhaps via relaxation of stretched filamin, or other mechanotransducers, and/or by inhibiting myosin.

4. TOOLS FOR MANIPULATION AND IMAGING

The study of mechanotransduction *in vivo* is accelerating, in part due to the development of new visualization tools.

4.1. FRET-based sensors

Fluorescence resonance energy transfer (FRET) has long been used to visualize protein interactions and protein aggregation in many systems,[204] including *C. elegans*, and is now being used as a valuable tool for measuring molecular-level deformations imparted by physical forces. FRET-based sensors embedded in various proteins, including PECAM-1, cadherin, collagen, actinin, filamin, spectrin and vinculin,[104,205–208] have been developed, mainly in cell culture, to image and quantify the magnitude of forces shear stress, stretch, and cell-generated contractile forces. FRET has also been used to detect molecular deformation (strain) in *C. elegans*. An stFRET module, consisting of the fluorescent proteins Cerulean and Venus, separated by an alpha helix derived from spectrin, has been incorporated into the *C. elegans* collagen COL-19.[206] As animals bearing the COL-19 stFRET construct are stretched, the fluorophores move apart and the FRET signal drops in a predictable and reproducible manner.[206]

A similar FRET-based tension sensor, TSmod, has been used in mammalian cell culture to measure forces experienced by focal adhesions in migrating mammalian cells[206,207] and to demonstrate that E-cadherin serves as a conduit for actomyosin-generated tension forces between cells.[104] In TSmod, the elastic linker is derived from spider silk, and the mathematical relationship between FRET and force has been determined. This module can be used to measure forces between 5 and 20 pN. Laser tweezer measurements of purified domains suggest that the conformations of mechanosensitive proteins, such as filamin, are susceptible to ~10 pN forces and that low pN forces can dramatically increase the affinity of these proteins for target peptides.[209,210] TSmod could be used in *C. elegans*, for example, to determine the mechanical stress on the cytoskeleton, or between cells, in tissues such as the spermatheca or the embryonic hypodermis.

FRET-based sensors for small molecules such as ATP[211,212] and cGMP[213] have been used successfully in *C. elegans*. FRET sensors that enable the visualization and quantification of signaling molecules such as IP_3,[214,215] cAMP,[216] and Ca[2217] and activation state of key regulators of cell contractility such as Rho and Rac[218] have been developed in mammalian cells and could be adapted for use in *C. elegans*. These sensors could be used in combination with other fluorescently labeled proteins localized to the actin cytoskeleton or cell membrane, enabling cotemporal investigation of stress and strain, cell shape changes, and signaling events. Such measurements would enable mechanical models of tissue response to mechanical input to be constructed.

4.2. Optogenetics

The field of optogenetics emerged from the discovery of the light-gated ion channels channelrhodopsin-1 (CHR1) and ChR2 in the microalgae *Chlamydomonas reinhardtii*[219,220] and the realization that light could be used to activate neurons engineered to express these proteins.[221] Channelrhodopsins are seven-transmembrane proteins, which when stimulated by blue light rapidly form a nonselective cation channel, leading to membrane depolarization.[222] Cell-specific expression of channelrhodopsin has been used extensively in *C. elegans* to study the neural circuitry controlling physiology and behavior.[223–227] *C. elegans* neurons also express LITE-1, a novel UV-activated G-protein-coupled receptor (GPCR). Similar to channelrhodopsins, heterologous expression of LITE-1 induces contraction of muscle cells, which are not normally light responsive.[228]

The recent use of photoactivatable channels to control muscle cells[229] opens up the possibility of using light-induced cell contraction as a mechanical manipulation in nonneuronal cells. In the *C. elegans* embryo, muscle contraction is required for maturation and strengthening of the FO in the hypodermal cells. Expression of channelrhodopsin in embryonic muscle would enable acute manipulation and a finer-grained and more quantitative analysis of the importance of frequency and amplitude of muscle contraction to embryonic elongation. This strategy could also be used to study internal cell response to contraction. For instance, expressing ChR2 or LITE-1 in the spermatheca would allow the effects of feedback on stretch-activated ion channels, phospholipases, or myosin regulatory proteins to be studied.

Pleiotropic effects or, conversely, genetic redundancy and compensation limit what can be learned from classical genetics and RNAi experiments. New optogenetic tools allow proteins to be activated, inhibited, moved to specific subcellular locations, or brought into complexes using light-activatable domains, and the acute effects of these manipulations can be studied.[230] For example, the light, oxygen, and voltage (LOV) domain found in plant and bacterial phototropins has been engineered into effective light-activated protein switches.[231]

The LOV domain is covalently bound to a flavin cofactor, usually a flavin mononucleotide (FMN).[232] When cells are illuminated with blue light, the FMN forms a covalent bond with a conserved Cys residue in the LOV domain. Bond formation drives conformational changes throughout the LOV domain including unwinding of an alpha helix.[233] The reaction is reversible, and dissociation occurs in seconds to minutes when the light stimulus is removed.[234] This conformational change in the LOV domain has enabled light-activatable protein activation, induced dimerization, and directed subcellular localization.

Enzymes such as the Rac family GTPases can be regulated by LOV fusions.[235,236] When the LOV domain helix unwinds, the domain no longer sterically inhibits Rac, allowing interaction with effectors.[236] Modified LOV domains fused to ePDZ-binding peptides have also been used to direct proteins to various subcellular locations, such as specific plasma membrane domains.[237] In addition, photoactivatable protein pairs have been developed that enable protein interaction and dissociation studies on various time scales. For example, the blue light-stimulated interaction of the *Arabidopsis* proteins FKF1 and GIGANTEA results in an interaction that lasts for around 90 min,[238] whereas stimulation of an interaction between Cryptochrome 2 and CIB1 results in a more transient (minute time scale) interaction.[239]

LOV-based protein interaction strategies have not yet been used in *C. elegans*, however, *C. elegans* cells naturally contain the FMN cofactor, and other LOV-based strategies have been successful. For example, a derivative of LOV engineered from *Arabidopsis* phototropin 2, the mini singlet oxygen generator (miniSOG), generates reactive oxygen species in response to blue light.[240] Originally, miniSOG was used to label cells for three-dimensional protein localization with electron tomography or electron microscopy.[241] Recently, miniSOG has been fused to the synaptic proteins VAMP2 or synaptophysin and used for chromophore-assisted light inactivation (CALI) of synaptic proteins.[242] Illumination of these cells results in the destruction of the fusion protein and reversible inhibition of synaptic transmission. Expression of miniSOG has also been used to selectively kill neurons in *C. elegans*.[243] Therefore, miniSOG enables destruction of specific proteins, or entire cells, in a spatially and temporally precise manner. Minimally invasive removal of specific cells, such as muscle cells or epithelial cells under strain, could be useful for studies of the mechanical properties of tissues in *C. elegans*.

C. elegans, with its transparent body and invariant cell lineages and locations, is an excellent model system for engineering control of cells and tissues in a living organism. Optogenetic strategies allow exquisite control over processes in single cells and the observation of acute responses to manipulations such as protein activation, inactivation, and localization. For example, light-triggered activation of myosin[244] could be induced in specific subcellular domains with cell response to contraction observed in real time. Coanalysis of photoactivatable proteins and the biosensors discussed in the previous section will enable analysis of how signals originating from subcellular locations sites are propagated within cells and how the cellular response to mechanical stimuli is integrated at the tissue level.

5. FUTURE PROSPECTS

Although protocols are being developed for mechanical manipulation of *C. elegans* cells using mechanical compression or magnetic nanoparticles, the physical barrier of the egg shell and/or cuticle can present a challenge to the delivery of precise and directed mechanical forces to *C. elegans* cells. Similarly, disruption of the cytoskeleton or mechanosensitive signaling pathways via delivery of pharmacological inhibitors to specific tissues is a formidable challenge. Because of these obstacles, the strengths of the *C. elegans* system are primarily in imaging, transgenesis, and genetic manipulations that can be designed to target both biochemical signaling and mechanical processes.

A key challenge in mechanotransduction is that the molecular mechanisms by which cells sense and respond to mechanical forces in the living animal remain largely unknown. For example, do a limited number of proteins or domains function as physiologically important mechanotransducers? How dynamic are structures such as the dynein regulatory cassette responsible for spindle positioning, integrin-based dense bodies, or CCCs? Among the many advantages of the *C. elegans* system for gene discovery are a complete genome sequence, an informative bioinformatics database, effective RNAi, and transgenesis strategies, including genome editing with CRISPR/Cas, and a freely available collection of mutant strains.[28,30,245] Biochemical approaches, such as identification of purified proteins by LC/MS, are also powerful tools in this system.[246] This chapter has highlighted a few notable examples of the successes of genetic strategies for identifying novel regulators of mechanotransduction in the zygote, in dendrite extension, and in cadherin function, but this area remains relatively unexplored.

Imaging tools, enabling visualization of the cytoskeleton, signaling processes, lipid constituents, and physical deformation of cells, will be important tools for building an integrated picture of how cells respond to mechanical forces *in vivo*. Many basic questions remain to be answered. For example, is the ability of cells to sense forces distributed uniformly, or are there specific subdomains? How do cells manage feedback to reset to a relaxed state after contracting? How are viscoelastic properties of cells established and how do these physical parameters couple with active processes such as actomyosin contraction and microtubule remodeling to determine the response of cells to mechanical perturbations? Quantitative imaging and mathematical modeling approaches, followed by experimental verification, will make *C. elegans* a powerful integrated system for the investigation of these basic questions.

How forces are coordinated across the cells that comprise organs remains a central question in biology.[10] For example, how do cells physically remodel themselves to drive macroscopic changes in tissue shape, mechanical compliance, etc.? How is immediate response to mechanical and biochemical cues integrated to produce long-term homeostasis and tissue function? Studies in experimentally tractable organisms such as *C. elegans* should improve our understanding of how mechanical inputs are coordinated between cells in living tissues and between tissues to regulate morphogenesis and organ function. The work discussed in this chapter and the development of biosensors for stress, strain, and mechanically activated signal transduction are the foundation for *C. elegans* to make significant contributions to this rapidly evolving field.

ACKNOWLEDGMENTS

Although all errors and omissions are my own, I would like to thank Anand Asthagiri, Jeff Hardin, Max Heiman, Jessica Weaver, and Douglas Finkbeiner for their helpful discussions and for critical reading of the chapter.

REFERENCES

1. Engler AJ, Sen S, Sweeney HL, Discher DE. Matrix elasticity directs stem cell lineage specification. *Cell*. 2006;126(4):677–689.
2. Kumar S, Weaver VM. Mechanics, malignancy, and metastasis: the force journey of a tumor cell. *Cancer Metastasis Rev*. 2009;28(1–2):113–127.
3. Khan S, Sheetz MP. Force effects on biochemical kinetics. *Annu Rev Biochem*. 1997;66:785–805.
4. Evans EA, Calderwood DA. Forces and bond dynamics in cell adhesion. *Science*. 2007;316(5828):1148–1153.

5. Discher DE, Janmey P, Wang YL. Tissue cells feel and respond to the stiffness of their substrate. *Science*. 2005;310(5751):1139–1143.
6. Harris AK, Wild P, Stopak D. Silicone rubber substrata: a new wrinkle in the study of cell locomotion. *Science*. 1980;208(4440):177–179.
7. Sukharev S, Sachs F. Molecular force transduction by ion channels: diversity and unifying principles. *J Cell Sci*. 2012;125(Pt 13):3075–3083.
8. Martinac B. The ion channels to cytoskeleton connection as potential mechanism of mechanosensitivity. *Biochim Biophys Acta*. 2014;1838(2):682–691.
9. Hoffman BD, Grashoff C, Schwartz MA. Dynamic molecular processes mediate cellular mechanotransduction. *Nature*. 2011;475(7356):316–323.
10. Eyckmans J, Boudou T, Yu X, Chen CS. A hitchhiker's guide to mechanobiology. *Dev Cell*. 2011;21(1):35–47.
11. Vogel V. Mechanotransduction involving multimodular proteins: converting force into biochemical signals. *Annu Rev Biophys Biomol Struct*. 2006;35:459–488.
12. Janmey PA, Miller RT. Mechanisms of mechanical signaling in development and disease. *J Cell Sci*. 2011;124(Pt 1):9–18.
13. Asparuhova MB, Gelman L, Chiquet M. Role of the actin cytoskeleton in tuning cellular responses to external mechanical stress. *Scand J Med Sci Sports*. 2009;19(4):490–499.
14. Wang N, Tytell JD, Ingber DE. Mechanotransduction at a distance: mechanically coupling the extracellular matrix with the nucleus. *Nat Rev Mol Cell Biol*. 2009;10(1):75–82.
15. Hardin J. Imaging embryonic morphogenesis in *C. elegans*. *Methods Cell Biol*. 2011;106:377–412.
16. Wuhr M, Obholzer ND, Megason SG, Detrich 3rd HW, Mitchison TJ. Live imaging of the cytoskeleton in early cleavage-stage zebrafish embryos. *Methods Cell Biol*. 2011;101:1–18.
17. Kasza KE, Zallen JA. Dynamics and regulation of contractile actin-myosin networks in morphogenesis. *Curr Opin Cell Biol*. 2011;23(1):30–38.
18. He L, Wang X, Montell DJ. Shining light on *Drosophila oogenesis*: live imaging of egg development. *Curr Opin Genet Dev*. 2011;21(5):612–619.
19. Farge E. Mechanotransduction in development. *Curr Top Dev Biol*. 2011;95:243–265.
20. Gonzalez-Rodriguez D, Guevorkian K, Douezan S, Brochard-Wyart F. Soft matter models of developing tissues and tumors. *Science*. 2012;338(6109):910–917.
21. Guillot C, Lecuit T. Mechanics of epithelial tissue homeostasis and morphogenesis. *Science*. 2013;340(6137):1185–1189.
22. Zhang H, Labouesse M. Signalling through mechanical inputs: a coordinated process. *J Cell Sci*. 2012;125(Pt 13):3039–3049.
23. Lumpkin EA, Marshall KL, Nelson AM. The cell biology of touch. *J Cell Biol*. 2010;191(2):237–248.
24. Chalfie M. Neurosensory mechanotransduction. *Nat Rev Mol Cell Biol*. 2009;10(1):44–52.
25. Bianchi L. Mechanotransduction: touch and feel at the molecular level as modeled in *Caenorhabditis elegans*. *Mol Neurobiol*. 2007;36(3):254–271.
26. Sulston JE, Horvitz HR. Post-embryonic cell lineages of the nematode, *Caenorhabditis elegans*. *Dev Biol*. 1977;56(1):110–156.
27. Brenner S. The genetics of *Caenorhabditis elegans*. *Genetics*. 1974;77(1):71–94.
28. Corsi AK. A biochemist's guide to *Caenorhabditis elegans*. *Anal Biochem*. 2006;359(1):1–17.
29. The *C. elegans* genome sequencing consortium. Genome sequence of the nematode *C. elegans*: a platform for investigating biology. *Science*. 1998;282(5396):2012–2018.

30. Harris TW, Antoshechkin I, Bieri T, et al. WormBase: a comprehensive resource for nematode research. *Nucleic Acids Res.* 2010;38(Database issue):D463–D467.
31. Chalfie M, Tu Y, Euskirchen G, Ward WW, Prasher DC. Green fluorescent protein as a marker for gene expression. *Science.* 1994;263(5148):802–805.
32. Dawes AT, Iron D. Cortical geometry may influence placement of interface between Par protein domains in early *Caenorhabditis elegans* embryos. *J Theor Biol.* 2013;333: 27–37.
33. Gonczy P. Mechanisms of asymmetric cell division: flies and worms pave the way. *Nat Rev Mol Cell Biol.* 2008;9(5):355–366.
34. St Johnston D, Ahringer J. Cell polarity in eggs and epithelia: parallels and diversity. *Cell.* 2010;141(5):757–774.
35. Munro E, Nance J, Priess JR. Cortical flows powered by asymmetrical contraction transport PAR proteins to establish and maintain anterior-posterior polarity in the early *C. elegans* embryo. *Dev Cell.* 2004;7(3):413–424.
36. Marston DJ, Goldstein B. Symmetry breaking in *C. elegans*: another gift from the sperm. *Dev Cell.* 2006;11(3):273–274.
37. Hird SN, White JG. Cortical and cytoplasmic flow polarity in early embryonic cells of *Caenorhabditis elegans*. *J Cell Biol.* 1993;121(6):1343–1355.
38. Mayer M, Depken M, Bois JS, Julicher F, Grill SW. Anisotropies in cortical tension reveal the physical basis of polarizing cortical flows. *Nature.* 2010;467(7315):617–621.
39. Sumiyoshi E, Sugimoto A. Cell polarity: centrosomes release signals for polarization. *Curr Biol.* 2012;22(8):R281–R283.
40. Boyd L, Guo S, Levitan D, Stinchcomb DT, Kemphues KJ. PAR-2 is asymmetrically distributed and promotes association of P granules and PAR-1 with the cortex in *C. elegans* embryos. *Development.* 1996;122(10):3075–3084.
41. Watts JL, Etemad-Moghadam B, Guo S, et al. par-6, a gene involved in the establishment of asymmetry in early *C. elegans* embryos, mediates the asymmetric localization of PAR-3. *Development.* 1996;122(10):3133–3140.
42. Goehring NW, Trong PK, Bois JS, et al. Polarization of PAR proteins by advective triggering of a pattern-forming system. *Science.* 2011;334(6059):1137–1141.
43. Cuenca AA, Schetter A, Aceto D, Kemphues K, Seydoux G. Polarization of the *C. elegans* zygote proceeds via distinct establishment and maintenance phases. *Development.* 2003;130(7):1255–1265.
44. Fievet BT, Rodriguez J, Naganathan S, et al. Systematic genetic interaction screens uncover cell polarity regulators and functional redundancy. *Nat Cell Biol.* 2013;15(1):103–112.
45. Oegema K, Hyman AA. Cell division. In: *WormBook*; 2006:1–40.
46. Galli M, van den Heuvel S. Determination of the cleavage plane in early *C. elegans* embryos. *Annu Rev Genet.* 2008;42:389–411.
47. Kotak S, Gonczy P. Mechanisms of spindle positioning: cortical force generators in the limelight. *Curr Opin Cell Biol.* 2013;25(6):741–748.
48. Harris TJ, Sawyer JK, Peifer M. How the cytoskeleton helps build the embryonic body plan: models of morphogenesis from *Drosophila*. *Curr Top Dev Biol.* 2009;89:55–85.
49. Dawes-Hoang RE, Parmar KM, Christiansen AE, et al. folded gastrulation, cell shape change and the control of myosin localization. *Development.* 2005;132(18):4165–4178.
50. Fletcher DA, Mullins RD. Cell mechanics and the cytoskeleton. *Nature.* 2010;463(7280):485–492.
51. Levayer R, Lecuit T. Biomechanical regulation of contractility: spatial control and dynamics. *Trends Cell Biol.* 2012;22(2):61–81.
52. Paluch E, Heisenberg CP. Biology and physics of cell shape changes in development. *Curr Biol.* 2009;19(17):R790–R799.

53. Pek YS, Wan AC, Ying JY. The effect of matrix stiffness on mesenchymal stem cell differentiation in a 3D thixotropic gel. *Biomaterials*. 2010;31(3):385–391.
54. Ruiz SA, Chen CS. Emergence of patterned stem cell differentiation within multicellular structures. *Stem Cells*. 2008;26(11):2921–2927.
55. Farge E. Mechanical induction of Twist in the *Drosophila* foregut/stomodeal primordium. *Curr Biol*. 2003;13(16):1365–1377.
56. Somogyi K, Rorth P. Evidence for tension-based regulation of *Drosophila* MAL and SRF during invasive cell migration. *Dev Cell*. 2004;7(1):85–93.
57. Guilak F, Cohen DM, Estes BT, et al. Control of stem cell fate by physical interactions with the extracellular matrix. *Cell Stem Cell*. 2009;5(1):17–26.
58. Meyers SG, Corsi AK. *C. elegans* twist gene expression in differentiated cell types is controlled by autoregulation through intron elements. *Dev Biol*. 2010;346(2):224–236.
59. Harfe BD, Vaz Gomes A, Kenyon C, et al. Analysis of a *Caenorhabditis elegans* Twist homolog identifies conserved and divergent aspects of mesodermal patterning. *Genes Dev*. 1998;12(16):2623–2635.
60. Philogene MC, Small SG, Wang P, Corsi AK. Distinct *Caenorhabditis elegans* HLH-8/twist-containing dimers function in the mesoderm. *Dev Dyn*. 2012;241(3):481–492.
61. Pan D. The hippo signaling pathway in development and cancer. *Dev Cell*. 2010;19(4):491–505.
62. Halder G, Johnson RL. Hippo signaling: growth control and beyond. *Development*. 2011;138(1):9–22.
63. Kim NG, Koh E, Chen X, Gumbiner BM. E-cadherin mediates contact inhibition of proliferation through Hippo signaling-pathway components. *Proc Natl Acad Sci USA*. 2011;108(29):11930–11935.
64. Chopra A, Tabdanov E, Patel H, Janmey PA, Kresh JY. Cardiac myocyte remodeling mediated by N-cadherin-dependent mechanosensing. *Am J Physiol Heart Circ Physiol*. 2011;300(4):H1252–H1266.
65. Mendez MG, Janmey PA. Transcription factor regulation by mechanical stress. *Int J Biochem Cell Biol*. 2012;44(5):728–732.
66. Dupont S, Morsut L, Aragona M, et al. Role of YAP/TAZ in mechanotransduction. *Nature*. 2011;474(7350):179–183.
67. Hilman D, Gat U. The evolutionary history of YAP and the hippo/YAP pathway. *Mol Biol Evol*. 2011;28(8):2403–2417.
68. Iwasa H, Maimaiti S, Kuroyanagi H, et al. Yes-associated protein homolog, YAP-1, is involved in the thermotolerance and aging in the nematode *Caenorhabditis elegans*. *Exp Cell Res*. 2013;319(7):931–945.
69. Ross TD, Coon BG, Yun S, et al. Integrins in mechanotransduction. *Curr Opin Cell Biol*. 2013;25(5):613–618.
70. Boettiger D. Mechanical control of integrin-mediated adhesion and signaling. *Curr Opin Cell Biol*. 2012;24(5):592–599.
71. Schiller HB, Fassler R. Mechanosensitivity and compositional dynamics of cell-matrix adhesions. *EMBO Rep*. 2013;14(6):509–519.
72. Zhang H, Labouesse M. The making of hemidesmosome structures in vivo. *Dev Dyn*. 2010;239(5):1465–1476.
73. Hu P, Luo BH. Integrin bi-directional signaling across the plasma membrane. *J Cell Physiol*. 2013;228(2):306–312.
74. Ye F, Kim C, Ginsberg MH. Molecular mechanism of inside-out integrin regulation. *J Thromb Haemost*. 2011;9(Suppl 1):20–25.
75. Gettner SN, Kenyon C, Reichardt LF. Characterization of beta pat-3 heterodimers, a family of essential integrin receptors in *C. elegans*. *J Cell Biol*. 1995;129(4):1127–1141.
76. Baum PD, Garriga G. Neuronal migrations and axon fasciculation are disrupted in ina-1 integrin mutants. *Neuron*. 1997;19(1):51–62.

77. Hresko MC, Williams BD, Waterston RH. Assembly of body wall muscle and muscle cell attachment structures in *Caenorhabditis elegans*. *J Cell Biol*. 1994;124(4):491–506.
78. Barstead RJ, Waterston RH. The basal component of the nematode dense-body is vinculin. *J Biol Chem*. 1989;264(17):10177–10185.
79. Moulder GL, Huang MM, Waterston RH, Barstead RJ. Talin requires beta-integrin, but not vinculin, for its assembly into focal adhesion-like structures in the nematode *Caenorhabditis elegans*. *Mol Biol Cell*. 1996;7(8):1181–1193.
80. Cram EJ, Clark SG, Schwarzbauer JE. Talin loss-of-function uncovers roles in cell contractility and migration in *C. elegans*. *J Cell Sci*. 2003;116(Pt 19):3871–3878.
81. Lecroisey C, Brouilly N, Qadota H, et al. ZYX-1, the unique zyxin protein of *Caenorhabditis elegans*, is involved in dystrophin-dependent muscle degeneration. *Mol Biol Cell*. 2013;24(8):1232–1249.
82. Hobert O, Moerman DG, Clark KA, Beckerle MC, Ruvkun G. A conserved LIM protein that affects muscular adherens junction integrity and mechanosensory function in *Caenorhabditis elegans*. *J Cell Biol*. 1999;144(1):45–57.
83. Williams BD, Waterston RH. Genes critical for muscle development and function in *Caenorhabditis elegans* identified through lethal mutations. *J Cell Biol*. 1994;124(4):475–490.
84. Moerman DG, Williams BD. Sarcomere assembly in *C. elegans* muscle. *WormBook*. 2006;1–16.
85. Brown NH. Cell-cell adhesion via the ECM: integrin genetics in fly and worm. *Matrix Biol*. 2000;19(3):191–201.
86. Labouesse M. Role of the extracellular matrix in epithelial morphogenesis: a view from *C. elegans*. *Organogenesis*. 2012;8(2):65–70.
87. Inglis PN, Ou G, Leroux MR, Scholey JM. The sensory cilia of *Caenorhabditis elegans*. *WormBook*. 2007;1–22.
88. Ward S, Thomson N, White JG, Brenner S. Electron microscopical reconstruction of the anterior sensory anatomy of the nematode *Caenorhabditis elegans*? *J Comp Neurol*. 1975;160(3):313–337.
89. Heiman MG, Shaham S. DEX-1 and DYF-7 establish sensory dendrite length by anchoring dendritic tips during cell migration. *Cell*. 2009;137(2):344–355.
90. Buffone MG, Foster JA, Gerton GL. The role of the acrosomal matrix in fertilization. *Int J Dev Biol*. 2008;52(5–6):511–522.
91. Legan PK, Rau A, Keen JN, Richardson GP. The mouse tectorins. Modular matrix proteins of the inner ear homologous to components of the sperm-egg adhesion system. *J Biol Chem*. 1997;272(13):8791–8801.
92. Monne M, Jovine L. A structural view of egg coat architecture and function in fertilization. *Biol Reprod*. 2011;85(4):661–669.
93. McLachlan IG, Heiman MG. Shaping dendrites with machinery borrowed from epithelia. *Curr Opin Neurobiol*. 2013;23(6):1005–1010.
94. Plaza S, Chanut-Delalande H, Fernandes I, Wassarman PM, Payre F. From A to Z: apical structures and zona pellucida-domain proteins. *Trends Cell Biol*. 2010;20(9): 524–532.
95. Jazwinska A, Ribeiro C, Affolter M. Epithelial tube morphogenesis during *Drosophila* tracheal development requires Piopio, a luminal ZP protein. *Nat Cell Biol*. 2003;5(10):895–901.
96. Kalamajski S, Oldberg A. The role of small leucine-rich proteoglycans in collagen fibrillogenesis. *Matrix Biol*. 2010;29(4):248–253.
97. de Wit J, Sylwestrak E, O'Sullivan ML, et al. LRRTM2 interacts with Neurexin1 and regulates excitatory synapse formation. *Neuron*. 2009;64(6):799–806.
98. Ko J, Fuccillo MV, Malenka RC, Sudhof TC. LRRTM2 functions as a neurexin ligand in promoting excitatory synapse formation. *Neuron*. 2009;64(6):791–798.

99. Siddiqui TJ, Pancaroglu R, Kang Y, Rooyakkers A, Craig AM. LRRTMs and neuroligins bind neurexins with a differential code to cooperate in glutamate synapse development. *J Neurosci*. 2010;30(22):7495–7506.
100. Mancuso VP, Parry JM, Storer L, et al. Extracellular leucine-rich repeat proteins are required to organize the apical extracellular matrix and maintain epithelial junction integrity in *C. elegans*. *Development*. 2012;139(5):979–990.
101. Cox EA, Hardin J. Sticky worms: adhesion complexes in *C. elegans*. *J Cell Sci*. 2004;117(Pt 10):1885–1897.
102. Segbert C, Johnson K, Theres C, van Furden D, Bossinger O. Molecular and functional analysis of apical junction formation in the gut epithelium of *Caenorhabditis elegans*. *Dev Biol*. 2004;266(1):17–26.
103. Dupin I, Camand E, Etienne-Manneville S. Classical cadherins control nucleus and centrosome position and cell polarity. *J Cell Biol*. 2009;185(5):779–786.
104. Borghi N, Sorokina M, Shcherbakova OG, et al. E-cadherin is under constitutive actomyosin-generated tension that is increased at cell-cell contacts upon externally applied stretch. *Proc Natl Acad Sci USA*. 2012;109(31):12568–12573.
105. Budnar S, Yap AS. A mechanobiological perspective on cadherins and the actin-myosin cytoskeleton. *F1000Prime Rep*. 2013;5:35.
106. Marcinkevicius E, Zallen JA. Regulation of cytoskeletal organization and junctional remodeling by the atypical cadherin Fat. *Development*. 2013;140(2):433–443.
107. Martin AC, Gelbart M, Fernandez-Gonzalez R, Kaschube M, Wieschaus EF. Integration of contractile forces during tissue invagination. *J Cell Biol*. 2010;188(5):735–749.
108. Loveless T, Hardin J. Cadherin complexity: recent insights into cadherin superfamily function in *C. elegans*. *Curr Opin Cell Biol*. 2012;24(5):695–701.
109. Collinet C, Lecuit T. Stability and dynamics of cell-cell junctions. *Prog Mol Biol Transl Sci*. 2013;116:25–47.
110. Kim JH, Dooling LJ, Asthagiri AR. Intercellular mechanotransduction during multicellular morphodynamics. *J R Soc Interface*. 2010;7(Suppl 3):S341–S350.
111. Yonemura S. Cadherin-actin interactions at adherens junctions. *Curr Opin Cell Biol*. 2011;23(5):515–522.
112. Pettitt J. The cadherin superfamily. *WormBook*. 2005;1–9.
113. Hardin J, Lynch A, Loveless T, Pettitt J. Cadherins and their partners in the nematode worm *Caenorhabditis elegans*. *Prog Mol Biol Transl Sci*. 2013;116:239–262.
114. Costa M, Raich W, Agbunag C, et al. A putative catenin-cadherin system mediates morphogenesis of the *Caenorhabditis elegans* embryo. *J Cell Biol*. 1998;141(1):297–308.
115. Raich WB, Agbunag C, Hardin J. Rapid epithelial-sheet sealing in the *Caenorhabditis elegans* embryo requires cadherin-dependent filopodial priming. *Curr Biol*. 1999;9(20):1139–1146.
116. Grana TM, Cox EA, Lynch AM, Hardin J. SAX-7/L1CAM and HMR-1/cadherin function redundantly in blastomere compaction and non-muscle myosin accumulation during *Caenorhabditis elegans* gastrulation. *Dev Biol*. 2010;344(2):731–744.
117. Firestein BL, Rongo C. DLG-1 is a MAGUK similar to SAP97 and is required for adherens junction formation. *Mol Biol Cell*. 2001;12(11):3465–3475.
118. Chisholm AD, Hardin J. Epidermal morphogenesis. *WormBook*. 2005;1–22.
119. Piekny AJ, Mains PE. Rho-binding kinase (LET-502) and myosin phosphatase (MEL-11) regulate cytokinesis in the early *Caenorhabditis elegans* embryo. *J Cell Sci*. 2002;115(Pt 11):2271–2282.
120. Diogon M, Wissler F, Quintin S, et al. The RhoGAP RGA-2 and LET-502/ROCK achieve a balance of actomyosin-dependent forces in *C. elegans* epidermis to control morphogenesis. *Development*. 2007;134(13):2469–2479.

121. Kwiatkowski AV, Maiden SL, Pokutta S, et al. In vitro and in vivo reconstitution of the cadherin-catenin-actin complex from *Caenorhabditis elegans*. *Proc Natl Acad Sci USA*. 2010;107(33):14591–14596.
122. Ladoux B, Anon E, Lambert M, et al. Strength dependence of cadherin-mediated adhesions. *Biophys J*. 2010;98(4):534–542.
123. Yamada S, Nelson WJ. Localized zones of Rho and Rac activities drive initiation and expansion of epithelial cell-cell adhesion. *J Cell Biol*. 2007;178(3):517–527.
124. Kametani Y, Takeichi M. Basal-to-apical cadherin flow at cell junctions. *Nat Cell Biol*. 2007;9(1):92–98.
125. Maddugoda MP, Crampton MS, Shewan AM, Yap AS. Myosin VI and vinculin cooperate during the morphogenesis of cadherin cell cell contacts in mammalian epithelial cells. *J Cell Biol*. 2007;178(3):529–540.
126. Bosveld F, Bonnet I, Guirao B, et al. Mechanical control of morphogenesis by Fat/Dachsous/Four-jointed planar cell polarity pathway. *Science*. 2012;336(6082):724–727.
127. Cox-Paulson EA, Walck-Shannon E, Lynch AM, et al. Tropomodulin protects alpha-catenin-dependent junctional-actin networks under stress during epithelial morphogenesis. *Curr Biol*. 2012;22(16):1500–1505.
128. Lockwood C, Zaidel-Bar R, Hardin J. The *C. elegans* zonula occludens ortholog cooperates with the cadherin complex to recruit actin during morphogenesis. *Curr Biol*. 2008;18(17):1333–1337.
129. Pettitt J, Cox EA, Broadbent ID, Flett A, Hardin J. The *Caenorhabditis elegans* p120 catenin homologue, JAC-1, modulates cadherin-catenin function during epidermal morphogenesis. *J Cell Biol*. 2003;162(1):15–22.
130. Zaidel-Bar R, Joyce MJ, Lynch AM, et al. The F-BAR domain of SRGP-1 facilitates cell-cell adhesion during *C. elegans* morphogenesis. *J Cell Biol*. 2010;191(4):761–769.
131. Yamashiro S, Gokhin DS, Kimura S, Nowak RB, Fowler VM. Tropomodulins: pointed-end capping proteins that regulate actin filament architecture in diverse cell types. *Cytoskeleton (Hoboken)*. 2012;69(6):337–370.
132. Yap AS, Niessen CM, Gumbiner BM. The juxtamembrane region of the cadherin cytoplasmic tail supports lateral clustering, adhesive strengthening, and interaction with p120ctn. *J Cell Biol*. 1998;141(3):779–789.
133. Thoreson MA, Anastasiadis PZ, Daniel JM, et al. Selective uncoupling of p120(ctn) from E-cadherin disrupts strong adhesion. *J Cell Biol*. 2000;148(1):189–202.
134. Anastasiadis PZ, Moon SY, Thoreson MA, et al. Inhibition of RhoA by p120 catenin. *Nat Cell Biol*. 2000;2(9):637–644.
135. Suetsugu S, Toyooka K, Senju Y. Subcellular membrane curvature mediated by the BAR domain superfamily proteins. *Semin Cell Dev Biol*. 2010;21(4):340–349.
136. Neukomm LJ, Frei AP, Cabello J, et al. Loss of the RhoGAP SRGP-1 promotes the clearance of dead and injured cells in *Caenorhabditis elegans*. *Nat Cell Biol*. 2011;13(1):79–86.
137. Papusheva E, Heisenberg CP. Spatial organization of adhesion: force-dependent regulation and function in tissue morphogenesis. *EMBO J*. 2010;29(16):2753–2768.
138. Gomez GA, McLachlan RW, Yap AS. Productive tension: force-sensing and homeostasis of cell-cell junctions. *Trends Cell Biol*. 2011;21(9):499–505.
139. Lecuit T, Lenne PF. Cell surface mechanics and the control of cell shape, tissue patterns and morphogenesis. *Nat Rev Mol Cell Biol*. 2007;8(8):633–644.
140. Zhang H, Landmann F, Zahreddine H, et al. A tension-induced mechanotransduction pathway promotes epithelial morphogenesis. *Nature*. 2011;471(7336):99–103.
141. Hardin J. Mechanotransduction: getting morphogenesis down pat. *Curr Biol*. 2011;21(9):R309–R311.
142. Hresko MC, Schriefer LA, Shrimankar P, Waterston RH. Myotactin, a novel hypodermal protein involved in muscle-cell adhesion in *Caenorhabditis elegans*. *J Cell Biol*. 1999;146(3):659–672.

143. Bosher JM, Hahn BS, Legouis R, et al. The *Caenorhabditis elegans* vab-10 spectraplakin isoforms protect the epidermis against internal and external forces. *J Cell Biol*. 2003;161(4):757–768.
144. Ding M, King RS, Berry EC, et al. The cell signaling adaptor protein EPS-8 is essential for *C. elegans* epidermal elongation and interacts with the ankyrin repeat protein VAB-19. *PLoS One*. 2008;3(10):e3346.
145. Zahreddine H, Zhang H, Diogon M, Nagamatsu Y, Labouesse M. CRT-1/calreticulin and the E3 ligase EEL-1/HUWE1 control hemidesmosome maturation in *C. elegans* development. *Curr Biol*. 2010;20(4):322–327.
146. Fotopoulos N, Wernike D, Chen Y, et al. *Caenorhabditis elegans* anillin (ani-1) regulates neuroblast cytokinesis and epidermal morphogenesis during embryonic development. *Dev Biol*. 2013;383(1):61–74.
147. Samakovlis C, Hacohen N, Manning G, et al. Development of the *Drosophila* tracheal system occurs by a series of morphologically distinct but genetically coupled branching events. *Development*. 1996;122(5):1395–1407.
148. Herwig L, Blum Y, Krudewig A, et al. Distinct cellular mechanisms of blood vessel fusion in the zebrafish embryo. *Curr Biol*. 2011;21(22):1942–1948.
149. Buechner M. Tubes and the single *C. elegans* excretory cell. *Trends Cell Biol*. 2002;12(10):479–484.
150. Nelson FK, Riddle DL. Functional study of the *Caenorhabditis elegans* secretory-excretory system using laser microsurgery. *J Exp Zool*. 1984;231(1):45–56.
151. Berry KL, Bulow HE, Hall DH, Hobert O. A *C. elegans* CLIC-like protein required for intracellular tube formation and maintenance. *Science*. 2003;302(5653): 2134–2137.
152. Kolotuev I, Hyenne V, Schwab Y, Rodriguez D, Labouesse M. A pathway for unicellular tube extension depending on the lymphatic vessel determinant Prox1 and on osmoregulation. *Nat Cell Biol*. 2013;15(2):157–168.
153. Khan LA, Zhang H, Abraham N, et al. Intracellular lumen extension requires ERM-1-dependent apical membrane expansion and AQP-8-mediated flux. *Nat Cell Biol*. 2013;15(2):143–156.
154. Gobel V, Barrett PL, Hall DH, Fleming JT. Lumen morphogenesis in *C. elegans* requires the membrane-cytoskeleton linker erm-1. *Dev Cell*. 2004;6(6):865–873.
155. Saotome I, Curto M, McClatchey AI. Ezrin is essential for epithelial organization and villus morphogenesis in the developing intestine. *Dev Cell*. 2004;6(6):855–864.
156. Maurel C, Verdoucq L, Luu DT, Santoni V. Plant aquaporins: membrane channels with multiple integrated functions. *Annu Rev Plant Biol*. 2008;59:595–624.
157. Stewart MP, Helenius J, Toyoda Y, et al. Hydrostatic pressure and the actomyosin cortex drive mitotic cell rounding. *Nature*. 2011;469(7329):226–230.
158. Brown D. The ins and outs of aquaporin-2 trafficking. *Am J Physiol Renal Physiol*. 2003;284(5):F893–F901.
159. Sugiya H, Matsuki-Fukushima M, Hashimoto S. Role of aquaporins and regulation of secretory vesicle volume in cell secretion. *J Cell Mol Med*. 2008;12(5A):1486–1494.
160. Kimble J, Hirsh D. The postembryonic cell lineages of the hermaphrodite and male gonads in *Caenorhabditis elegans*. *Dev Biol*. 1979;70(2):396–417.
161. McCarter J, Bartlett B, Dang T, Schedl T. Soma-germ cell interactions in *Caenorhabditis elegans*: multiple events of hermaphrodite germline development require the somatic sheath and spermathecal lineages. *Dev Biol*. 1997;181(2):121–143.
162. McCarter J, Bartlett B, Dang T, Schedl T. On the control of oocyte meiotic maturation and ovulation in *Caenorhabditis elegans*. *Dev Biol*. 1999;205(1):111–128.
163. Miller MA, Nguyen VQ, Lee MH, et al. A sperm cytoskeletal protein that signals oocyte meiotic maturation and ovulation. *Science*. 2001;291(5511):2144–2147.

164. Clandinin TR, DeModena JA, Sternberg PW. Inositol trisphosphate mediates a RAS-independent response to LET-23 receptor tyrosine kinase activation in *C. elegans*. *Cell*. 1998;92(4):523–533.
165. Yin X, Gower NJ, Baylis HA, Strange K. Inositol 1,4,5-trisphosphate signaling regulates rhythmic contractile activity of myoepithelial sheath cells in *Caenorhabditis elegans*. *Mol Biol Cell*. 2004;15(8):3938–3949.
166. Kariya K, Bui YK, Gao X, Sternberg PW, Kataoka T. Phospholipase Cepsilon regulates ovulation in *Caenorhabditis elegans*. *Dev Biol*. 2004;274(1):201–210.
167. Kim H, McCulloch CA. Filamin A mediates interactions between cytoskeletal proteins that control cell adhesion. *FEBS Lett*. 2011;585(1):18–22.
168. Djinovic-Carugo K, Carugo O. Structural portrait of filamin interaction mechanisms. *Curr Protein Pept Sci*. 2010;11(7):639–650.
169. Gorlin JB, Yamin R, Egan S, et al. Human endothelial actin-binding protein (ABP-280, nonmuscle filamin): a molecular leaf spring. *J Cell Biol*. 1990;111(3):1089–1105.
170. Stossel TP, Condeelis J, Cooley L, et al. Filamins as integrators of cell mechanics and signalling. *Nat Rev Mol Cell Biol*. 2001;2(2):138–145.
171. Himmel M, Van Der Ven PF, Stocklein W, Furst DO. The limits of promiscuity: isoform-specific dimerization of filamins. *Biochemistry*. 2003;42(2):430–439.
172. Kovacevic I, Cram EJ. FLN-1/filamin is required for maintenance of actin and exit of fertilized oocytes from the spermatheca in *C. elegans*. *Dev Biol*. 2010;347(2):247–257.
173. Kovacevic I, Orozco JM, Cram EJ. Filamin and phospholipase C-epsilon are required for calcium signaling in the *Caenorhabditis elegans* spermatheca. *PLoS Genet*. 2013;9(5): e1003510.
174. Razinia Z, Makela T, Ylanne J, Calderwood DA. Filamins in mechanosensing and signaling. *Annu Rev Biophys*. 2012;41:227–246.
175. Rhee SG. Regulation of phosphoinositide-specific phospholipase C. *Annu Rev Biochem*. 2001;70:281–312.
176. Bunney TD, Katan M. Phospholipase C epsilon: linking second messengers and small GTPases. *Trends Cell Biol*. 2006;16(12):640–648.
177. Bunney TD, Harris R, Gandarillas NL, et al. Structural and mechanistic insights into ras association domains of phospholipase C epsilon. *Mol Cell*. 2006;21(4):495–507.
178. Kelley GG, Reks SE, Ondrako JM, Smrcka AV. Phospholipase C(epsilon): a novel Ras effector. *EMBO J*. 2001;20(4):743–754.
179. Seifert JP, Zhou Y, Hicks SN, Sondek J, Harden TK. Dual activation of phospholipase C-epsilon by Rho and Ras GTPases. *J Biol Chem*. 2008;283(44):29690–29698.
180. Shibatohge M, Kariya K, Liao Y, et al. Identification of PLC210, a *Caenorhabditis elegans* phospholipase C, as a putative effector of Ras. *J Biol Chem*. 1998;273(11):6218–6222.
181. Nakamura F, Osborn E, Janmey PA, Stossel TP. Comparison of filamin A-induced cross-linking and Arp2/3 complex-mediated branching on the mechanics of actin filaments. *J Biol Chem*. 2002;277(11):9148–9154.
182. Nakamura F, Stossel TP, Hartwig JH. The filamins: organizers of cell structure and function. *Cell Adh Migr*. 2011;5(2):160–169.
183. Calderwood DA, Huttenlocher A, Kiosses WB, et al. Increased filamin binding to beta-integrin cytoplasmic domains inhibits cell migration. *Nat Cell Biol*. 2001;3(12): 1060–1068.
184. Popowicz GM, Schleicher M, Noegel AA, Holak TA. Filamins: promiscuous organizers of the cytoskeleton. *Trends Biochem Sci*. 2006;31(7):411–419.
185. Feng Y, Walsh CA. The many faces of filamin: a versatile molecular scaffold for cell motility and signalling. *Nat Cell Biol*. 2004;6(11):1034–1038.
186. Zhou AX, Hartwig JH, Akyurek LM. Filamins in cell signaling, transcription and organ development. *Trends Cell Biol*. 2010;20(2):113–123.

187. Gehler S, Baldassarre M, Lad Y, et al. Filamin A-beta1 integrin complex tunes epithelial cell response to matrix tension. *Mol Biol Cell.* 2009;20(14):3224–3238.
188. Kainulainen T, Pender A, D'Addario M, et al. Cell death and mechanoprotection by filamin a in connective tissues after challenge by applied tensile forces. *J Biol Chem.* 2002;277(24):21998–22009.
189. Shifrin Y, Arora PD, Ohta Y, Calderwood DA, McCulloch CA. The role of FilGAP-filamin A interactions in mechanoprotection. *Mol Biol Cell.* 2009;20(5):1269–1279.
190. Ehrlicher AJ, Nakamura F, Hartwig JH, Weitz DA, Stossel TP. Mechanical strain in actin networks regulates FilGAP and integrin binding to filamin A. *Nature.* 2011;478(7368):260–263.
191. Nakamura F, Osborn TM, Hartemink CA, Hartwig JH, Stossel TP. Structural basis of filamin A functions. *J Cell Biol.* 2007;179(5):1011–1025.
192. Ithychanda SS, Qin J. Evidence for multisite ligand binding and stretching of filamin by integrin and migfilin. *Biochemistry.* 2011;50(20):4229–4231.
193. Mao T, O'Connor DH, Scheuss V, Nakai J, Svoboda K. Characterization and subcellular targeting of GCaMP-type genetically-encoded calcium indicators. *PLoS One.* 2008;3(3):e1796.
194. Tian L, Hires SA, Mao T, et al. Imaging neural activity in worms, flies and mice with improved GCaMP calcium indicators. *Nat Methods.* 2009;6(12):875–881.
195. Suh PG, Park JI, Manzoli L, et al. Multiple roles of phosphoinositide-specific phospholipase C isozymes. *BMB Rep.* 2008;41(6):415–434.
196. Berridge MJ. Inositol trisphosphate and calcium signalling mechanisms. *Biochim Biophys Acta.* 2009;1793(6):933–940.
197. Wissmann A, Ingles J, Mains PE. The *Caenorhabditis elegans* mel-11 myosin phosphatase regulatory subunit affects tissue contraction in the somatic gonad and the embryonic epidermis and genetically interacts with the Rac signaling pathway. *Dev Biol.* 1999;209(1):111–127.
198. Piekny AJ, Johnson JL, Cham GD, Mains PE. The *Caenorhabditis elegans* nonmuscle myosin genes nmy-1 and nmy-2 function as redundant components of the let-502/Rho-binding kinase and mel-11/myosin phosphatase pathway during embryonic morphogenesis. *Development.* 2003;130(23):5695–5704.
199. Piekny AJ, Wissmann A, Mains PE. Embryonic morphogenesis in *Caenorhabditis elegans* integrates the activity of LET-502 Rho-binding kinase, MEL-11 myosin phosphatase, DAF-2 insulin receptor and FEM-2 PP2c phosphatase. *Genetics.* 2000;156(4):1671–1689.
200. Gally C, Wissler F, Zahreddine H, et al. Myosin II regulation during *C. elegans* embryonic elongation: LET-502/ROCK, MRCK-1 and PAK-1, three kinases with different roles. *Development.* 2009;136(18):3109–3119.
201. Herve JC, Derangeon M. Gap-junction-mediated cell-to-cell communication. *Cell Tissue Res.* 2013;352(1):21–31.
202. Delmar M, Makita N. Cardiac connexins, mutations and arrhythmias. *Curr Opin Cardiol.* 2012;27(3):236–241.
203. Burridge K, Wittchen ES. The tension mounts: stress fibers as force-generating mechanotransducers. *J Cell Biol.* 2013;200(1):9–19.
204. Fernandez-Duenas V, Llorente J, Gandia J, et al. Fluorescence resonance energy transfer-based technologies in the study of protein-protein interactions at the cell surface. *Methods.* 2012;57(4):467–472.
205. Conway DE, Breckenridge MT, Hinde E, et al. Fluid shear stress on endothelial cells modulates mechanical tension across VE-cadherin and PECAM-1. *Curr Biol.* 2013;23(11):1024–1030.
206. Meng F, Suchyna TM, Lazakovitch E, Gronostajski RM, Sachs F. Real time FRET based detection of mechanical stress in cytoskeletal and extracellular matrix proteins. *Cell Mol Bioeng.* 2011;4(2):148–159.

207. Grashoff C, Hoffman BD, Brenner MD, et al. Measuring mechanical tension across vinculin reveals regulation of focal adhesion dynamics. *Nature*. 2010;466:263–267.
208. Austen K, Kluger C, Freikamp A, Chrostek-Grashoff A, Grashoff C. Generation and analysis of biosensors to measure mechanical forces within cells. *Methods Mol Biol*. 2013;1066:169–184.
209. Rognoni L, Stigler J, Pelz B, Ylanne J, Rief M. Dynamic force sensing of filamin revealed in single-molecule experiments. *Proc Natl Acad Sci USA*. 2012;109(48): 19679–19684.
210. Chen H, Zhu X, Cong P, et al. Differential mechanical stability of filamin A rod segments. *Biophys J*. 2011;101(5):1231–1237.
211. Kishikawa J, Fujikawa M, Imamura H, et al. MRT letter: expression of ATP sensor protein in *Caenorhabditis elegans*. *Microsc Res Tech*. 2012;75(1):15–19.
212. Tsuyama T, Kishikawa J, Han YW, et al. In vivo fluorescent adenosine 5′-triphosphate (ATP) imaging of *Drosophila* melanogaster and *Caenorhabditis elegans* by using a genetically encoded fluorescent ATP biosensor optimized for low temperatures. *Anal Chem*. 2013;85(16):7889–7896.
213. Couto A, Oda S, Nikolaev VO, Soltesz Z, de Bono M. In vivo genetic dissection of O2-evoked cGMP dynamics in a *Caenorhabditis elegans* gas sensor. *Proc Natl Acad Sci USA*. 2013;110(35):E3301–E3310.
214. Tanimura A, Morita T, Nezu A, Tojyo Y. Monitoring of IP3 dynamics during Ca2+ oscillations in HSY human parotid cell line with FRET-based IP3 biosensors. *J Med Investig*. 2009;56(Suppl):357–361.
215. Nezu A, Tanimura A, Morita T, Tojyo Y. Visualization of Ins(1,4,5)P3 dynamics in living cells: two distinct pathways for Ins(1,4,5)P3 generation following mechanical stimulation of HSY-EA1 cells. *J Cell Sci*. 2010;123(Pt 13):2292–2298.
216. Borner S, Schwede F, Schlipp A, et al. FRET measurements of intracellular cAMP concentrations and cAMP analog permeability in intact cells. *Nat Protoc*. 2011;6(4): 427–438.
217. Miyawaki A, Llopis J, Heim R, et al. Fluorescent indicators for Ca2+ based on green fluorescent proteins and calmodulin. *Nature*. 1997;388(6645):882–887.
218. Yoshizaki H, Ohba Y, Kurokawa K, et al. Activity of Rho-family GTPases during cell division as visualized with FRET-based probes. *J Cell Biol*. 2003;162(2):223–232.
219. Nagel G, Ollig D, Fuhrmann M, et al. Channelrhodopsin-1: a light-gated proton channel in green algae. *Science*. 2002;296(5577):2395–2398.
220. Nagel G, Szellas T, Huhn W, et al. Channelrhodopsin-2, a directly light-gated cation-selective membrane channel. *Proc Natl Acad Sci USA*. 2003;100(24):13940–13945.
221. Boyden ES, Zhang F, Bamberg E, Nagel G, Deisseroth K. Millisecond-timescale, genetically targeted optical control of neural activity. *Nat Neurosci*. 2005;8(9): 1263–1268.
222. Kato HE, Zhang F, Yizhar O, et al. Crystal structure of the channelrhodopsin light-gated cation channel. *Nature*. 2012;482(7385):369–374.
223. Xu X, Kim SK. The early bird catches the worm: new technologies for the *Caenorhabditis elegans* toolkit. *Nat Rev Genet*. 2011;12(11):793–801.
224. Goodman MB, Lindsay TH, Lockery SR, Richmond JE. Electrophysiological methods for *Caenorhabditis elegans* neurobiology. *Methods Cell Biol*. 2012;107:409–436.
225. Franks CJ, Murray C, Ogden D, O'Connor V, Holden-Dye L. A comparison of electrically evoked and channel rhodopsin-evoked postsynaptic potentials in the pharyngeal system of *Caenorhabditis elegans*. *Invert Neurosci*. 2009;9(1):43–56.
226. Liu Q, Hollopeter G, Jorgensen EM. Graded synaptic transmission at the *Caenorhabditis elegans* neuromuscular junction. *Proc Natl Acad Sci USA*. 2009;106(26):10823–10828.
227. Husson SJ, Gottschalk A, Leifer AM. Optogenetic manipulation of neural activity in *C. elegans*: from synapse to circuits and behaviour. *Biol Cell*. 2013;105(6):235–250.

228. Edwards SL, Charlie NK, Milfort MC, et al. A novel molecular solution for ultraviolet light detection in *Caenorhabditis elegans*. *PLoS Biol*. 2008;6(8):e198.
229. Stirman JN, Crane MM, Husson SJ, et al. Real-time multimodal optical control of neurons and muscles in freely behaving *Caenorhabditis elegans*. *Nat Methods*. 2011;8(2):153–158.
230. Yin T, Wu YI. Guiding lights: recent developments in optogenetic control of biochemical signals. *Pflugers Arch*. 2013;465(3):397–408.
231. Krauss U, Lee J, Benkovic SJ, Jaeger KE. LOVely enzymes—towards engineering light-controllable biocatalysts. *Microb Biotechnol*. 2010;3(1):15–23.
232. Moglich A, Ayers RA, Moffat K. Structure and signaling mechanism of Per-ARNT-Sim domains. *Structure*. 2009;17(10):1282–1294.
233. Harper SM, Christie JM, Gardner KH. Disruption of the LOV-Jalpha helix interaction activates phototropin kinase activity. *Biochemistry*. 2004;43(51):16184–16192.
234. Swartz TE, Corchnoy SB, Christie JM, et al. The photocycle of a flavin-binding domain of the blue light photoreceptor phototropin. *J Biol Chem*. 2001;276(39): 36493–36500.
235. Wu YI, Wang X, He L, Montell D, Hahn KM. Spatiotemporal control of small GTPases with light using the LOV domain. *Methods Enzymol*. 2011;497:393–407.
236. Wu YI, Frey D, Lungu OI, et al. A genetically encoded photoactivatable Rac controls the motility of living cells. *Nature*. 2009;461(7260):104–108.
237. Strickland D, Lin Y, Wagner E, et al. TULIPs: tunable, light-controlled interacting protein tags for cell biology. *Nat Methods*. 2012;9(4):379–384.
238. Yazawa M, Sadaghiani AM, Hsueh B, Dolmetsch RE. Induction of protein-protein interactions in live cells using light. *Nat Biotechnol*. 2009;27(10):941–945.
239. Kennedy MJ, Hughes RM, Peteya LA, et al. Rapid blue-light-mediated induction of protein interactions in living cells. *Nat Methods*. 2010;7(12):973–975.
240. Yan D, Noma K, Jin Y. Expanding views of presynaptic terminals: new findings from *Caenorhabditis elegans*. *Curr Opin Neurobiol*. 2012;22(3):431–437.
241. Shu X, Lev-Ram V, Deerinck TJ, et al. A genetically encoded tag for correlated light and electron microscopy of intact cells, tissues, and organisms. *PLoS Biol*. 2011;9(4): e1001041.
242. Lin JY, Sann SB, Zhou K, et al. Optogenetic inhibition of synaptic release with chromophore-assisted light inactivation (CALI). *Neuron*. 2013;79(2):241–253.
243. Qi YB, Garren EJ, Shu X, Tsien RY, Jin Y. Photo-inducible cell ablation in *Caenorhabditis elegans* using the genetically encoded singlet oxygen generating protein miniSOG. *Proc Natl Acad Sci USA*. 2012;109(19):7499–7504.
244. Goguen BN, Hoffman BD, Sellers JR, Schwartz MA, Imperiali B. Light-triggered myosin activation for probing dynamic cellular processes. *Angew Chem Int Ed Engl*. 2011;50(25):5667–5670.
245. Friedland AE, Tzur YB, Esvelt KM, et al. Heritable genome editing in *C. elegans* via a CRISPR-Cas9 system. *Nat Methods*. 2013;10(8):741–743.
246. Volkel P, Le Faou P, Angrand PO. Interaction proteomics: characterization of protein complexes using tandem affinity purification-mass spectrometry. *Biochem Soc Trans*. 2010;38(4):883–887.

CHAPTER THIRTEEN

Mechanical Force Sensing in Tissues

Soline Chanet, Adam C. Martin
Department of Biology, Massachusetts Institute of Technology, Cambridge, Massachusetts, USA

Contents

Abstract

Tissue size, shape, and organization reflect individual cell behaviors such as proliferation, shape change, and movement. Evidence suggests that mechanical signals operate in tandem with biochemical cues to properly coordinate cell behavior and pattern tissues. The objective of this chapter is to present recent evidence demonstrating that forces transmitted between cells act as signals that coordinate cell behavior across tissues. We first briefly summarize molecular and cellular mechanisms by which forces are sensed by cells with an emphasis on forces generated and transmitted by cytoskeletal networks. We then discuss evidence for these mechanisms operating in multicellular contexts to coordinate complex cell and tissue behaviors that occur during embryonic development: specifically growth and morphogenesis.

Progress in Molecular Biology and Translational Science, Volume 126
ISSN 1877-1173
http://dx.doi.org/10.1016/B978-0-12-394624-9.00013-0

1. INTRODUCTION: MOLECULAR MECHANISMS OF MULTICELLULAR FORCE SENSING

Tissues are dynamic multicellular structures that are extensively remodeled, particularly during embryonic development when tissues of all different shapes and sizes are formed. Precise control of cell behaviors, such as growth, death, shape change, and movement within a tissue is crucial to generate and maintain the characteristic shape, size, and function of embryos and organs. Thus, understanding tissue organization and function requires knowledge of the mechanisms responsible for coordinating cell behaviors between the different cells. One way for cells to communicate is to exchange biochemical cues such as secreted signaling ligands. In addition to biochemical signals, cells also sense and respond to mechanical cues. Because cells in tissues (e.g., epithelia) are physically coupled to each other through intercellular junctions, forces are transmitted between the cells of a tissue and also between neighboring connected tissues. Such forces can rapidly and globally impact cell behavior in a tissue.[1] Thus, mechanical forces transmitted between cells provide a critical complement to biochemical signals to coordinate multicellular behavior.

Animal cells exert mechanical forces on their environment largely through the action of the actin cytoskeleton. Actin networks that vary in network architecture can generate different types of force, such as protrusive and contractile force. Forces that are transmitted between cells and result in mechanical signals often rely on the contractile activity of actin networks that contain the molecular motor myosin II (Myo-II).[2,3] Actomyosin networks can be organized into fibers made of bundles of antiparallel actin filaments (F-actin) that are cross-linked by Myo-II, such as cytoplasmic stress fibers. Alternatively, F-actin and Myo-II can form interconnected two-dimensional contractile meshworks, such as the actomyosin cortex that underlies the plasma membrane. These different network types are coupled to the cell membrane and to neighboring cells and/or the extracellular matrix (ECM) by adhesion complexes, transmitting tension between cells via cell–cell junctions or to the ECM via focal adhesions.[3] The magnitude and direction of transmitted forces depend on the connectivity of the network to adhesion complexes.[4–7]

In addition to actively generating force, actomyosin networks also provide cells with mechanical properties such as elasticity and viscoelasticity,[8] therefore conferring mechanical resistance to deformation by increasing cell and tissue stiffness.[9–13] The actin cortex as well as stress fibers resist external forces and exert traction forces at adhesion sites against the surrounding cells or the underlying ECM.[14,15] Elasticity occurs over short time scales where stretch or compression of actin networks leads to a restoration force that is proportional to the strain. Strains occurring over longer time scales can result in a viscoelastic response due to the turnover (assembly and disassembly) of F-actin within the network and binding/unbinding of F-actin cross-linkers.[16] In addition to resisting external forces, the actin cortex also resists the hydrostatic pressure from the cell cytoplasm (in plant cells, this turgor pressure is resisted by the cell wall). These mechanical properties are important in multicellular contexts for transmitting and sensing mechanical signals.

To effectively use force as a signal to coordinate cell behavior in tissues, cells must sense different types of stress or strain, such as compression, tension, or shear.[17] How do cells sense forces transmitted through a tissue? Transduction of a mechanical signal (mechanotransduction) resembles classical biochemical signal transduction in many ways. A specific mechanical force, which can be distinguished by its magnitude, orientation, and/or frequency, must be recognized by specific mechanosensing machinery. Several molecules or molecular complexes can directly respond to physical stress or strain by changing conformation or macromolecular assemblies. Classic examples are the unfolding or stretching of molecules or the opening of ion channels under mechanical forces that would transduce a signal to downstream-signaling pathways.[18] In addition, rather than a single molecule or molecular complex responding to force, mechanical constraints that alter cell geometry can lead to rearrangements of the cytoskeleton due to the self-organizing properties of such cellular systems.[19] We first describe several molecular- and systems-level mechanisms by which cells respond to forces. We then discuss evidence that suggests roles for these mechanisms of multicellular sensing during tissue growth and morphogenesis.

1.1. Force sensing by adhesion complexes

As cells interact with each other and their physical environment through adhesion complexes, these complexes serve as the cell's front line for receiving mechanical stimuli. The classic example of an adhesive complex serving as a mechanosensor is focal adhesions: complexes of greater than 100 different proteins including integrin receptors that link the cell's actin cytoskeleton and plasma membrane to the underlying ECM.[20,21] Focal adhesions grow in size and exhibit different protein composition and dynamics in response to tension.[22,23] Stretching of several focal adhesion proteins (e.g., Talin, Vinculin, and p130-Cas) can result in the exposure of cryptic-binding sites for cytoskeletal proteins or kinases, resulting in increased adhesion assembly.[24–26] While focal adhesions connect isolated cells, such as fibroblasts, to the ECM, integrin-containing adhesions can also mediate intercellular connections through an intervening ECM as in the case of muscle–tendon cell attachment.[27]

More common examples of intercellular adhesions include adherens junctions, tight junctions, and desmosomes. Apical adherens junctions are the principal structures mediating mechanical coupling between the cells of epithelial tissues. Adherens junctions contain adhesion receptors called Cadherins (e.g., E-Cadherin in epithelial cells) that mediate homophilic adhesion between cells in the epithelium. The intracellular domain of E-Cadherin binds to the adaptors β-catenin and α-catenin that link adhesion receptors to the underlying actin cortex.[28] β-Catenin forms a stable linkage to the intracellular tail of E-Cadherin, while the actin-binding protein α-catenin binds dynamically to β-catenin.[29–31] Recent studies have suggested that, like focal adhesions, adherens junctions can respond to mechanical forces. Using microfabricated force sensor arrays on which endothelial cells were plated, Liu *et al.* demonstrated that adherens junctions undergo mechanically induced growth in response to loading.[32] Pulling forces exerted by the contractile cytoskeleton on adherens junctions or direct application of mechanical tugging force tend to cluster and develop adhesion complexes, increasing the size of adherens junctions.[30,32] The mechanism of mechanotransduction is still not clear, but could involve protein stretching that reveals cryptic-binding sites for additional actin-binding proteins, similar to focal adhesions. It has been proposed that force-induced stretching of α-catenin exposes a binding site for the actin-binding protein Vinculin[33] (Fig. 13.1A). Recruitment of Vinculin to adherens junctions seems to play a central role in anchoring adhesions to actin networks as disruption of α-catenin–Vinculin binding can lead to reduced epithelial integrity between cells undergoing force-dependent junction remodeling.[14,34] Thus, for both focal adhesions and adherens junctions, the primary cellular response to applied tension is to strengthen adhesion complexes that bear tension and thus reinforce cell–cell or cell–matrix adhesion.

1.2. Force sensing by actomyosin networks

Forces transmitted through adhesion receptors are subsequently transmitted to the cytoskeleton. In animal cells, adhesion complexes are coupled to

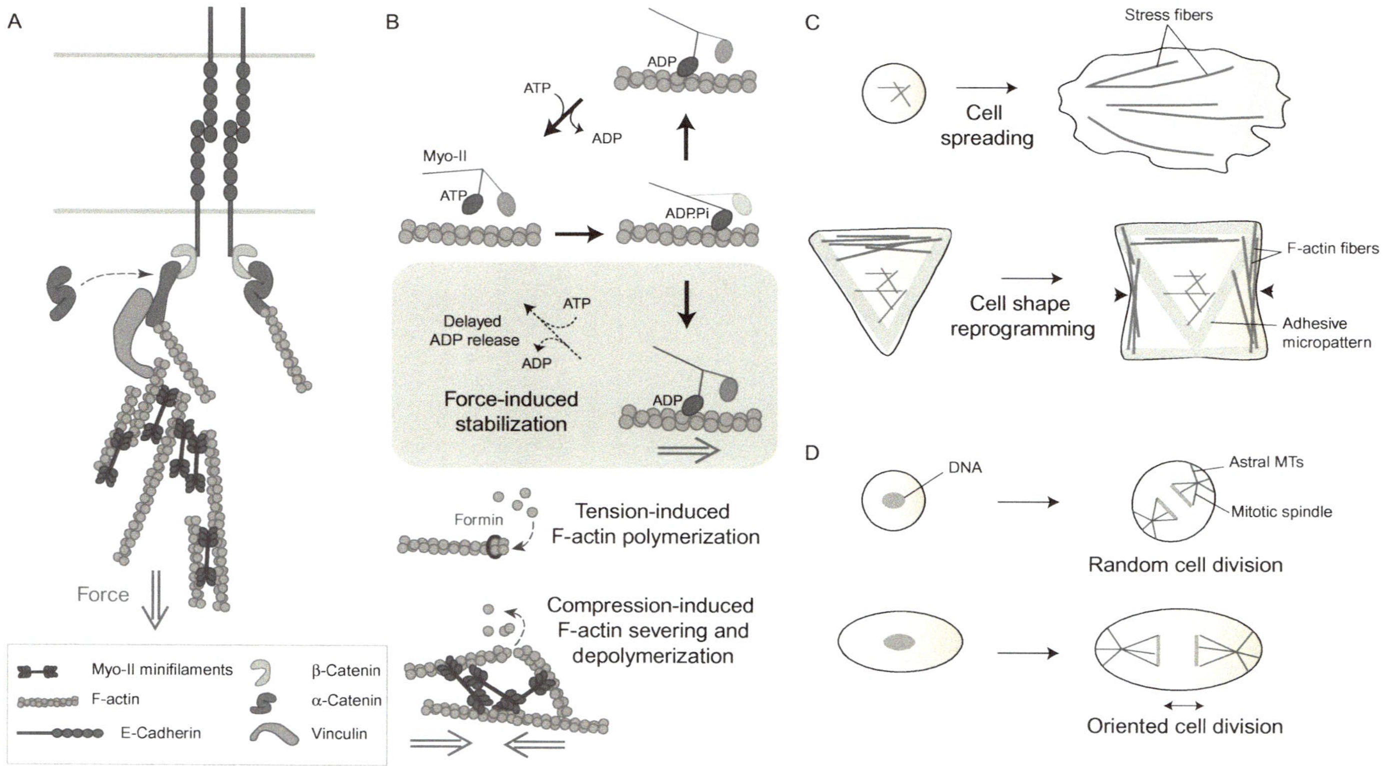
A
B
C
D
Force
Myo-II minifilaments
F-actin
E-Cadherin
β-Catenin
α-Catenin
Vinculin
Myo-II
ATP
ADP
ADP.Pi
Delayed
ADP release
Force-induced
stabilization
Formin
Tension-induced
F-actin polymerization
Compression-induced
F-actin severing and
depolymerization
Stress fibers
Cell
spreading
Cell shape
reprogramming
F-actin fibers
Adhesive
micropattern
DNA
Astral MTs
Mitotic spindle
Random cell division
Oriented cell division

(*Continued*)

either the actin cytoskeleton or intermediate filaments. While we will discuss later how force can alter intermediate filaments attachment to adhesion sites, we will focus here on how mechanical forces alter the dynamics and force generation by actomyosin networks. Spatial regulation of Myo-II motor activity and F-actin assembly–disassembly are determinants of the magnitude and direction of the contractile forces generated. Evidence suggests that mechanical forces can influence both Myo-II motors and F-actin networks. Myo-II motors convert the energy from ATP hydrolysis into mechanical work, undergoing a conformational "power-stroke" that is linked to the ATP hydrolysis cycle[40] (Fig. 13.1B). Myo-II motors differ from motors such as Kinesin in that the fraction of the time during the ATP hydrolysis cycle in which they are bound to F-actin (the duty ratio) is small, necessitating that Myo-II motors assemble into antiparallel oligomers called minifilaments to processively move along F-actin. Because myosins are mechanochemical enzymes, the duty ratio is influenced by mechanical force. Single-molecule measurements have shown that load stabilizes myosins, including Myo-II, in the ADP-bound state, favoring their association with F-actin in a state that sustains tension, with low levels of ATP cycling.[41–43] Interestingly, Myo-II is ectopically recruited to the cortex in response to micropipette aspiration,[44,45] potentially due to the force

---

Figure 13.1—Cont'd Molecular and cellular mechanisms of mechanosensing. (A) Mechanosensing at intercellular junctions. Cell–cell adhesion in epithelial tissues is mediated by homophilic interaction of E-Cadherin in adjacent cells. E-Cadherin molecules are bound to the actomyosin cytoskeleton through adaptor proteins such as β-catenin and α-catenin. Pulling forces generated by actomyosin contraction can induce conformational changes in adhesion proteins, possibly α-catenin, causing the recruitment of proteins such as Vinculin. Recruitment of additional adaptor proteins and growth of adhesion size can reinforce adhesion complex attachment to actin filaments.[32–34] (B) Forces in actin filament networks regulate Myo-II dynamics and actin turnover. Tension resisting the myosin power-stroke can decrease the rate of ADP release and cause myosin to persist in an F-actin-bound conformation. In addition, mechanical stimulation can increase actin monomer availability and induce formin-mediated F-actin assembly. Compression of actin filaments during myosin-mediated contraction, and potentially external forces applied to cells, causes filament bucking and increased severing.[35–37] (C) Response of actomyosin networks to changes in cell geometry. Cell spreading is associated with actomyosin stress fiber assembly (upper panels).[38] Reprogramming cell shape by changing the pattern of cell-substrate adhesion (gray pattern) leads to stress fiber assembly between well-separated adhesion sites (lower panels).[39] (D) Response of cell division to changes in cell geometry. During cell division, the mitotic spindle preferentially aligns with the longest axis of the cell, resulting in oriented cell division even in absence of extrinsic biochemical polarizing cues. (See the color plate.)

sensitivity of the Myo-II motor. Moreover, Myo-II has a higher affinity for stretched actin filaments, which might be another mechanism stabilizing the F-actin–Myo-II interaction under tension.[46]

In addition to regulating Myo-II motor dynamics, mechanical forces could also impact the activity of proteins involved in F-actin assembly and disassembly (turnover). Mechanical forces modulate the activity of formins that are actin nucleation and elongation factors[35] (Fig. 13.1B). Recent results demonstrate that application of force on the cell cortex triggers actin monomer release from filaments. This transient increase in actin monomer concentration in turn stimulates formin-dependent actin polymerization.[36] Furthermore, Myo-II in some settings accelerates F-actin turnover, suggesting that intrinsic and possibly extrinsic mechanical forces could modulate actin network depolymerization and remodeling.[47,48] A potential mechanism for force regulated F-actin disassembly could be that compression of actin filaments results in buckling, where bends in the filament are more prone to severing[37] (Fig. 13.1B). Thus, mechanical forces have the potential to influence force generation by altering Myo-II and F-actin dynamics.

1.3. Force sensing by the cell plasma membrane

Because actomyosin networks are linked to the plasma membrane through adhesion sites, intrinsic forces generated by cortical actin networks and their response to extrinsic forces from neighboring cells can be sensed through changes in plasma membrane tension.[49] Plasma membrane tension can modulate cellular processes and activate signals that would change cell behavior in several different ways, of which we will discuss two. First, membrane tension can induce conformational changes in transmembrane channels called mechanosensitive ion channels.[50] Mechanosensitive channels can be activated by increased membrane tension, which increases the probability for the channel to adopt an open conformation. Other mechanosensitive channels are linked to the internal cytoskeleton and/or the external ECM, which could allow these channels to be regulated by strain resulting from movement of the cytoskeleton relative to the cell exterior.[51] By mediating flux of a specific ion, opening of mechanosensitive channels can induce a variety of cell responses. For example, signaling through Ca^{2+}-stretched-sensitive channels, such as those in the transient receptor potential channel M (TRPM) family, results in actin cytoskeleton remodeling.[52] Increased plasma membrane tension can also modify plasma membrane removal and insertion by modulating rates or mechanisms of endocytosis and exocytosis. Increased plasma membrane tension counteracts endocytosis, increasing the dependence of internalization on the actin cytoskeleton.[53] Conversely, increased membrane tension activates exocytosis.[54] Modulating the rate of endocytosis and exocytosis will in turn regulate plasma membrane tension and can directly impact on biochemical signaling by modulating the number of secreted ligands and/or transmembrane receptors present at the cell surface.

1.4. Force sensing by change in cell geometry

Cellular structures, such as the cytoskeleton, exhibit self-organization that depends on the geometry of the cell and the arrangement of anchor points at the cell periphery.[19,55] Because forces transmitted between cells in a tissue are likely to impact cell shape, cells in multicellular tissues could also sense forces via changes in the organization of supramolecular structures that depend on cell geometry.[19] Exposure of confluent layers of endothelial cells to laminar flow (and thus shear stress) leads to cell elongation in the direction of flow and the assembly or maintenance of stress fibers aligned with the long axis of the cell.[56] There is a positive correlation between the extent of cell spreading (cell-substrate contact area) and stress fiber assembly, such that cells confined to a small ECM substrate contact area display fewer focal adhesions and stress fibers that are also thinner compared to spread cells[38,57,58] (Fig. 13.1C, upper panels). Recent studies using laser nanopatterning illustrated how dynamically changing cell shape and the geometry of adhesive connections to the substrate changes the organization and orientation of actomyosin stress fibers. Generating new adhesion sites separated by non-adhesive regions promotes actin stress fiber assembly at the expense of other actin network types, suggesting that the distance between adhesive contacts is a critical determinant of actomyosin fiber assembly and contraction[39] (Fig. 13.1C, lower panels). Thus, reorientation of actomyosin networks in response to altering cell shape and adhesion geometry will change how these cells generate and resist forces.

Cell shape is also a critical determinant for the orientation of the microtubule (MT) apparatus that segregates chromosomes during cell division, the mitotic spindle. Mitotic spindles tend to align with the longest axis of the cell, resulting in a division plane that is perpendicular to the long axis[59–61] (Fig. 13.1D). The position and orientation of the mitotic spindle are set by forces exerted on astral MTs that extend from the spindle poles toward the cell cortex. Recent studies in sea urchin eggs (nonadherent cells with no external cues to direct spindle orientation) suggested a model where the length-dependent pulling forces exerted on astral MTs link cell geometry to the alignment of the mitotic spindle.[62] Cell division in adherent cells is linked to the geometry of matrix-adhesion sites, which affect the location of cortical cues that orient the spindle.[63,64] Interestingly, a recent study demonstrated that forces exerted on adhesive sites can reorient the spindle independently of cell shape.[65] Thus, the geometry of cells, and also the geometry of externally applied forces, can regulate the organization of cytoskeletal networks, providing a cellular mechanism of responding to mechanical cues during development.[66,67]

Described above are some of the mechanisms for mechanotransduction that have been documented in cells. An important next step is to determine how molecular mechanisms for mechanosensing operate in tissues to coordinate cell behavior during development. Connecting specific mechanical forces to distinct molecular mechanisms in embryonic tissues presents several challenges. Cells in tissues are exposed to a combination of biochemical and mechanical signals that often cannot be easily disentangled. In addition, tissues experience combinations of compression, tension, and shear stress along different dimensions, making it difficult to determine what type of stress or strain is sensed and what cellular structure(s) are responding. Finally, it is often difficult to find a handle on a tissue to apply a specific type of force to individual cells or populations of cells. Despite these limitations, research on developmental systems has identified key roles for mechanical forces in cellular coordination. We will discuss evidence for multicellular force sensing in tissues, focusing on tissue growth and morphogenesis.

2. MULTICELLULAR SENSING DURING TISSUE GROWTH

During development and homeostasis, tissue growth is tightly regulated to generate, repair, or renew organs with precise size, structure, and function.[68,69] Thus, individual cells in a tissue must somehow sense the overall size or growth state of a tissue to prevent overproliferation that would lead to inappropriate tissue size or tumor formation. Much effort has focused on secreted signals that regulate tissue growth, but tissues also experience mechanical stress during growth that could, in principle, serve to globally coordinate cell growth across a tissue. Tissues can grow either by increasing their cell volume or by increasing cell number through cell division/proliferation. In both cases, growth might engender local stress in the context of a tissue where cells exhibit intercellular adhesion and are thus not free to immediately rearrange in a manner that would relieve stress. Here, we discuss evidence that suggests a role for mechanical forces in modulating the magnitude and orientation of cell growth and proliferation within developing tissues.

2.1. Mechanical regulation of tissue growth

Tissue growth and patterning during animal development relies on biochemical signals called morphogens. Morphogens are signaling molecules that diffuse from a localized source and form a concentration gradient across the developing tissue, providing a series of concentration thresholds that control the behavior of surrounding cells depending on their distance from the source.[70–72] Therefore, morphogen gradients serve as instructive signals that pattern tissues and also serve as growth factors that orchestrate the growth and division of cells that constitute the tissue.[73,74] One of the clearest examples where morphogen gradients regulate tissue growth is the wing imaginal disc in the fruit fly *Drosophila melanogaster*. Wing imaginal discs are epithelial layers that grow from about 50 cells to 50,000 cells within 5 days during *Drosophila* larval development. The growth of the wing disc is primarily controlled by the morphogens Decapentaplegic (Dpp), which is a bone morphogenetic protein homolog and Wingless (Wg), which is a Wnt homolog. Dpp and Wg are produced in anterior–posterior (AP) and dorsal–ventral (DV) stripes, respectively, which intersect in the middle of the disc (Fig. 13.2A).[69,73,74,80] Both Dpp and Wg are secreted ligands that are distributed in a graded manner in the disc from their site of production. However, the occurrence of cell divisions is uniform across the disc,[81–83] indicating that growth rate does not simply follow the concentration of these secreted growth factors. Moreover, divisions cease when the disc reaches its critical size corresponding to the time when the larvae enter pupation. Cessation of growth can be controlled by extrinsic signals such as insulin-like growth hormones, but tissue-intrinsic factors, including signaling by morphogens also play critical role in determining organ final size. While models using changes in morphogen concentration and gradient shape as regulators of tissue size can account for growth regulation, these models do not accurately predict experimental results where morphogen expression is perturbed.[69,75,84–86] Therefore, additional signals must spatially and temporally regulate growth in the wing disc.

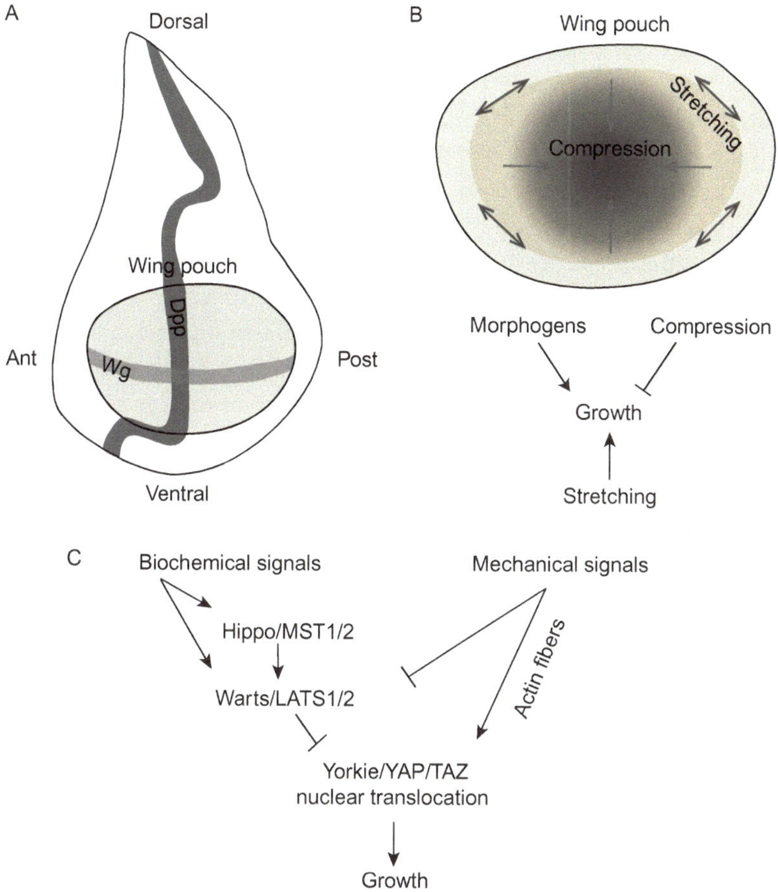

Figure 13.2 Mechanical forces influence growth rate. (A) Expression pattern of the morphogens *Dpp* and *Wg* in the developing wing imaginal disc. The central region of the disc (wing pouch, gray) will give rise to the adult wing. (B) Model of wing disc growth regulation by a combination of biochemical and mechanical signals. The secreted morphogens form a concentration gradient from the center to the periphery of the wing pouch (green gradient), providing a positive signal for cell proliferation in the center. Localized growth in the center causes stretching forces (blue) in the periphery promoting proliferation, and compressive forces (red) in the center inhibiting proliferation. Combination of morphogen signaling and mechanical feedback could promote uniform growth in the wing disc.[75] (C) The signaling activity of the Hippo pathway integrates biochemical and mechanical signals to regulate growth. Biochemical activation of the Hippo kinase leads to activation of the Warts/LATS1/2 kinase, cytoplasmic retention of the transcription factors Yorkie/YAP and TAZ, and inhibition of growth. Yorkie/YAP/TAZ activity is also regulated by mechanical signals. The mechanism linking force sensing to nuclear translocation of Yorkie/YAP/TAZ remains unknown and may or may not involve Warts/LATS1/2 inhibition, but does involve the formation of actin fibers.[38,76–79] (See the color plate.)

Mechanical models of tissue growth have predicted that the buildup of compressive stress at the center of the wing disc that results from inhomogeneous growth in the tissue could provide an inhibitory growth signal that stabilizes uniform cell division patterns.[87] Models of tissue growth combining positive growth signals provided by morphogen concentration in the middle of the disc (represented in Fig. 13.2B by the green gradient) with mechanical signals, whereby compression inhibits and stretching promotes cell proliferation, recapitulates uniform cell division and the cessation of cell division at a critical tissue size[75,86] (Fig. 13.2B). Quantitative analyses of cell behavior in growing imaginal discs verified several important assumptions of theoretical models predicting a role for forces. First, very few cell rearrangement events (such as neighbor exchanges) have been observed during wing disc growth, suggesting that tissue stress that results from inhomogeneous growth is not immediately dissipated.[88,89] In addition, cells in the center of the disc exhibit smaller apical areas, consistent with increased compression in the disc center as it grows. In contrast, cells at the periphery of the disc become stretched in the direction tangential to the boundary of the tissue, consistent with peripheral cells serving as a barrier to growth that would result in the compression of more central cells (Fig. 13.2B).[88,90] Moreover, the distribution of mechanical stress has been measured in the wing disc using photoelasticity, showing that cells are compressed in the center and stretched in the periphery.[91,92] Although a direct demonstration of inhibition of proliferation by compression remains a technical challenge, it has been recently reported that mechanical stretching increases growth rate in the wing the disc.[92] An outstanding question is how mechanical signals are sensed by the cells and transduced to influence growth rate.

A possible pathway that could link mechanical forces to growth control is the Hippo pathway: a signaling pathway implicated in tissue growth in *Drosophila* and vertebrates, including a role in imaginal disc growth.[93–98] The Hippo signaling pathway includes a kinase called Hippo (MST1 and MST2 in mammals), which regulates a downstream kinase called Warts (LATS1 and LATS2 in mammals). Activation of the Hippo pathway suppresses cell proliferation and promotes apoptosis through inhibitory phosphorylation and nuclear exclusion of a transcriptional coactivator called Yorkie (YAP and TAZ in mammals). Inhibition of the Hippo pathway leads to Yorkie/YAP/TAZ nuclear translocation and activation that ultimately induces proliferation and differentiation[93,94,99,100] (Fig. 13.2C). The signal(s) required for spatial and temporal activation of the Hippo pathway within a tissue remains a debated and controversial question. Several biochemical signals can regulate the activity of the Hippo/MST1/2 or Warts/LATS1/2 kinases, including the protocadherins Fat and Dachsous, the membrane-associated FERM-proteins Merlin and Expanded, the scaffolding protein Salvador or the apical polarity protein Crumbs.[101] Recent evidence indicates that mechanical forces may also serve as inputs that regulate YAP/TAZ and Yorkie activity.[76,102] Studies examining YAP/TAZ nuclear localization and transcriptional activity in mammalian cells cultured on micropatterned substrates have demonstrated that YAP/TAZ activity is positively correlated with ECM stiffness and the area of cell spreading.[38,76] Importantly, YAP/TAZ activity is induced on polylysine-coated substrates, suggesting that the signal to YAP/TAZ does not involve focal adhesion signaling, but rather changes in cell geometry/cytoskeletal organization.[103] Substrate stiffness and cell area are correlated with actomyosin stress fiber formation and traction forces exerted by cells on the substrate.[38,102] Correspondingly, YAP/TAZ activity and their nuclear localization are inhibited by treatments that disrupt actomyosin stress fiber assembly.[38,76–78,103,104] Experiments using cultured cells support a model where mechanical forces and/or stress fiber formation activate YAP/TAZ activity. F-actin networks might directly regulate Hippo pathway components, either by inactivating LATS1/2[38,103] or by sequestering a LATS-independent inhibitor of YAP/TAZ.[76,78] In accordance with the first model, a recent study has found that the activation of the LATS kinase and its interaction with the plasma membrane-associated protein Merlin is regulated by the status of the actin cytoskeleton, where disruption of actin cytoskeleton enhances Warts/LATS1/2 activity.[105] While there is not yet consensus on whether regulation of YAP/TAZ by mechanical forces depends on LATS1/2 activity, a LATS1/2-independent pathway, or both, cell culture studies support a link between mechanical forces and resulting changes in cell geometry and cytoskeletal organization to the activity of a key regulator of growth, YAP/TAZ (Fig. 13.2C).

Does mechanical regulation of cell shape and/or cytoskeletal organization adaptively regulate YAP/TAZ activity in a multicellular context? When confluent monolayers of cells are grown in culture, YAP/TAZ becomes excluded from nuclei in dense monolayers where cells have a small contact area with the substrate.[78] Furthermore, YAP/TAZ nuclear translocation and cell proliferation are induced by locally stretching cultured epithelial monolayers.[78] Importantly, increasing actin filament assembly and actin stress fiber formation by depleting actin-capping/severing proteins can restore YAP/TAZ activity in dense monolayers, resulting in constitutive cell proliferation. Therefore, uncoupling stress fiber assembly from cell shape can make proliferation independent of tensile or compressive stress.[78] Experiments in *Drosophila* imaginal discs also support a crucial role for actin network organization in regulating Yorkie activity.[77,79] Increased actin assembly resulting form either depletion of inhibitors of actin assembly (e.g., capping protein) or by overexpressing the formin *Diaphanous* induces overgrowth in the wing disc and promote Yorkie nuclear accumulation.[77,79]

While a definitive mechanism for mechanosensing is yet to be established in developing tissues, the current data suggest that mechanical stress in the disc tissue modulates cellular forces and/or cytoskeletal assembly, which feeds back to regulate cell division during wing disc development. Overall, a possible model for the mechanical regulation of cell proliferation is that tissue growth and resulting cell compression at the center of the tissue reduces apical and basal cell area and stress fiber assembly, resulting in cytoplasmic/inactive Yorkie. Conversely, mechanical stretching of peripheral cells could enhance actin fiber formation, by analogy to cell culture, promoting Yorkie nuclear translocation and transcriptional activity. Indeed, actomyosin fibers are observed to form along cell interfaces in stretched peripheral cells[88] (Fig. 13.3A, lower panels), although it is not yet clear whether induction of cell division by stretching is associated with Yorkie activity and whether these actomyosin fibers are required for nuclear translocation of Yorkie. Tension could induce formin-dependent actin polymerization or stabilize cortical Myo-II providing possible pathways for tension-sensitive actomyosin fiber assembly.[35,36,45,107] How these candidate mechanical signals are integrated with biochemical signals to regulate, YAP/TAZ and Yorkie nuclear localization will be a very interesting area of research in the coming years.

2.2. Mechanical regulation of cell growth orientation

In addition to growth rate, the orientation of cell growth and division is a critical determinant of how tissue growth affects tissue mechanics. For instance, oriented cell division along the principal axis of tension has been shown to release tension and might therefore be an adaptive mechanism used in growing epithelia to adapt to their mechanical environment.[108] It is therefore possible that mechanosensitive pathways regulating the orientation of growth and division in a tissue will also impact the growth rate, final size, and shape of tissues.

Different parameters regulate cell division orientation. Biochemical signals can polarize cortical proteins that interact with astral MTs and regulate

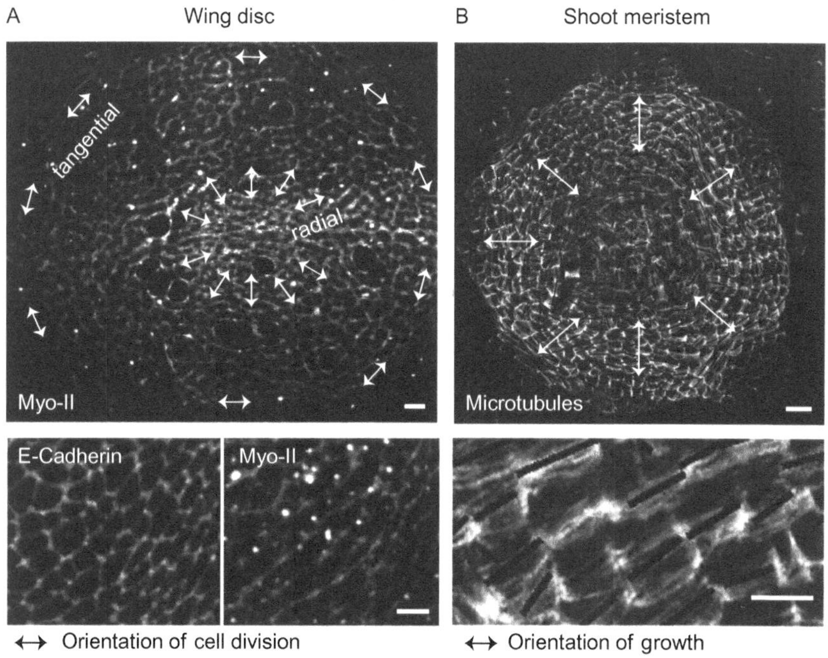

Figure 13.3 Mechanical forces influence growth orientation. (A) The pattern of mechanical stress influences cell division orientation in the wing disc possibly by regulating cell shape. Wing disc marked for Myo-II (upper panel) or E-Cadherin and Myo-II (lower panels). Cells are compressed in the center leading to small apical areas, whereas cells are stretched in the periphery. Blue arrows illustrate the orientation of cell divisions, which follows the patterns of stress. Divisions are radial in the center and tangential in the periphery. At the disc periphery, Myo-II forms polarized supracellular fibers lining the direction of maximal stress, which resist further stretching. Scale bar: 5 μm. (B) The pattern of mechanical stress influences growth orientation in the root apical meristem. Apical meristem marked for MTs (green). The red lines indicate the orientation of the MT array for each cell. MTs align with the pattern of stress in the tissue. MT orientation confers greater resistance to cell deformation along the maximal stress axis. Blue arrows indicate orientation bias of growth, toward the center of the meristem. Scales bars: 10 μm (upper panel) and 2.5 μm (lower panel). *(A) Adapted from Ref. 88 and (B) adapted from Ref. 106.* (See the color plate.)

spindle orientation, as occurs in cells that undergo asymmetric cell divisions.[109] In addition, oriented cell division can occur in the absence of an extrinsic biochemical cue, depending instead on cell geometry and/or mechanical forces.[62,65] Cell division orientation in the wing disc appears to depend on the interplay between biochemical and mechanical cues. Cell divisions are preferentially aligned radially from the center of the disc to the periphery in the central region and tangential to the tissue boundary in the periphery (Fig. 13.3A). Radially oriented cell divisions depend on the activity of the protocadherin Fat, which acts as a receptor for its ligand Dachsous. Wings in *fat* mutants are less elongated than wild-type wings and the division planes are random during wing disc development in the central region, demonstrating that cell division orientation affects tissue shape.[110] In the wing disc, activation of the Fat–Dachous pathway leads to the planar polarization of the atypical myosin Dachs, which is enriched in tangential junctions (oriented perpendicular to the radial axis of the tissue) and depleted from radial junctions.[111] Planar polarization of Dachs could localize cortical cues that directly interact with astral MTs to orient the mitotic spindle. However, since Dachs is a myosin, planar polarization of Dachs might indirectly orient the spindle by controlling cell shape. In accordance with this latter view, Dachs has been shown to define local patterns of tension and contraction in tissues,[112] and Dachs planar polarization leads to increased tension along the tangential cellular junctions where it accumulates.[110] This polarized tension is predicted to cause radial cell elongation, which is correlated with the axis of cell division. Thus, biochemical signals could polarize local patterns of tissue tension to orient cell and tissue growth by altering cell shape.[110]

The planar polarization of the myosin Dachs does not explain why cell divisions are oriented tangentially in the periphery of the wing disc. In this case, the orientation of cell division may rely on the global pattern of tissue stress that results from tissue growth. Cells at the disc periphery that are elongated exhibit higher tension tangential to the boundary of the tissue.[88,90] The higher tangential tension is associated with tangential Myo-II fibers at junctions, suggesting that cells at the periphery adapt to balance cell stretching by assembling Myo-II fibers[88] (Fig. 13.3A, lower panels). Similar to more central cells, peripheral cells also tend to divide along the axis of elongation.[88,90] Importantly, local induction of cell growth can reorient cell division plane and Myo-II fibers assembly in surrounding cells (occurring tangentially to the source of growth), suggesting that mechanical stress that results from tissue growth is the cue that reorients cell division, possibly providing a means to dissipate tension at the tissue boundary and allow for further tissue growth. A critical question will be to determine how epithelial cells sense mechanical forces to orient the mitotic spindle: Do they sense cell geometry[62] or could mechanical forces polarize cortical cues[65]?

During animal development, tissue growth generates and adapts to mechanical forces. In plant tissues, mechanical forces are also thought to guide tissue growth; however, the nature of the forces and mechanotransduction mechanisms are different.[113] In plants, cell and tissue growth is counteracted by stiff extracellular walls that also prevent cell rearrangements. Increase in plant cell volume results from internal turgor pressure that induces tension in the surrounding cell wall. Anisotropy in the cell wall stiffness dictates the direction of cell growth, defined as growth anisotropy: the cell will grow in a preferred direction where it encounters less resistance from the extracellular wall. Because cellulose fibrils are thought to bear most of the cell wall's tensile stress, the orientation of cellulose fibrils is a critical determinant of growth in plants.[114,115] Importantly, the orientation of parallel MT arrays in the apical cortex of the cells correlates with anisotropic deposition of stiff cellulose microfibrils in the cell wall because MTs guide cellulose synthase.[114,116–118] Recent studies of the growth at the shoot apical meristem have elucidated mechanisms that allow plant cells to sense and respond to forces to orient plant growth. The shoot apical meristem consists in a population of stem cells that can continuously grow (by increasing cell volume) and divide. The meristem adopts the shape of a dome at the periphery of which flower primordia emerge forming small buds (Fig. 13.3B). Interestingly, cortical MTs are aligned with theoretically predicted directions of principal stress in the different regions of the meristem.[119] MTs are poorly aligned in the central region where there is isotropic stress, whereas MTs are arranged into circumferential arrays of parallel bundles at the periphery of the dome where there is highly anisotropic stress (Fig. 13.3B). Artificially changing the pattern of stress in the tissue through laser ablation or tissue compression is sufficient to reorient MT array.[119] Thus, MT array orientation and consequently anisotropic growth are responses of mechanical stress in plants.

The mechanotransduction pathway allowing cells of the shoot apical meristem to sense and respond to mechanical forces involves MT dynamics and their self-organization into parallel arrays. Compressive forces tend to induce MT shrinking, whereas tensile forces favor MT elongation.[120–122] Thus, the direction of maximal tension may bias MT self-organization in a preferred orientation. The MT-severing protein Katanin is an important regulator of MT dynamics and has been implicated in the force-sensing mechanism. *katanin* mutants exhibit more disorganized and less dynamic cortical MTs in the whole meristem. Interestingly, *katanin* mutants are not able to efficiently respond to changes in mechanical stress pattern: cells do not reorient cortical MT arrays after laser ablation or compression.[106] Katanin has been shown to cut MTs where they intersect each other,[123,124] disfavoring MT crossover, and increasing the number of free MTs amenable to self-organize according to stress directions.[113] This suggests that MT dynamics is required to allow cells to sense and respond to mechanical forces. In addition, in *katanin* mutants, the meristem adopts an aberrant morphology, which is correlated with a more homogeneous growth rate. Thus, growth orientation that results from mechanical feedback is critical to regulate the morphology of the growing meristem. This feedback has been proposed to be important for emergence of the floral primordial buds at the periphery of the meristem[106] (Fig. 13.3B).

3. MULTICELLULAR SENSING DURING TISSUE MORPHOGENESIS

During tissue morphogenesis, individual cells have to constantly adapt and coordinate cell shape changes and movements relative to their neighbors to maintain tissue integrity and collectively drive coherent movements or deformations of the overall tissue. Forces critical for tissue morphogenesis can be generated within cells changing shape (cell autonomous). Mechanosensitive signaling could potentially impact how cells generate forces, possibly helping coordination between cells or propagation of cell shape changes among cells. In addition, forces extrinsic to a tissue (cell non-autonomous) can influence cell and tissue shape by compressing/stretching a neighboring tissue or serving as a mechanical constraint to deformation. Here, we discuss how multicellular sensing plays a role in shaping tissues.

3.1. Mechanical coordination of actomyosin contractility

Many morphogenetic events that shape embryos and organs take place within epithelial tissues, which undergo highly coordinated movements and deformations.[2] Actomyosin contractility within epithelial cells can promote tissue remodeling such as tissue bending, elongation, contraction, or tube formation. During morphogenesis, cells in epithelia might coordinate force generation to engender a specific tissue response and the appropriate deformation or movement. The magnitude, direction and timing of cellular forces depend on the organization of the contractile actomyosin meshworks and fibers at the level of the cell and how these networks are connected between cells at the level of the tissue.

Recent live imaging and quantitative analyses of actomyosin contraction and associated cell shape changes have highlighted dynamic behaviors of cortical actomyosin meshworks, where actomyosin contractility is not continuous, but occurs as discrete contractile events or pulses. In these cases, Myo-II structures appear and often coalesce or condense in the apical (and sometime basal) cell cortex, before being remodeled.[125–131] In epithelia, Myo-II pulses have been shown to correlate with transient fluctuations of cell shape and area. Contraction pulses are not only characterized by a temporal cycle of actomyosin appearance and remodeling but can also exhibit spatial dynamics such as directional flow toward one edge of the cell.[7,127,132,133] Contraction pulses have been observed for example in the amnioserosa, an epithelial tissue that covers the dorsal surface of the *Drosophila* embryo. During a process called dorsal closure, the amnioserosa cells constrict and eventually disappear inside the embryo, pulling the neighboring tissue, the lateral epidermis, toward the dorsal midline of the embryo in a process that resembles wound healing[134,135] (Fig. 13.4A). At the beginning of dorsal closure, amnioserosa cells exhibit pulsatile constrictions, where pulses in neighboring cells are predominantly out of phase, resulting in a back-and-forth contraction between neighbors[127,136] (Fig. 13.4A and B). This begs the question whether amnioserosa cells sense pulses in neighboring cells and whether there is a mechanosensitive mechanism to stagger contraction pulses.

When tension is locally released by laser ablation of a single cell–cell junction or by laser-induced wounding of a single amnioserosa cell, cells surrounding the wound instantaneously stop pulsing.[136,138,139] In addition, inhibiting contractility in individual cells by driving the expression of dominant negative RhoA not only inhibits contractile pulses cell autonomously but also inhibits contraction pulses and Myo-II accumulation in adjacent cells[138] (Fig. 13.3C). Inducing the arrest of pulsed contractions by locally wounding a cell is accompanied by a change in Myo-II dynamics in the nearest neighbor of the perturbed cell. Instead of flowing toward the center of the cell apex, Myo-II flows toward the cellular junction with wounded cell.[138] These data suggest that the transmission of tension between neighboring cells is required for the collective pulsing behavior and that mechanical signals sensed by neighboring cells leads to staggered contraction pulses.[136]

In contrast to the pulsatile contraction of cortical actomyosin meshworks, other morphogenetic processes are associated with more stable actomyosin structures that resemble stress fibers, such as actomyosin cables that assemble at intercellular junctions. Junctional actomyosin cables are often connected between cells, leading to multicellular actomyosin cables that are critical for tissue morphogenesis such as during axis elongation in *Drosophila* or during neural tube closure in vertebrate.[137,140,141] The assembly of multicellular actomyosin cables requires coordination to properly

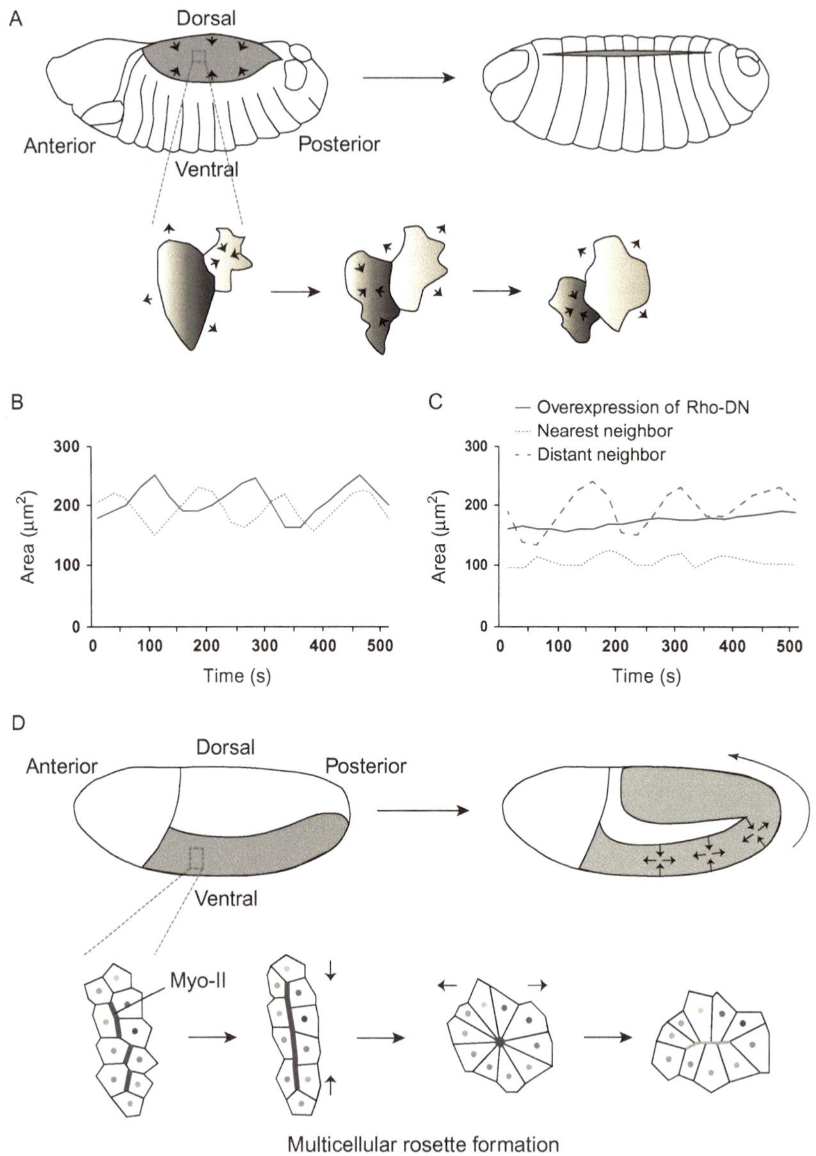

Figure 13.4 Mechanical feedback during tissue morphogenesis. (A–C) Mechanical feedback is required for collective pulsed contraction of amnioserosa cells. (A) Schematic representation of dorsal closure in *Drosophila* embryo, the amnioserosa epithelium is highlighted in orange. Amnioserosa cells display pulsed apical contraction during dorsal closure such that neighboring cells pulse out of phase[136] (B). (C) Overexpression of a dominant negative version of Rho (Rho-DN) in a single cell abolishes constriction pulses cell autonomously (pink curve) as well as in the nearest neighbor (green dotted curve).

link actomyosin structures between cells. An attractive hypothesis is that cells utilize mechanical forces to coordinate the assembly of multicellular actomyosin cables. In the early *Drosophila* embryo, actomyosin cables are assembled to drive directional rearrangement among epithelial cells that leads to the elongation of the embryonic epithelium[142] (Fig. 13.4D). Directed cell–cell intercalation is achieved by the shrinkage and loss of cell–cell interfaces followed by the elongation of new interfaces in the orthogonal direction, leading to convergence and extension of the tissue.[126,143] Shrinkage is associated with the formation of actomyosin cables along the interface of convergent cells. Actomyosin cables along a single interface bring four cells together; however, interfaces and their associated actomyosin cables can become aligned across multiple cells whose contraction results in the formation of multicellular rosettes, with 5–11 cell converging at a single point[137] (Fig. 13.3D, lower panel). Evidence suggests that a positive mechanical feedback coordinates actomyosin cable assembly to align multiple interfaces. Analysis of tissue recoil velocity after laser ablation of cell–cell junctions revealed that interfaces that are part of a multicellular actomyosin cables support higher levels of tension than isolated interfaces enriched in Myo-II. In addition, fluorescence recovery after photobleaching (FRAP) experiments demonstrate that Myo-II is more stable in these multicellular cables, indicating that tension is inversely correlated with Myo-II dynamics. Importantly, laser ablation of multicellular actomyosin cables results in an acute loss of Myo-II intensity in adjacent unablated interfaces. Induction of local stress by micropipette aspiration of the apical membrane of these cells induces Myo-II recruitment. Because tension appears to stabilize Myo-II at junctions, cell–cell interfaces under tension might recruit or stabilize Myo-II in neighboring junctions, resulting higher order organization of actomyosin into multicellular cables.[107]

Pulsatile constriction is not affected however in distant neighbor (gray dashed curve). (D) Mechanical feedback regulates convergence and extension of the *Drosophila* embryonic epithelium. Schematic representation of the elongation of the embryonic epithelium (blue) in *Drosophila* and multicellular rosette formation.[137] Myo-II (red) becomes enriched at junctions oriented with the dorsal–ventral axis. Mechanical feedback recruits or stabilizes Myo-II in neighboring junctions resulting in supracellular cables assembly and alignment of cells interfaces.[107] Contraction of myosin cables leads to rosette formation, which is resolved by the growing of new junction in the orthogonal direction (light blue) and convergence and extension of the tissue. *(C) Adapted from Ref. 138.* (See the color plate.)

While the evidence suggests that both actomyosin meshworks and cables are influenced by mechanical signals transmitted between cells, a future challenge is to determine how cells sense mechanical signals and transduce these signals to cytoskeletal regulation. Because Myo-II itself is a mechanosensitive motor whose attachment to actin filaments is influenced by force, a possibility is that tension directly modulates the mechanochemical cycle of the Myo-II motor leading to stabilization of actomyosin cables or meshworks.[41,45] However, tension could also modulate upstream signaling pathways, such as the activation of the small GTPase RhoA.[144] Activation of RhoA leads to the activation of its downstream effector Rho-associated Kinase, which phosphorylates and activates Myo-II. During *Drosophila* gastrulation, RhoA signaling functions to promote apical constriction and tissue invagination.[5,145,146] Interestingly, mechanical compression of *Drosophila* embryos at this stage can rescue Myo-II accumulation and contraction in mutants that are normally defective in these processes, potentially by mechanically inducing RhoA signaling activity.[147] RhoA signaling could be activated by elevating the activity of upstream G-protein-coupled receptors that localize to the plasma membrane. Increased membrane tension due to mechanical perturbation could inhibit endocytosis of G-protein-coupled receptors, which would increase the signaling activity (more receptors can be activated by an extracellular ligand) and therefore elevate RhoA signaling and Myo-II accumulation. Mechanical compression has also been shown to upregulate expression levels of transcription factors known to activate RhoA signaling, providing a positive transcriptional feedback reinforcing Myo-II accumulation and/or activity.[5,148–151] Interestingly, this model assumes that morphogenetic movements that profoundly reshape the embryo can in turn regulate developmental genes expression through mechanosensitive pathways. Compressive forces exerted by the extension of the underlying tissue on cells that will form the anterior gut have been shown to induce gene expression further required for gut formation in the *Drosophila* embryo.[149] Moreover, a similar mechanosensitive pathway has recently been described in the zebrafish embryo to induce mesoderm identity in a subset of cells that are compressed by epiboly movements.[152] In both cases, mechanical strains developed by morphogenetic movements induce β-catenin nuclear translocation where it regulates targets genes expression. How mechanical stimulus activates β-catenin pathway remains to be fully understood but evidence supports a model where mechanical forces stimulate β-catenin phosphorylation by the kinase Src, leading to β-catenin nuclear translocation.[152]

Actomyosin activity in cells of a tissue may also be altered by the action of mechanosensitive ion channels. The stress-activated calcium channel

TRPM has been implicated in epithelial response to wounding both in *Caenorhabditis elegans* and *Drosophila*.[153,154] Wounding induces the formation of a supracellular actomyosin cable at the margin of the wound acting as a purse string to close the hole. Interestingly, downregulation of TRPM channel impairs actomyosin cable formation and compromise wound closure. Epithelial wounding results in the stretching of the cells surrounding the wound, which may induce the opening of TRPM. In accordance with this view, an increase of intracellular calcium in cells that surround the wound has been observed in both *Drosophila* and *C. elegans*. Calcium fluxes and several ion channels have also recently been implicated in Myo-II accumulation in unwounded embryos, specifically during dorsal closure.[155] Intracellular calcium signaling might regulate actomyosin remodeling through the activation of Gelsolin, a calcium-activated actin filament-severing protein.[154]

3.2. Mechanical reinforcement of junctions

Cell–cell adhesion complexes are critical to transmit tension between cells. The absence of strong adhesion, especially in the presence of tension, can result in catastrophic loss of adhesion and the loss of mechanical integrity in the tissue.[156–158] Given the importance of maintaining tissue integrity during morphogenesis, there could be mechanisms that strengthen adhesion in response to tension to prevent loss of cell–cell adhesion during morphogenetic events. In addition, cell–cell adhesions are well positioned to sense forces transmitted between cells, which could activate signaling pathways that coordinate cell behavior within a tissue. Although recent cell culture studies have suggested that the size and protein composition of E-Cadherin-containing adherens junctions can be modulated by mechanical forces, evidence for a role of adherens junction strengthening has not been yet observed during tissue morphogenesis. However, there is evidence for mechanotransduction playing a role to reinforce other types of cell–cell adhesion between neighboring tissues during morphogenesis.

Because muscle contraction can generate tensile forces of high magnitude, adhesion sites linking muscles to tendons or to an overlying epithelium need to be tightly regulated and reinforced, such that muscle contraction is transduced to surrounding tissues without impairing tissue integrity. In *Drosophila*, muscle–tendon cell attachment is mediated by an integrin-based intercellular adhesion complex that links the two tissues via an intervening ECM. FRAP experiments have been used to measure the stability of proteins within these intercellular adhesions in response to increased or decreased muscle contraction. Inducing muscle hypercontraction leads to greater stability of integrin and associated proteins at adhesion sites, whereas integrin turnover is increased when muscle contraction is blocked.[159] This property may have important implications in the context of development to maintain muscle–tendon connectivity and integrity of both tissues; the authors show that adhesive complexes become more stable over time from late embryogenesis to larval stages as muscle develop and generate increasing levels of tensile force.

Adhesion reinforcement by mechanical forces is not only important for long-term tissue maintenance but also participates in tissue morphogenesis. In *C. elegans*, mechanical stimuli applied by the muscles on the developing epidermis are required for the full elongation of the worm epidermis. Communication between muscles and the epidermis relies on hemidesmosome-like junctions in the epidermis that attach the epidermis to the muscle at the basal side and connect the epidermis to the exoskeleton at the apical side. Intermediate filaments span the apical-basal attachment sites, ensuring mechanical continuity between muscles and the exoskeleton. A signaling pathway has been identified at these junctions that is activated by mechanical tension and stimulates phosphorylation of intermediate filaments and their recruitment to hemidesmosome junctions, hence strengthening the adhesion sites.[160] In addition, genetic interaction studies suggested that muscle tension also activates Myo-II in the epidermis. Together, these results describe a mechanosensitive pathway that reinforces muscle–epidermis attachment sites and their linkage to the intracellular cytoskeleton. In this context, the mechanotransduction pathway is required for tissue elongation and coordination of epidermal and muscle tissue morphogenesis.[160]

In addition to reinforcing stable intercellular junctions, recent studies have demonstrated that forces transmitted through Cadherin-containing junctions can polarize cell migration. Collective cell migration requires maintenance of adhesive contacts between cells and the coordination of cellular protrusions and movements among cells. Cell culture systems of collective cell migration have shown that the speed, persistence, directionality, and coordination of cell movements is increased on stiff substrates, indicating that migrating cells can collectively sense and respond to their mechanical environment.[161] Importantly, this response depends on Cadherin-mediated adhesion between migrating cells as reducing cell–cell adhesion decreases directional migration and the polarization of cells in the direction of migration.[161–163] The importance of cell–cell adhesion in collective cell migration and a potential mechanism for mechanotransduction has been demonstrated during *Xenopus* gastrulation. In *Xenopus*, mesendodermal cells collectively migrate along the ectoderm using the ECM as substrate. Every cell displays leading edge protrusions and a rear retracting edge, but when cells from this tissue are dissociated from one another, they become multipolar, protrude randomly, and fail to effectively migrate toward a source of guidance cues. Strikingly, polarized protrusion in a single cell can be restored when a pulling force is applied to Cadherin adhesion proteins using magnetic tweezers, with protrusions forming on the opposite side of the cell as the pulled bead.[164] The mechanotransduction pathway involves the catenin Plakoglobin (PG) that links the cytoplasmic tail of Cadherin to intermediate filaments. Local forces applied to Cadherin result in PG-dependent recruitment of intermediate filaments and reorganization of the intermediate filament network.[164] PG might then act as a mechanosensor, possibly changing conformation under force as it as been shown for α-catenin.[33] The model posits that intercellular adhesion integrates traction forces arising from every migrating cell, leading to higher cell–cell tension at the rear of migrating cells. This polarized mechanical signal is transduced into polarized localization of intermediate filaments and polarized protrusive activity of cells.[164]

3.3. Extrinsic mechanical constraints influencing tissue shape

Morphogenesis is an integrated process and different tissue movements and deformations often occur at the same time. Cells and tissues undergoing shape changes in concert may actively apply forces to each other. In addition, cells and tissues have viscoelastic properties, such that neighboring cells or tissues can give rise to opposing forces that must be overcome to reshape a tissue.[13] Forces that result from mechanical constraints on tissues could be sensed by cells and impact both cell and tissue shape. A classic example of the impact of surrounding tissues on cell shape change is the apical constriction of bottle cells during *Xenopus* gastrulation. *In vivo*, bottle cells display anisotropic constriction, however, when explanted and cultured *in vitro* bottle cells constrict isotropically.[165] This observation suggests that the mechanics of the surrounding tissue can influence individual cell shape changes.

Apical constriction is also anisotropic during *Drosophila* gastrulation. *Drosophila* gastrulation involves collective apical constriction of cells in an epithelial sheet that forms a furrow along the ventral midline where cells are pulled inside the embryo along the DV axis, but the tissue remains extended along the AP axis[166] (Fig. 13.5A). Because of this anisotropy in tissue movement, it is not surprising that the underlying cell shape changes

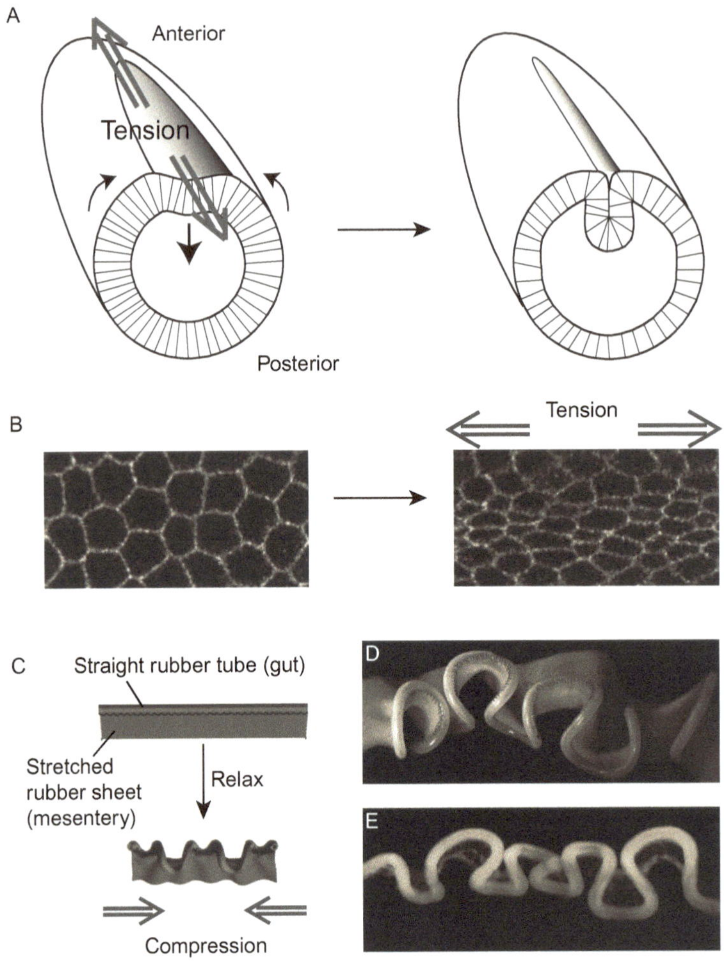

Figure 13.5 Extrinsic mechanical constraints influencing tissue shape. (A) Tissue-level tension influences individual cell shape change and regulates tissue invagination. Schematic representation of ventral furrow formation in *Drosophila* embryo. During invagination, tension is highest along the anterior–posterior axis (red arrows).[157] Anisotropic epithelial tension results in anisotropic apical constriction (B) and the formation of a narrow furrow. (C) Gut-looping morphogenesis is driven by mechanical constraints exerted by the connected mesentery to the gut tube. Looping pattern of the gut can be modeled using a stretched rubber sheet (pink, mesentery) attached to a straight, non-stretched rubber tube (blue, gut). Relaxation leads to generation of compression forces and looping of the tube. The rubber model (D) deforms into a structure similar to the chicken gut (E). *Adapted from Ref. 167.* (See the color plate.)

are anisotropic: ventral furrow cells undergo anisotropic apical constriction such that cells constrict predominantly in the DV direction and remain longer along the AP axis (Fig. 13.5B). This anisotropy is not a cell autonomous property but appears to rely on the global tissue force pattern. Laser ablation experiments and the resulting recoil of the surrounding tissue have suggested that epithelial tension is highest along the AP axis, the axis along which cells are most elongated. Epithelial tension can be locally released by reducing cell–cell adhesion, which causes a loss in tissue integrity in the form of tissue-wide tearing events where clusters of cells separate. Importantly, individual cells that undergo apical constriction after a neighboring tissue tear constrict isotropically, suggesting that integration of tensile forces across the tissue causes cell anisotropy.[157] The mechanisms that generate this asymmetry in epithelial tension are still not clear. It is possible that there is greater resistance to pulling forces oriented in the AP direction at the poles of the embryo than to pulling lateral cells. Alternatively, a gradient of Myo-II contractility along the DV axis could be responsible for directing movement toward the ventral midline.[168] The importance of this anisotropy for tissue invagination and how this tension might alter the activity of the force generating actomyosin cytoskeleton are future areas of investigation.

The ability of tissue-level forces to impact morphogenesis extends beyond morphogenetic processes that require active force generation by cytoskeletal networks. A series of recent studies of the vertebrate gut have demonstrated that differences in the growth rate and mechanics of connected tissues can sculpt tissues into different shapes. The embryonic gut is initially a straight tube, but rapidly grows and folds in the intestinal cavity forming a looping pattern that is stereotypical for different species of animals. However, gut looping neither does not result from differential cell proliferation within the gut tissue, nor does looping result from the spatial constraint of the body cavity in which it grows. Instead, gut looping results from the coupling of the gut along its entire length to a tissue called the mesentery. The gut tube grows faster than the mesentery, resulting in an elastic stretch of the mesentery and resulting compression and buckling of the gut tube, giving it a wavy shape (Fig. 13.5C–E). Quantitative measurement of the mechanical properties of the gut and mesentery combined with a theoretical model of the gut–mesentery composite was able to accurately predict the looping pattern in several species[167] (Fig. 13.5C–E). Thus, differential growth between coupled tissues can drive morphogenesis.

In addition to looping, the gut tissue forms internal folds that result in villi. Interestingly, the formation of villi can also be explained by mechanical interactions between two tissues, in this case the gut epithelium and surrounding muscle cells.[169] During villification, the gut epithelium first buckles inward to form longitudinal ridges that are parallel to the axis of the gut tube. Longitudinal ridges subsequently buckle to form parallel zigzags and finally bulges arise from the zigzag pattern and individual finger-like villi are produced. Importantly, every step in this morphogenetic process leading to a new geometrical conformation of the epithelium coincides with the differentiation of a new layer of smooth muscle, which are all required for proper villification. Given that the formation of longitudinal ridges can be rescued by replacing muscle cells with a synthetic silk tube, the muscle layer appears to act as a stiff barrier that prevents expansion of the growing epithelium and promotes buckling.[169] Mathematical modeling considering the gut epithelial layer as flat elastic sheet that is compressed during growth first in the axial then in the longitudinal axes mimic the role of the three muscle layers in the stepwise process of villification. Although the two above models of gut looping and villification both describe purely mechanical interpretations of gut morphogenesis during growth, suggesting that the epithelium fold passively, it is not known whether epithelial cells actively respond to mechanical stress by adjusting their growth rate or by regulating their stiffness through remodeling of the actomyosin cytoskeleton. These examples illustrate how forces across different spatial scales, molecular, cellular to the tissue level can serve as signals in multicellular contexts to coordinate cell and tissue behavior.

ACKNOWLEDGMENTS

We thank Loic LeGoff, Thomas Lecuit, Olivier Hamant, Maithreyi Narasimha, Amy Syher, Clifford Tabin, and L. Mahadevan for kindly providing us with original figures and allow us to reproduce their data. In addition, we would like to thank the members of the Martin lab for fruitful discussions. We are supported by NIH grant GM105984 (A. C. M.) and EMBO fellowship ALTF 1082-2012 (S. C.).

REFERENCES

1. Na S, Collin O, Chowdhury F, et al. Rapid signal transduction in living cells is a unique feature of mechanotransduction. *Proc Natl Acad Sci*. 2008;105(18):6626–6631.
2. Quintin S, Gally C, Labouesse M. Epithelial morphogenesis in embryos: asymmetries, motors and brakes. *Trends Genet*. 2008;24(5):221–230.
3. Lecuit T, Lenne P-F, Munro E. Force generation, transmission, and integration during cell and tissue morphogenesis. *Annu Rev Cell Dev Biol*. 2011;27(1):157–184.
4. Reymann AC, Boujemaa-Paterski R, Martiel JL, et al. Actin network architecture can determine myosin motor activity. *Science*. 2012;336(6086):1310–1314.

5. Dawes-Hoang RE, Parmar KM, Christiansen AE, Phelps CB, Brand AH, Wieschaus EF. Folded gastrulation, cell shape change and the control of myosin localization. *Development*. 2007;134(24):4507.
6. Roh-Johnson M, Shemer G, Higgins CD, et al. Triggering a cell shape change by exploiting preexisting actomyosin contractions. *Science*. 2012;335(6073):1232–1235.
7. Levayer R, Lecuit T. Oscillation and polarity of E-Cadherin asymmetries control actomyosin flow patterns during morphogenesis. *Dev Cell*. 2013;26(2):162–175.
8. Salbreux G, Charras G, Paluch E. Actin cortex mechanics and cellular morphogenesis. *Trends Cell Biol*. 2012;22(10):536–545.
9. Storm C, Pastore JJ, MacKintosh FC, Lubensky TC, Janmey PA. Nonlinear elasticity in biological gels. *Nature*. 2005;435(7039):191–194.
10. Zhou J, Kim HY, Davidson LA. Actomyosin stiffens the vertebrate embryo during crucial stages of elongation and neural tube closure. *Development*. 2009;136(4):677–688.
11. Gardel ML, Nakamura F, Hartwig JH, Crocker JC, Stossel TP, Weitz DA. Prestressed F-actin networks cross-linked by hinged filamins replicate mechanical properties of cells. *Proc Natl Acad Sci*. 2006;103(6):1762–1767.
12. Kasza KE, Rowat AC, Liu J, et al. The cell as a material. *Curr Opin Cell Biol*. 2007;19(1):101–107.
13. Davidson LA. Embryo mechanics: balancing force production with elastic resistance during morphogenesis. *Curr Top Dev Biol*. 2011;95:215–241.
14. Huveneers S, Oldenburg J, Spanjaard E, et al. Vinculin associates with endothelial VE-cadherin junctions to control force-dependent remodeling. *J Cell Biol*. 2012;196(5):641–652.
15. Naumanen P, Lappalainen P, Hotulainen P. Mechanisms of actin stress fibre assembly. *J Microsc*. 2008;231(3):446–454.
16. Stricker J, Falzone T, Gardel ML. Mechanics of the F-actin cytoskeleton. *J Biomech*. 2010;43(1):9–14.
17. Davidson L, Dassow von M, Zhou J. Multi-scale mechanics from molecules to morphogenesis. *Int J Biochem Cell Biol*. 2009;41(11):2147–2162.
18. Hoffman BD, Grashoff C, Schwartz MA. Dynamic molecular processes mediate cellular mechanotransduction. *Nature*. 2011;475(7356):316–323.
19. Vignaud T, Blanchoin L, Théry M. Directed cytoskeleton self-organization. *Trends Cell Biol*. 2012;22(12):671–682.
20. Zaidel-Bar R, Geiger B. The switchable integrin adhesome. *J Cell Sci*. 2010;123(Pt 9):1385–1388.
21. Parsons JT, Horwitz AR, Schwartz MA. Cell adhesion: integrating cytoskeletal dynamics and cellular tension. *Nat Rev Mol Cell Biol*. 2010;11(9):633–643.
22. Riveline D, Zamir E, Balaban NQ, et al. Focal contacts as mechanosensors: externally applied local mechanical force induces growth of focal contacts by an mDia1-dependent and ROCK-independent mechanism. *J Cell Biol*. 2001;153(6):1175–1186.
23. Schiller HB, Fässler R. Mechanosensitivity and compositional dynamics of cell–matrix adhesions. *EMBO Rep*. 2013;14(6):509–519.
24. del Rio A, Perez-Jimenez R, Liu R, Roca-Cusachs P, Fernandez JM, Sheetz MP. Stretching single talin rod molecules activates vinculin binding. *Science*. 2009;323(5914):638–641.
25. Grashoff C, Hoffman BD, Brenner MD, et al. Measuring mechanical tension across vinculin reveals regulation of focal adhesion dynamics. *Nature*. 2010;466(7303):263–266.
26. Sawada Y, Tamada M, Dubin-Thaler BJ, et al. Force sensing by mechanical extension of the Src family kinase substrate p130Cas. *Cell*. 2006;127(5):1015–1026.
27. Brown NH. Cell–cell adhesion via the ECM: integrin genetics in fly and worm. *Matrix Biol*. 2000;19(3):191–201.

28. Niessen CM, Leckband D, Yap AS. Tissue organization by cadherin adhesion molecules: dynamic molecular and cellular mechanisms of morphogenetic regulation. *Physiol Rev*. 2011;91(2):691–731.
29. Yamada S, Pokutta S, Drees F, Weis WI, Nelson WJ. Deconstructing the cadherin–catenin–actin complex. *Cell*. 2005;123(5):889–901.
30. Cavey M, Rauzi M, Lenne P-F, Lecuit T. A two-tiered mechanism for stabilization and immobilization of E-cadherin. *Nature*. 2008;453(7196):751–756.
31. Huber AH, Stewart DB, Laurents DV, Nelson WJ, Weis WI. The cadherin cytoplasmic domain is unstructured in the absence of beta-catenin. A possible mechanism for regulating cadherin turnover. *J Biol Chem*. 2001;276(15):12301–12309.
32. Liu Z, Tan JL, Cohen DM, et al. Mechanical tugging force regulates the size of cell–cell junctions. *Proc Natl Acad Sci*. 2010;107(22):9944–9949.
33. Yonemura S, Wada Y, Watanabe T, Nagafuchi A, Shibata M. α-Catenin as a tension transducer that induces adherens junction development. *Nat Cell Biol*. 2010;12(6): 533–542.
34. Twiss F, le Duc Q, Van Der Horst S, et al. Vinculin-dependent Cadherin mechanosensing regulates efficient epithelial barrier formation. *Biol Open*. 2012;1(11): 1128–1140.
35. Bershadsky AD, Ballestrem C, Carramusa L, et al. Assembly and mechanosensory function of focal adhesions: experiments and models. *Eur J Cell Biol*. 2006;85(3–4):165–173.
36. Higashida C, Kiuchi T, Akiba Y, et al. F- and G-actin homeostasis regulates mechanosensitive actin nucleation by formins. *Nat Cell Biol*. 2013;15(4):395–405.
37. Murrell MP, Gardel ML. F-actin buckling coordinates contractility and severing in a biomimetic actomyosin cortex. *Proc Natl Acad Sci*. 2012;109(51):20820–20825.
38. Wada KI, Itoga K, Okano T, Yonemura S, Sasaki H. Hippo pathway regulation by cell morphology and stress fibers. *Development*. 2011;138(18):3907–3914.
39. Vignaud T, Galland R, Tseng Q, Blanchoin L, Colombelli J, Thery M. Reprogramming cell shape with laser nano-patterning. *J Cell Sci*. 2012;125(9):2134–2140.
40. Howard J. *Mechanics of Motor Proteins and the Cytoskeleton*. Sunderland, MA: Sinauer Associates, Inc.; 2001.
41. Kovács M, Thirumurugan K, Knight PJ, Sellers JR. Load-dependent mechanism of nonmuscle myosin 2. *Proc Natl Acad Sci*. 2007;104(24):9994–9999.
42. Veigel C, Molloy JE, Schmitz S, Kendrick-Jones J. Load-dependent kinetics of force production by smooth muscle myosin measured with optical tweezers. *Nat Cell Biol*. 2003;5(11):980–986.
43. Laakso JM, Lewis JH, Shuman H, Ostap EM. Myosin I can act as a molecular force sensor. *Science*. 2008;321(5885):133–136.
44. Merkel R, Simson R, Simson DA, et al. A micromechanic study of cell polarity and plasma membrane cell body coupling in Dictyostelium. *Biophys J*. 2000;79(2):707–719.
45. Ren Y, Effler JC, Norstrom M, et al. Mechanosensing through cooperative interactions between myosin II and the actin crosslinker cortexillin I. *Curr Biol*. 2009;19(17): 1421–1428.
46. Uyeda TQP, Iwadate Y, Umeki N, Nagasaki A, Yumura S. Stretching actin filaments within cells enhances their affinity for the myosin II motor domain. *PLoS One*. 2011;6(10):e26200.
47. Wilson CA, Tsuchida MA, Allen GM, et al. Myosin II contributes to cell-scale actin network treadmilling through network disassembly. *Nature*. 2010;465(7296):373–377.
48. Murthy K, Wadsworth P. Myosin-II-dependent localization and dynamics of F-actin during cytokinesis. *Curr Biol*. 2005;15(8):724–731.
49. Diz-Muñoz A, Fletcher DA, Weiner OD. Use the force: membrane tension as an organizer of cell shape and motility. *Trends Cell Biol*. 2013;23(2):47–53.

50. Arnadóttir J, Chalfie M. Eukaryotic mechanosensitive channels. *Annu Rev Biophys.* 2010;39(1):111–137.
51. Clark K, Middelbeek J, van Leeuwen FN. Interplay between TRP channels and the cytoskeleton in health and disease. *Eur J Cell Biol.* 2008;87(8–9):631–640.
52. Kuipers AJ, Middelbeek J, van Leeuwen FN, Eur J. *Cell Biol.* 2012;91(11–12):834–846.
53. Boulant S, Kural C, Zeeh J-C, Ubelmann F, Kirchhausen T. Actin dynamics counteract membrane tension during clathrin-mediated endocytosis. *Nat Cell Biol.* 2011;13(9):1124–1131.
54. Gauthier NC, Fardin MA, Roca-Cusachs P, Sheetz MP. Temporary increase in plasma membrane tension coordinates the activation of exocytosis and contraction during cell spreading. *Proc Natl Acad Sci.* 2011;108(35):14467–14472.
55. Théry M, Pépin A, Dressaire E, Chen Y, Bornens M. Cell distribution of stress fibres in response to the geometry of the adhesive environment. *Cell Motil Cytoskeleton.* 2006;63(6):341–355.
56. Davies PF, Robotewskyj A, Griem ML. Quantitative studies of endothelial cell adhesion. Directional remodeling of focal adhesion sites in response to flow forces. *J Clin Invest.* 1994;93(5):2031–2038.
57. Galbraith CG, Yamada KM, Sheetz MP. The relationship between force and focal complex development. *J Cell Biol.* 2002;159(4):695–705.
58. Vogel V, Sheetz M. Local force and geometry sensing regulate cell functions. *Nat Rev Mol Cell Biol.* 2006;7(4):265–275.
59. Hertwig O. The problem of fertilization and of isotropy of the egg, a theory of heredity. *Jena Z Naturwiss.* 1884;18:21–23.
60. Wilson EB. *The Cell in Development and Inheritance.* Macmillan; 1925.
61. Théry M, Bornens M. Cell shape and cell division. *Curr Opin Cell Biol.* 2006;18(6):648–657.
62. Minc N, Burgess D, Chang F. Influence of cell geometry on division-plane positioning. *Cell.* 2011;144(3):414–426.
63. Théry M, Racine V, Pépin A, et al. The extracellular matrix guides the orientation of the cell division axis. *Nat Cell Biol.* 2005;7(10):947–953.
64. Théry M, Jiménez-Dalmaroni A, Racine V, Bornens M, Jülicher F. Experimental and theoretical study of mitotic spindle orientation. *Nature.* 2007;447(7143):493–496.
65. Fink J, Carpi N, Betz T, et al. External forces control mitotic spindle positioning. *Nat Cell Biol.* 2011;13(7):771–778.
66. Louveaux M, Hamant O. The mechanics behind cell division. *Curr Opin Plant Biol.* 2013;16(6):774–779.
67. Minc N, Piel M. Predicting division plane position and orientation. *Trends Cell Biol.* 2012;22(4):193–200.
68. Lecuit T, Le Goff L. Orchestrating size and shape during morphogenesis. *Nature.* 2007;450(7167):189–192.
69. Schwank G, Basler K. Regulation of organ growth by morphogen gradients. *Cold Spring Harb Perspect Biol.* 2010;2(1):a001669.
70. Wartlick O, Kicheva A, Gonzalez-Gaitan M. Morphogen gradient formation. *Cold Spring Harb Perspect Biol.* 2009;1(3):a001255.
71. Wolpert L. Positional information revisited. *Development.* 1989;107(Suppl):3–12.
72. Wolpert L. Positional information and patterning revisited. *J Theor Biol.* 2011;269(1):359–365.
73. Wartlick O, Mumcu P, Jülicher F, Gonzalez-Gaitan M. Understanding morphogenetic growth control—lessons from flies. *Nat Rev Mol Cell Biol.* 2011;12(9):594–604.
74. Affolter M, Basler K. The Decapentaplegic morphogen gradient: from pattern formation to growth regulation. *Nat Rev Genet.* 2007;8(9):663–674.

75. Aegerter-Wilmsen T, Aegerter CM, Hafen E, Basler K. Model for the regulation of size in the wing imaginal disc of Drosophila. *Mech Dev*. 2007;124(4):318–326.
76. Dupont S, Morsut L, Aragona M, et al. Role of YAP/TAZ in mechanotransduction. *Nature*. 2012;474(7350):179–183.
77. Sansores-Garcia L, Bossuyt W, Wada K-I, et al. Modulating F-actin organization induces organ growth by affecting the Hippo pathway. *EMBO J*. 2011;30(12):2325–2335.
78. Aragona M, Panciera T, Manfrin A, et al. A mechanical checkpoint controls multicellular growth through YAP/TAZ regulation by actin-processing factors. *Cell*. 2013;154(5):1047–1059.
79. Fernandez BG, Gaspar P, Bras-Pereira C, Jezowska B, Rebelo SR, Janody F. Actin-capping protein and the Hippo pathway regulate F-actin and tissue growth in Drosophila. *Development*. 2011;138(11):2337–2346.
80. Wartlick O, Mumcu P, Kicheva A, et al. Dynamics of Dpp signaling and proliferation control. *Science*. 2011;331(6021):1154–1159.
81. Milán M, Campuzano S, García-Bellido A. Cell cycling and patterned cell proliferation in the wing primordium of Drosophila. *Proc Natl Acad Sci*. 1996;93(2):640–645.
82. Martín-Castellanos C, Edgar BA. A characterization of the effects of Dpp signaling on cell growth and proliferation in the Drosophila wing. *Development*. 2002;129(4):1003–1013.
83. García-Bellido A, Merriam JR. Parameters of the wing imaginal disc development of Drosophila melanogaster. *Dev Biol*. 1971;24(1):61–87.
84. Day SJ, Lawrence PA. Measuring dimensions: the regulation of size and shape. *Development*. 2000;127(14):2977–2987.
85. Hufnagel L, Teleman AA, Rouault H, Cohen SM, Shraiman BI. On the mechanism of wing size determination in fly development. *Proc Natl Acad Sci*. 2007;104(10):3835–3840.
86. Aegerter-Wilmsen T, Heimlicher MB, Smith AC, et al. Integrating force-sensing and signaling pathways in a model for the regulation of wing imaginal disc size. *Development*. 2012;139(17):3221–3231.
87. Shraiman BI. Mechanical feedback as a possible regulator of tissue growth. *Proc Natl Acad Sci*. 2005;102(9):3318–3323.
88. LeGoff L, Rouault H, Lecuit T. A global pattern of mechanical stress polarizes cell divisions and cell shape in the growing Drosophila wing disc. *Development*. 2013;140(19):4051–4059.
89. Gibson MC, Patel AB, Nagpal R, Perrimon N. The emergence of geometric order in proliferating metazoan epithelia. *Nature*. 2006;442(7106):1038–1041.
90. Mao Y, Tournier AL, Hoppe A, Kester L, Thompson BJ, Tapon N. Differential proliferation rates generate patterns of mechanical tension that orient tissue growth. *EMBO J*. 2013;32(21):2790–2803.
91. Nienhaus U, Aegerter-Wilmsen T, Aegerter CM. Determination of mechanical stress distribution in Drosophila wing discs using photoelasticity. *Mech Dev*. 2009;126(11–12):942–949.
92. Schluck T, Nienhaus U, Aegerter-Wilmsen T, Aegerter CM. Mechanical control of organ size in the development of the Drosophila wing disc. *PLoS One*. 2013;8(10):e76171.
93. Saucedo LJ, Edgar BA. Filling out the Hippo pathway. *Nat Rev Mol Cell Biol*. 2007;8(8):613–621.
94. Pan D. The hippo signaling pathway in development and cancer. *Dev Cell*. 2010;19(4):491–505.
95. Harvey KF, Pfleger CM, Hariharan IK. The Drosophila Mst ortholog, hippo, restricts growth and cell proliferation and promotes apoptosis. *Cell*. 2003;114(4):457–467.
96. Oh H, Irvine KD. In vivo regulation of Yorkie phosphorylation and localization. *Development*. 2008;135(6):1081–1088.

97. Oh H, Irvine KD. In vivo analysis of Yorkie phosphorylation sites. *Oncogene*. 2009; 28(17):1916–1927.
98. Rogulja D, Rauskolb C, Irvine KD. Morphogen control of wing growth through the Fat signaling pathway. *Dev Cell*. 2008;15(2):309–321.
99. Yu F-X, Guan K-L. The Hippo pathway: regulators and regulations. *Genes Dev*. 2013;27(4):355–371.
100. Halder G, Johnson RL. Hippo signaling: growth control and beyond. *Development*. 2011;138(1):9–22.
101. Boggiano JC, Fehon RG. Growth control by committee: intercellular junctions, cell polarity, and the cytoskeleton regulate Hippo signaling. *Dev Cell*. 2012;22(4): 695–702.
102. Halder G, Dupont S, Piccolo S. Transduction of mechanical and cytoskeletal cues by YAP and TAZ. *Nat Rev Mol Cell Biol*. 2012;13(9):591–600.
103. Zhao B, Li L, Wang L, Wang C-Y, Yu J, Guan K-L. Cell detachment activates the Hippo pathway via cytoskeleton reorganization to induce anoikis. *Genes Dev*. 2012;26(1):54–68.
104. Calvo F, Ege N, Grande-Garcia A, et al. Mechanotransduction and YAP-dependent matrix remodelling is required for the generation and maintenance of cancer-associated fibroblasts. *Nat Cell Biol*. 2013;15(6):637–646.
105. Yin F, Yu J, Zheng Y, Chen Q, Zhang N, Pan D. Spatial organization of Hippo signaling at the plasma membrane mediated by the tumor suppressor merlin/NF2. *Cell*. 2013;154(6):1342–1355.
106. Uyttewaal M, Burian A, Alim K, et al. Mechanical stress acts via katanin to amplify differences in growth rate between adjacent cells in Arabidopsis. *Cell*. 2012;149(2): 439–451.
107. Fernandez-Gonzalez R, Simoes S de M, Röper J-C, Eaton S, Zallen JA. Myosin II dynamics are regulated by tension in intercalating cells. *Dev Cell*. 2009;17(5):736–743.
108. Campinho P, Behrndt M, Ranft J, Risler T, Minc N, Heisenberg C-P. Tension-oriented cell divisions limit anisotropic tissue tension in epithelial spreading during zebrafish epiboly. *Nat Cell Biol*. 2013;5(11):1–13.
109. Morin X, Bellaïche Y. Mitotic spindle orientation in asymmetric and symmetric cell divisions during animal development. *Dev Cell*. 2011;21(1):102–119.
110. Mao Y, Tournier AL, Bates PA, Gale JE, Tapon N, Thompson BJ. Planar polarization of the atypical myosin Dachs orients cell divisions in Drosophila. *Genes Dev*. 2011;25(2):131–136.
111. Mao Y, Rauskolb C, Cho E, et al. Dachs: an unconventional myosin that functions downstream of Fat to regulate growth, affinity and gene expression in Drosophila. *Development*. 2006;133(13):2539–2551.
112. Bosveld F, Bonnet I, Guirao B, et al. Mechanical control of morphogenesis by Fat/Dachsous/Four-jointed planar cell polarity pathway. *Science*. 2012;336(6082):724–727.
113. Landrein B, Hamant O. How mechanical stress controls microtubule behavior and morphogenesis in plants: history, experiments and revisited theories. *Plant J*. 2013; 75(2):324–338.
114. Green PB. Mechanism for plant cellular morphogenesis. *Science*. 1962;138(3548): 1404–1405.
115. Baskin T, Meekes H, Liang B, Sharp R. Regulation of growth anisotropy in well-watered and water-stressed maize roots. II. Role of cortical microtubules and cellulose microfibrils. *Plant Physiol*. 1999;119(2):681–692.
116. Paredez AR, Somerville CR, Ehrhardt DW. Visualization of cellulose synthase demonstrates functional association with microtubules. *Science*. 2006;312(5779): 1491–1495.

117. Bringmann M, Landrein B, Schudoma C, Hamant O, Hauser M-T, Persson S. Cracking the elusive alignment hypothesis: the microtubule-cellulose synthase nexus unraveled. *Trends Plant Sci.* 2012;17(11):666–674.
118. Baskin TI, Gu Y. Making parallel lines meet: transferring information from microtubules to extracellular matrix. *Cell Adh Migr.* 2012;6(5):404–408.
119. Hamant O, Heisler MG, Jönsson H, et al. Developmental patterning by mechanical signals in Arabidopsis. *Science.* 2008;322(5908):1650–1655.
120. Putnam AJ, Cunningham JJ, Dennis RG, Linderman JJ, Mooney DJ. Microtubule assembly is regulated by externally applied strain in cultured smooth muscle cells. *J Cell Sci.* 1998;111(Pt 22):3379–3387.
121. Janson ME, Dogterom M. Scaling of microtubule force–velocity curves obtained at different tubulin concentrations. *Phys Rev Lett.* 2004;92(24):248101.
122. Zheng J, Buxbaum RE, Heidemann SR. Investigation of microtubule assembly and organization accompanying tension-induced neurite initiation. *J Cell Sci.* 1993; 104(Pt 4):1239–1250.
123. Wightman R, Turner SR. Severing at sites of microtubule crossover contributes to microtubule alignment in cortical arrays. *Plant J.* 2007;52(4):742–751.
124. Zhang Q, Fishel E, Bertroche T, Dixit R. Microtubule severing at crossover sites by katanin generates ordered cortical microtubule arrays in Arabidopsis. *Curr Biol.* 2013;23(21):2191–2195.
125. Martin AC, Kaschube M, Wieschaus EF. Pulsed contractions of an actin. *Nature.* 2009;457(7228):495–501.
126. Rauzi M, Verant P, Lecuit T, Lenne P-F. Nature and anisotropy of cortical forces orienting Drosophila tissue morphogenesis. *Nat Cell Biol.* 2008;10(12):1401–1410.
127. Blanchard GB, Murugesu S, Adams RJ, Martinez-Arias A, Gorfinkiel N. Cytoskeletal dynamics and supracellular organisation of cell shape fluctuations during dorsal closure. *Development.* 2010;137(16):2743–2752.
128. David DJV, Tishkina A, Harris TJC. The PAR complex regulates pulsed actomyosin contractions during amnioserosa apical constriction in Drosophila. *Development.* 2010;137(10):1645–1655.
129. He L, Wang X, Tang HL, Montell DJ. Tissue elongation requires oscillating contractions of a basal actomyosin network. *Nat Cell Biol.* 2010;12(12):1133–1142.
130. Skoglund P, Rolo A, Chen X, Gumbiner BM, Keller R. *Development.* 2008;135(14): 2435–2444, 18550716.
131. Kim HY, Davidson LA. Punctuated actin contractions during convergent extension and their permissive regulation by the non-canonical Wnt-signaling pathway. *J Cell Sci.* 2011;124(Pt 4):635–646.
132. Rauzi M, Lenne P-F, Lecuit T. Planar polarized actomyosin contractile flows control epithelial junction remodelling. *Nature.* 2011;468(7327):1110–1114.
133. Gorfinkiel N, Blanchard GB. Dynamics of actomyosin contractile activity during epithelial morphogenesis. *Curr Opin Cell Biol.* 2011;23(5):531–539.
134. Jacinto A, Martin P. Morphogenesis: unravelling the cell biology of hole closure. *Curr Biol.* 2001;11(17):R705–R707.
135. Jacinto A, Woolner S, Martin P. Dynamic analysis of dorsal closure in Drosophila: from genetics to cell biology. *Dev Cell.* 2002;3(1):9–19.
136. Solon J, Kaya-Copur A, Colombelli J, Brunner D. Pulsed forces timed by a ratchet-like mechanism drive directed tissue movement during dorsal closure. *Cell.* 2009;137(7): 1331–1342.
137. Blankenship JT, Backovic ST, Sanny JSP, Weitz O, Zallen JA. Multicellular rosette formation links planar cell polarity to tissue morphogenesis. *Dev Cell.* 2006;11(4): 459–470.

138. Saravanan S, Meghana C, Narasimha M. Local, cell-nonautonomous feedback regulation of myosin dynamics patterns transitions in cell behavior: a role for tension and geometry? *Mol Biol Cell.* 2013;24(15):2350–2361.
139. Jayasinghe AK, Crews SM, Mashburn DN, Hutson MS. Apical oscillations in amnioserosa cells: basolateral coupling and mechanical autonomy. *Biophys J.* 2013;105(1):255–265.
140. Nishimura T, Honda H, Takeichi M. Planar cell polarity links axes of spatial dynamics in neural-tube closure. *Cell.* 2012;149(5):1084–1097.
141. Jacinto A, Wood W, Woolner S, et al. Dynamic analysis of actin cable function during Drosophila dorsal closure. *Curr Biol.* 2002;12(14):1245–1250.
142. Zallen JA, Wieschaus E. Patterned gene expression directs bipolar planar polarity in Drosophila. *Dev Cell.* 2004;6(3):343–355.
143. Bertet C, Sulak L, Lecuit T. Myosin-dependent junction remodelling controls planar cell intercalation and axis elongation. *Nature.* 2004;429(6992):667–671.
144. Guilluy C, Swaminathan V, Garcia-Mata R, O'Brien ET, Superfine R, Burridge K. The Rho GEFs LARG and GEF-H1 regulate the mechanical response to force on integrins. *Nat Cell Biol.* 2011;13(6):724–729.
145. Mason FM, Tworoger M, Martin AC. Apical domain polarization localizes actin-myosin activity to drive ratchet-like apical constriction. *Nat Cell Biol.* 2013;15(8):926–936.
146. Kolsch V, Seher T, Fernandez-Ballester GJ, Serrano L, Leptin M. Control of Drosophila gastrulation by apical localization of adherens junctions and RhoGEF2. *Science.* 2007;315(5810):384–386.
147. Pouille PA, Ahmadi P, Brunet AC, Farge E. Mechanical signals trigger myosin II redistribution and mesoderm invagination in Drosophila embryos. *Sci Signal.* 2009;2(66): ra16.
148. Farge E. Mechanical induction of twist in the Drosophila foregut/stomodeal primordium. *Curr Biol.* 2003;13(16):1365–1377.
149. Desprat N, Supatto W, Pouille P-A, Beaurepaire E, Farge E. Tissue deformation modulates twist expression to determine anterior midgut differentiation in Drosophila embryos. *Dev Cell.* 2008;15(3):470–477.
150. Costa M, Wilson ET, Wieschaus E. A putative cell signal encoded by the folded gastrulation gene coordinates cell shape changes during Drosophila gastrulation. *Cell.* 1994;76(6):1075–1089.
151. Morize P, Christiansen AE, Costa M, Parks S, Wieschaus E. Hyperactivation of the folded gastrulation pathway induces specific cell shape changes. *Development.* 1998;125(4):589–597.
152. Brunet T, Bouclet A, Ahmadi P, et al. Evolutionary conservation of early mesoderm specification by mechanotransduction in Bilateria. *Nat Commun.* 2013;14:2821.
153. Xu S, Chisholm AD. Signaling pathway promotes actin-mediated epidermal wound closure in *C. elegans*. *Curr Biol.* 2011;21(23):1960–1967.
154. Antunes M, Pereira T, Cordeiro JV, Almeida L, Jacinto A. Coordinated waves of actomyosin flow and apical cell constriction immediately after wounding. *J Cell Biol.* 2013;202(2):365–379.
155. Hunter GL, Crawford JM, Genkins JZ, Kiehart DP. Ion channels contribute to the regulation of cell sheet forces during Drosophila dorsal closure. *Development.* 2014;141(2):325–334.
156. Gorfinkiel N, Blanchard GB, Adams RJ, Martinez-Arias A. Mechanical control of global cell behaviour during dorsal closure in Drosophila. *Development.* 2009;136(11):1889–1898.
157. Martin AC, Gelbart M, Fernandez-Gonzalez R, Kaschube M, Wieschaus EF. Integration of contractile forces during tissue invagination. *J Cell Biol.* 2010;188(5):735–749.

158. Narasimha M, Brown NH. Novel functions for integrins in epithelial morphogenesis. *Curr Biol.* 2004;14(5):381–385.
159. Pines M, Das R, Ellis SJ, et al. Mechanical force regulates integrin turnover *in Drosophila in vivo*. *Nat Cell Biol.* 2012;14(9):935–943.
160. Zhang H, Landmann F, Zahreddine H, Rodriguez D, Koch M, Labouesse M. A tension-induced mechanotransduction pathway promotes epithelial morphogenesis. *Nature.* 2012;470(7336):99–103.
161. Ng MR, Besser A, Danuser G, Brugge JS. Substrate stiffness regulates cadherin-dependent collective migration through myosin-II contractility. *J Cell Biol.* 2012;199(3):545–563.
162. Tambe DT, Hardin CC, Angelini TE, et al. Collective cell guidance by cooperative intercellular forces. *Nat Mater.* 2011;10(6):469–475.
163. Trepat X, Chen Z, Jacobson K. Cell migration. *Compr Physiol.* 2012;2(4):2369–2392.
164. Weber GF, Bjerke MA, DeSimone DW. A mechanoresponsive cadherin–keratin complex directs polarized protrusive behavior and collective cell migration. *Dev Cell.* 2012;22(1):104–115.
165. Hardin J, Keller R. The behaviour and function of bottle cells during gastrulation of Xenopus laevis. *Development.* 1988;103(1):211–230.
166. Leptin M. Gastrulation movements: the logic and the nuts and bolts. *Dev Cell.* 2005;8(3):305–320.
167. Savin T, Kurpios NA, Shyer AE, et al. On the growth and form of the gut. *Nature.* 2012;476(7358):57–62.
168. Spahn P, Reuter R. A vertex model of Drosophila ventral furrow formation. *PLoS One.* 2013;8(9):e75051.
169. Shyer AE, Tallinen T, Nerurkar NL, et al. Villification: how the gut gets its villi. *Science.* 2013;342(6155):212–218.

INDEX

Note: Page numbers followed by "*f*" indicate figures and "*t*" indicate tables.

A

D

N

O

P

Q

R

S

T

V

W

X

Y

Z

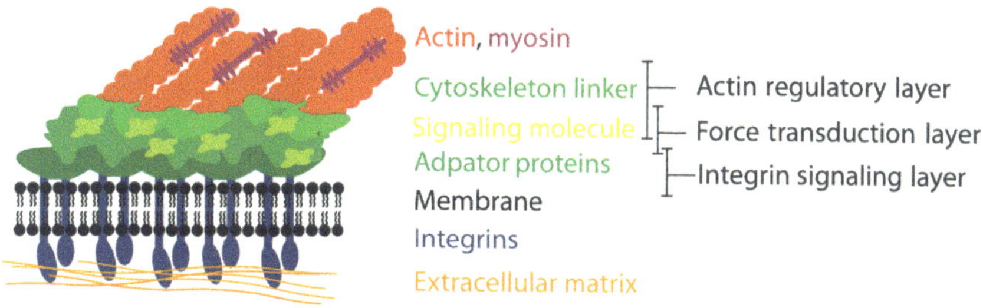

Brenton D. Hoffman, Figure 1.1

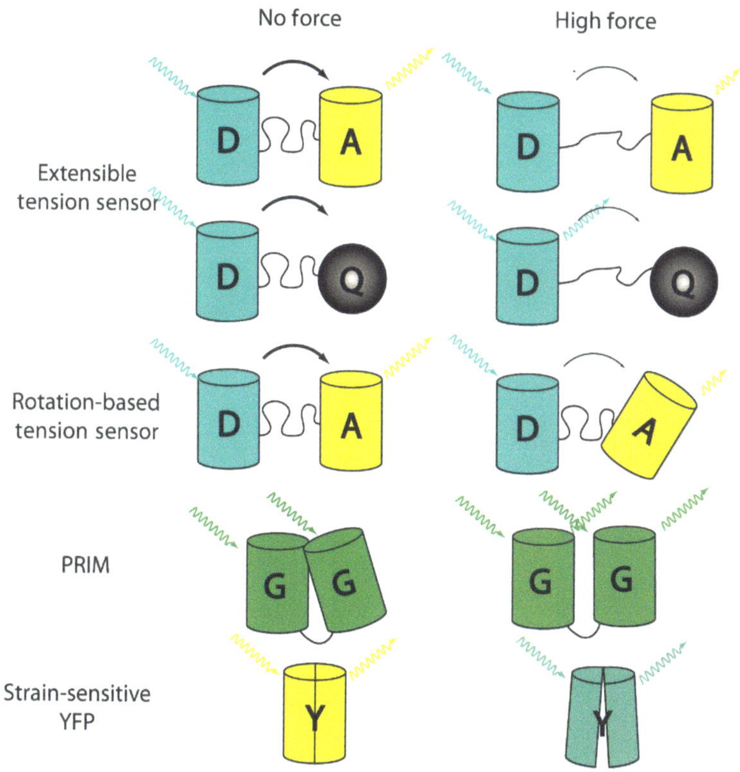

Brenton D. Hoffman, Figure 1.4

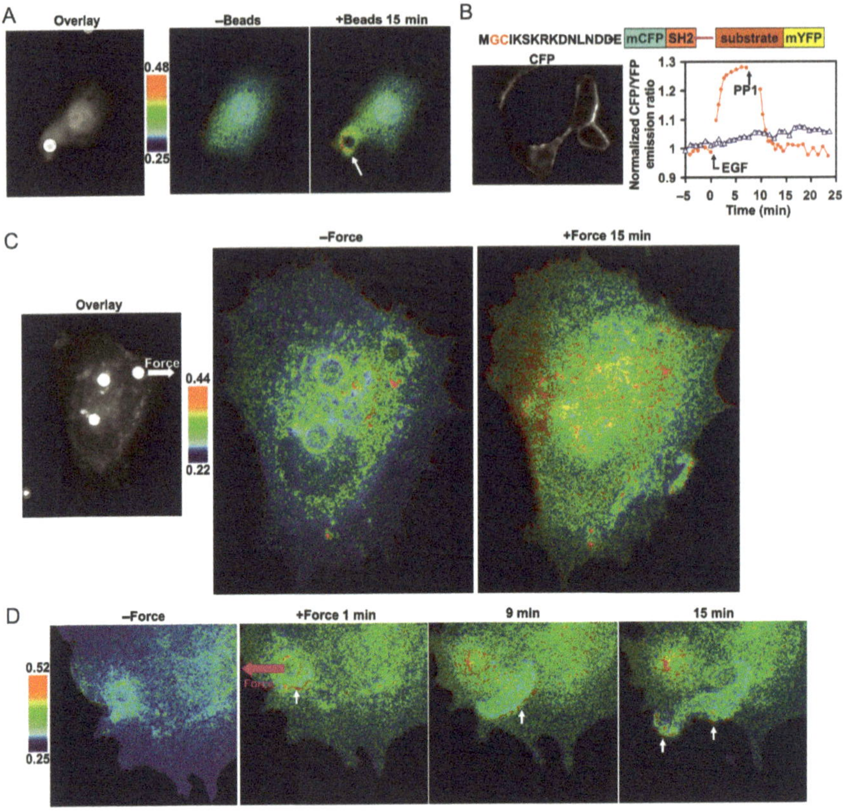

Shaoying Lu and Yingxiao Wang, Figure 2.1

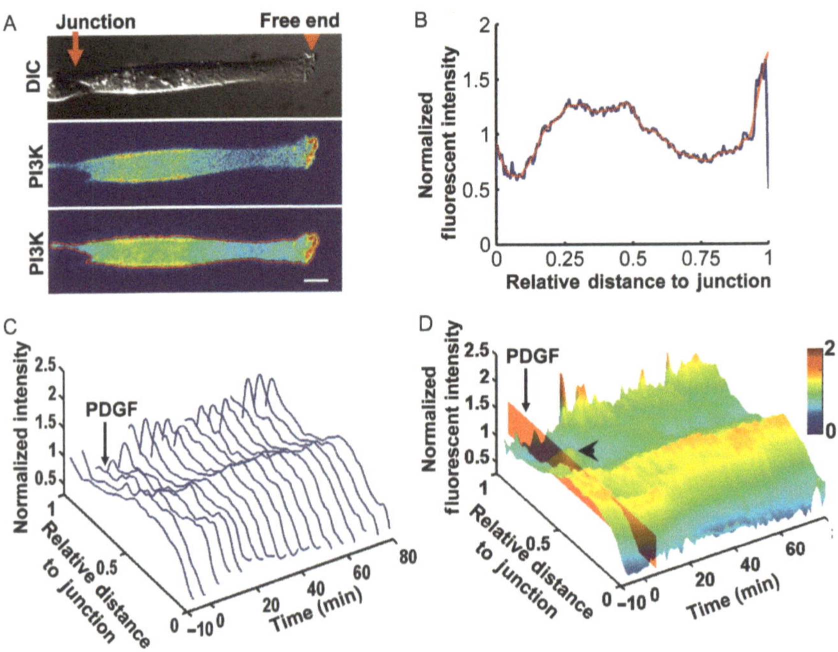

Shaoying Lu and Yingxiao Wang, Figure 2.3

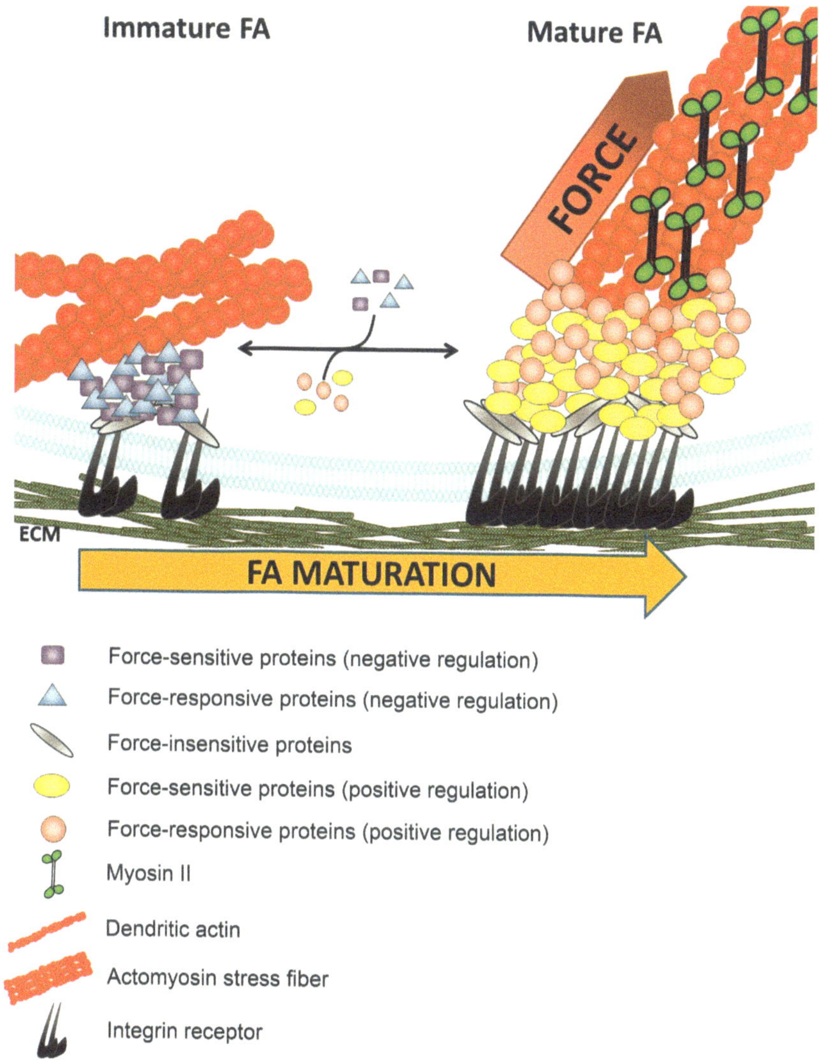

Jean-Cheng Kuo, Figure 3.2

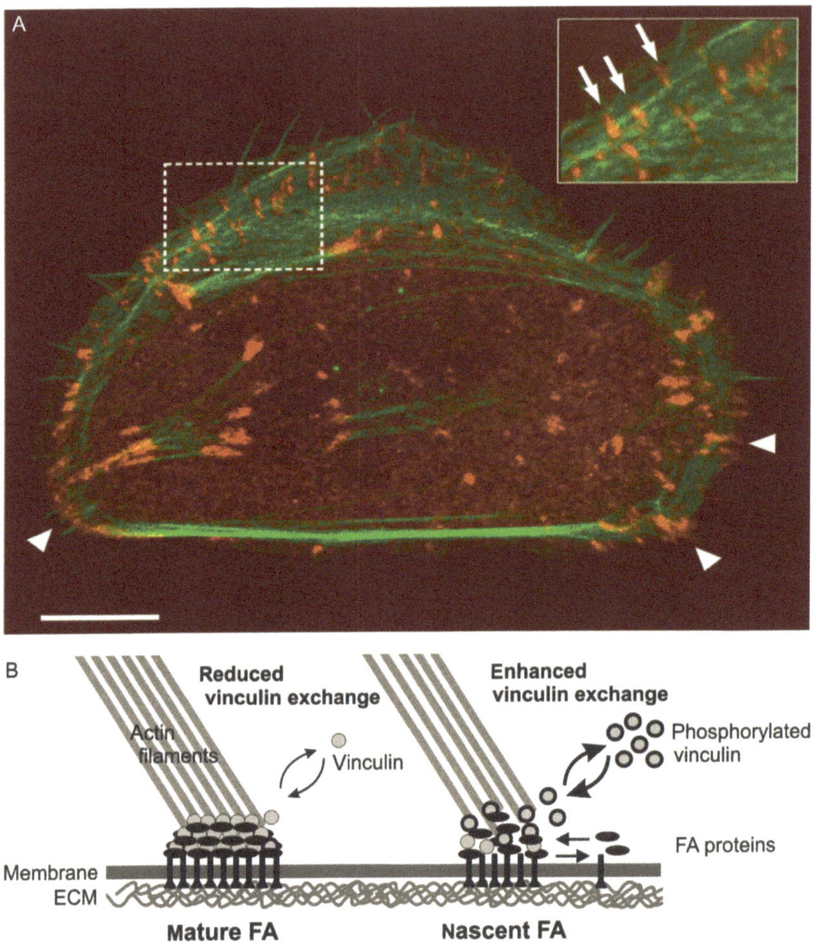

Wolfgang H. Goldmann, Figure 4.2

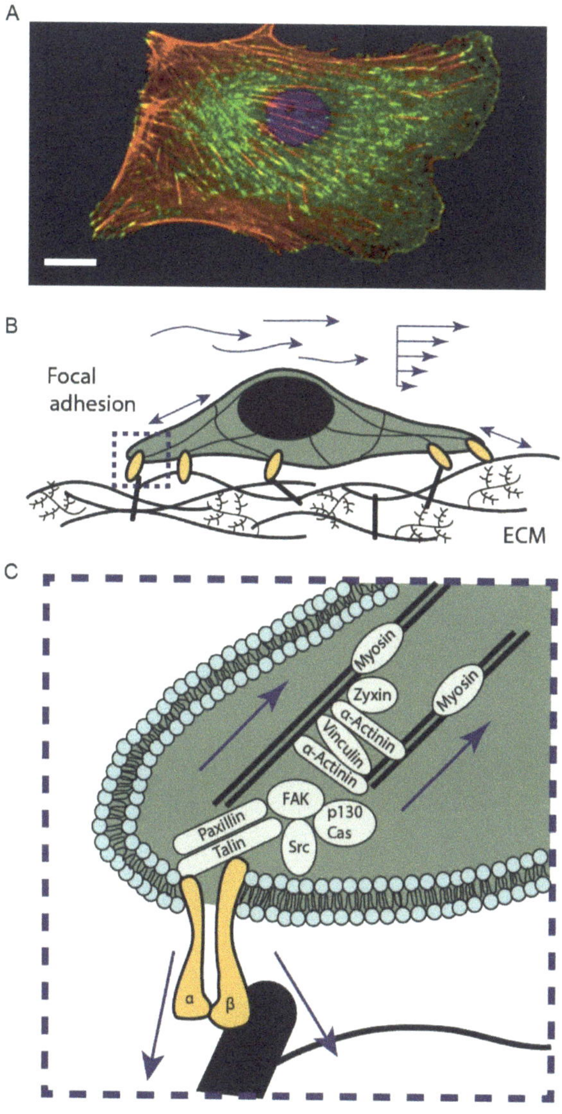

KRISTINA HAASE *ET AL.*, FIGURE 5.1

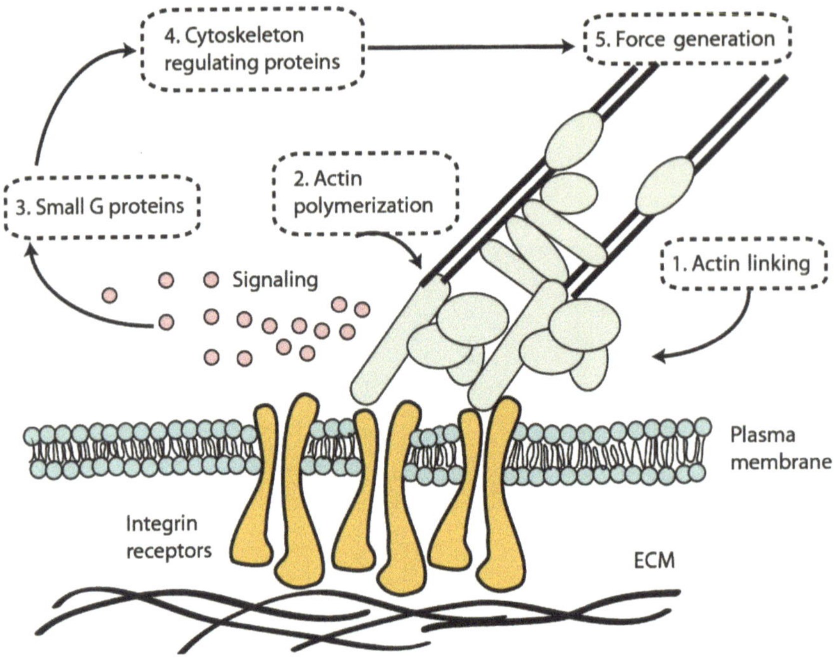

Kristina Haase *et al.*, Figure 5.4

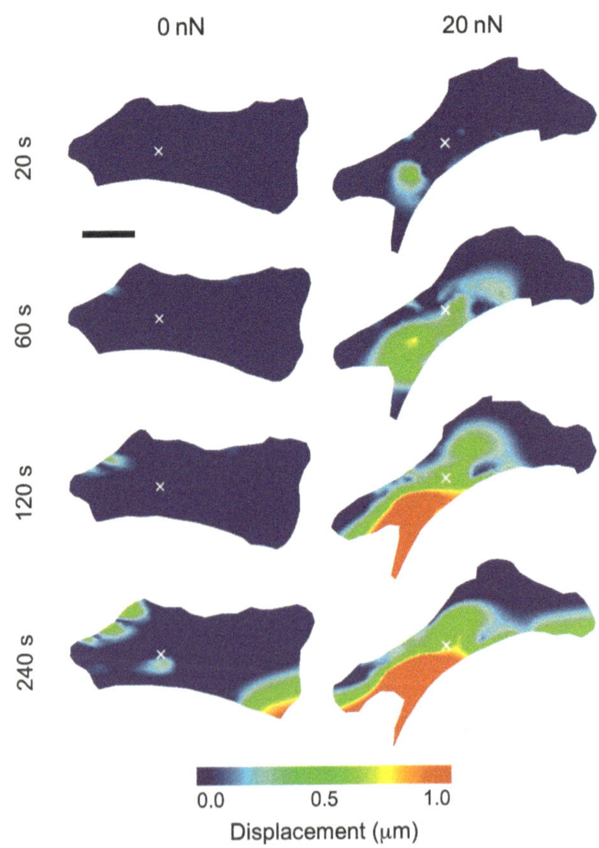

Kristina Haase *et al.*, Figure 5.6

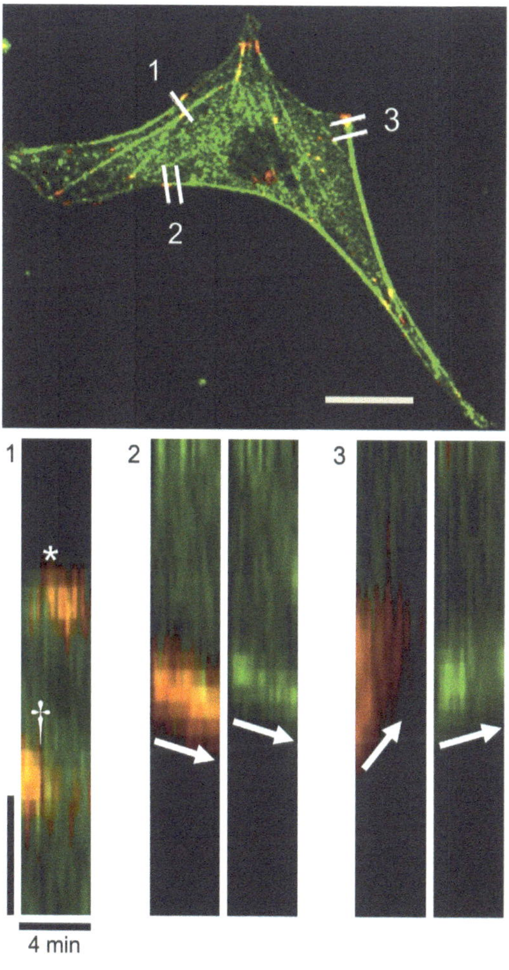

KRISTINA HAASE *ET AL.*, FIGURE 5.7

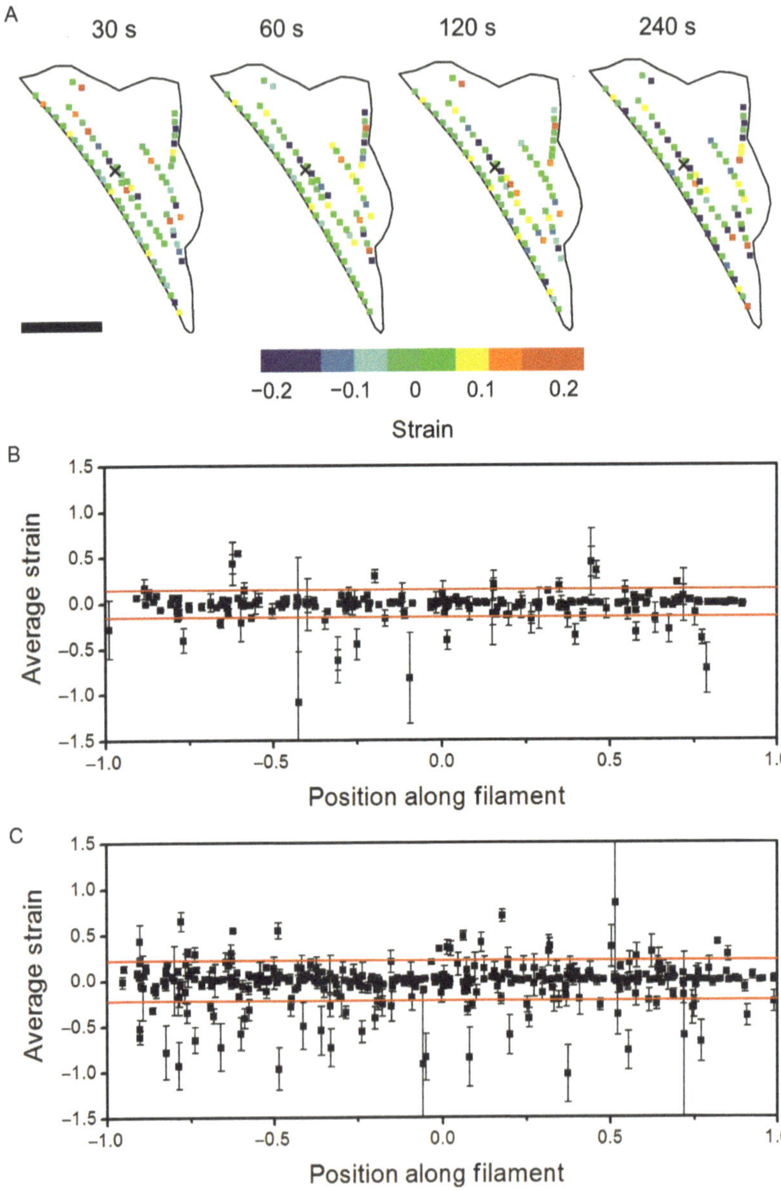

KRISTINA HAASE *ET AL.*, FIGURE 5.8

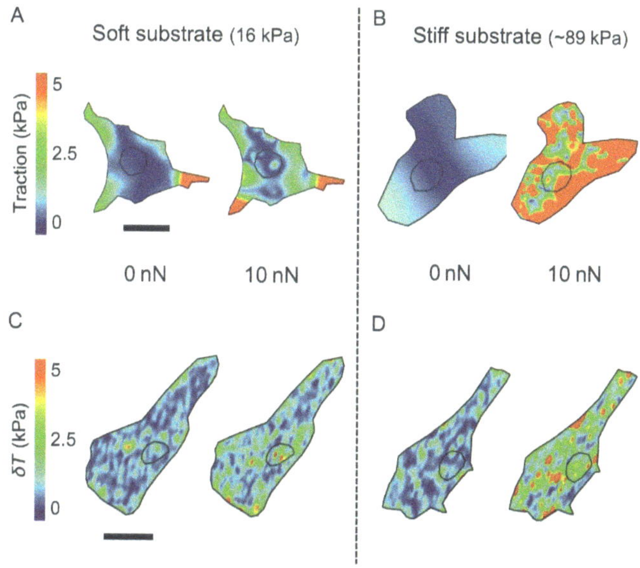

Kristina Haase *et al.*, Figure 5.11

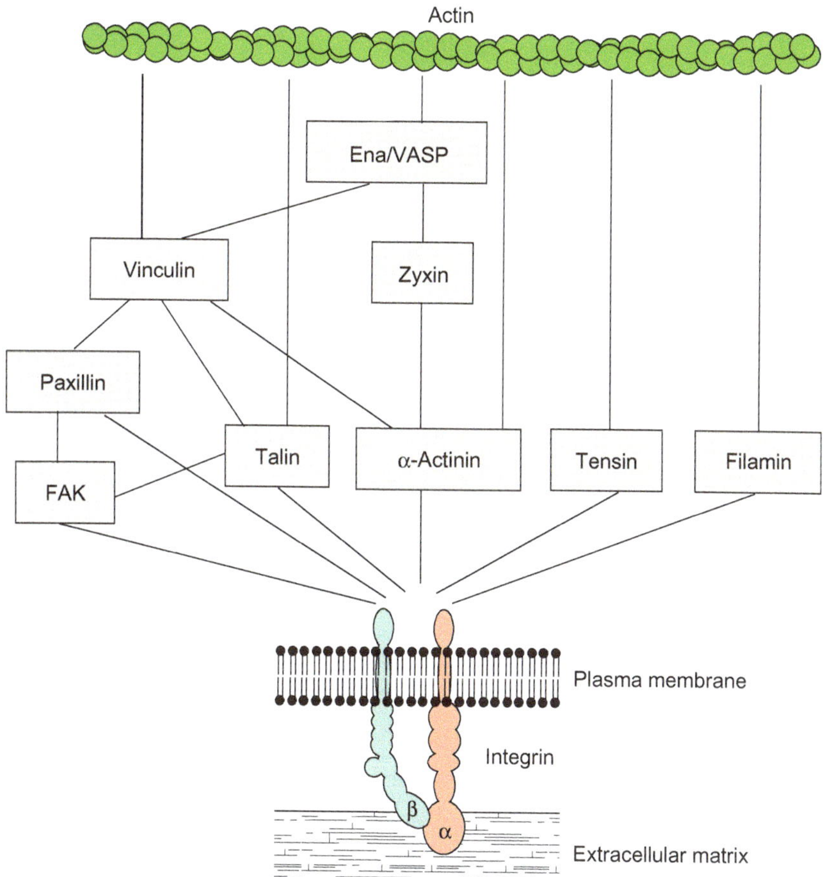

Hiroaki Hirata *et al.*, Figure 6.1

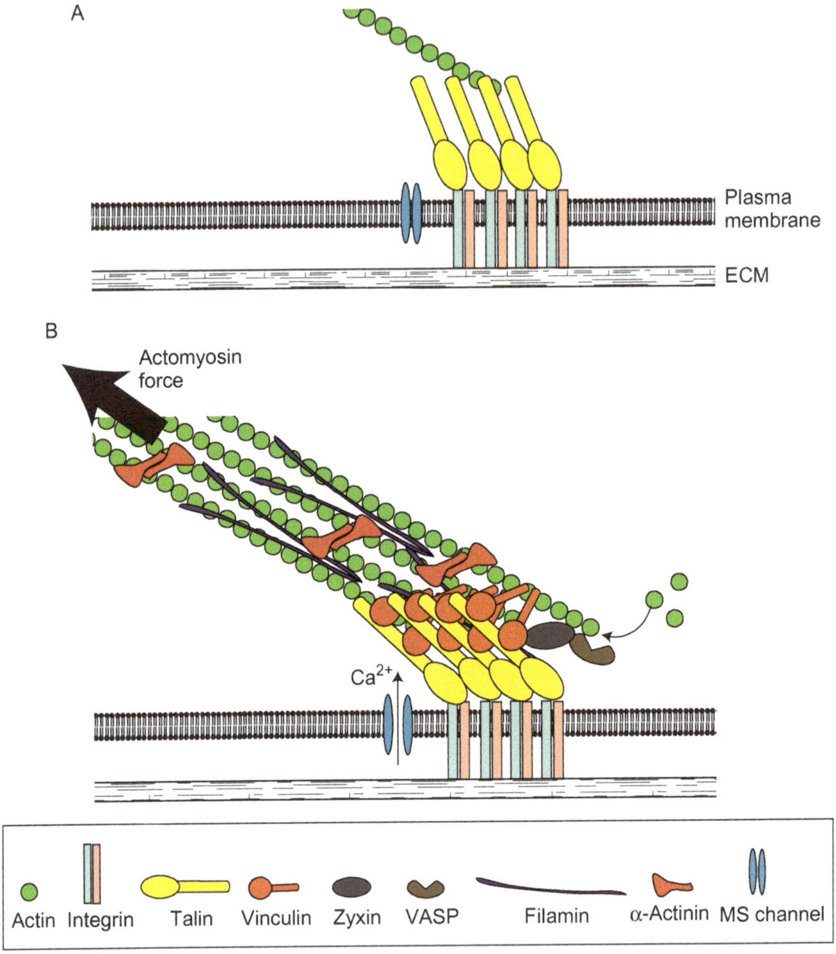

Hiroaki Hirata *et al.*, Figure 6.2

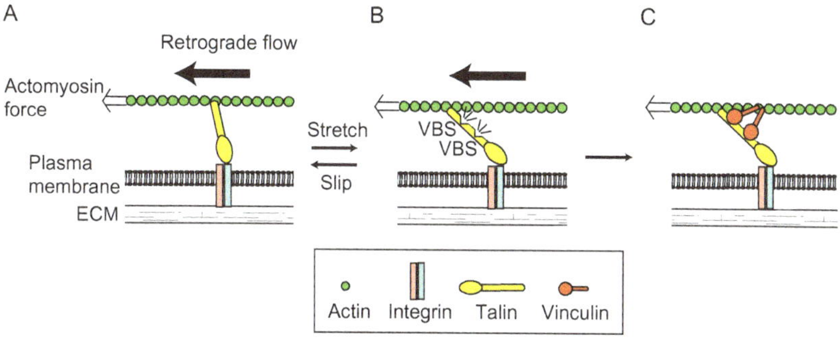

HIROAKI HIRATA *ET AL.*, FIGURE 6.3

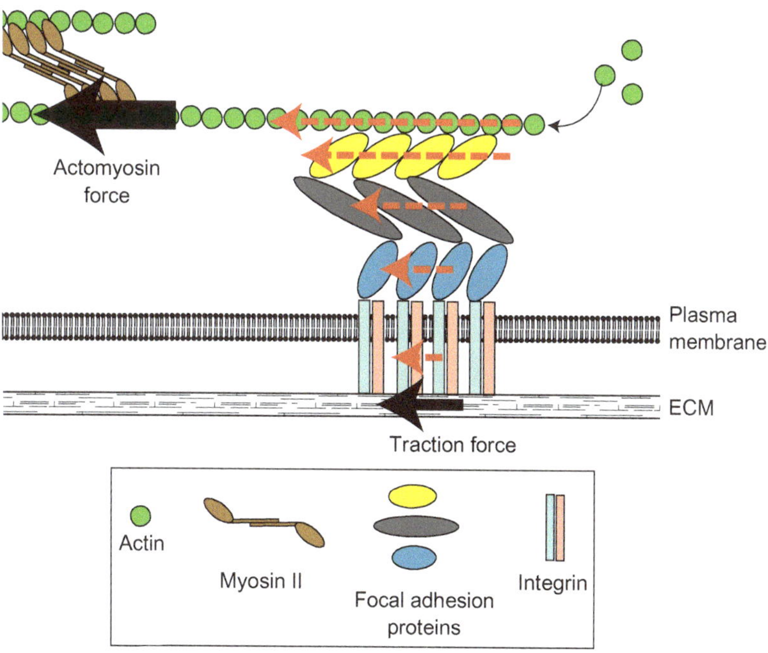

HIROAKI HIRATA *ET AL.*, FIGURE 6.4

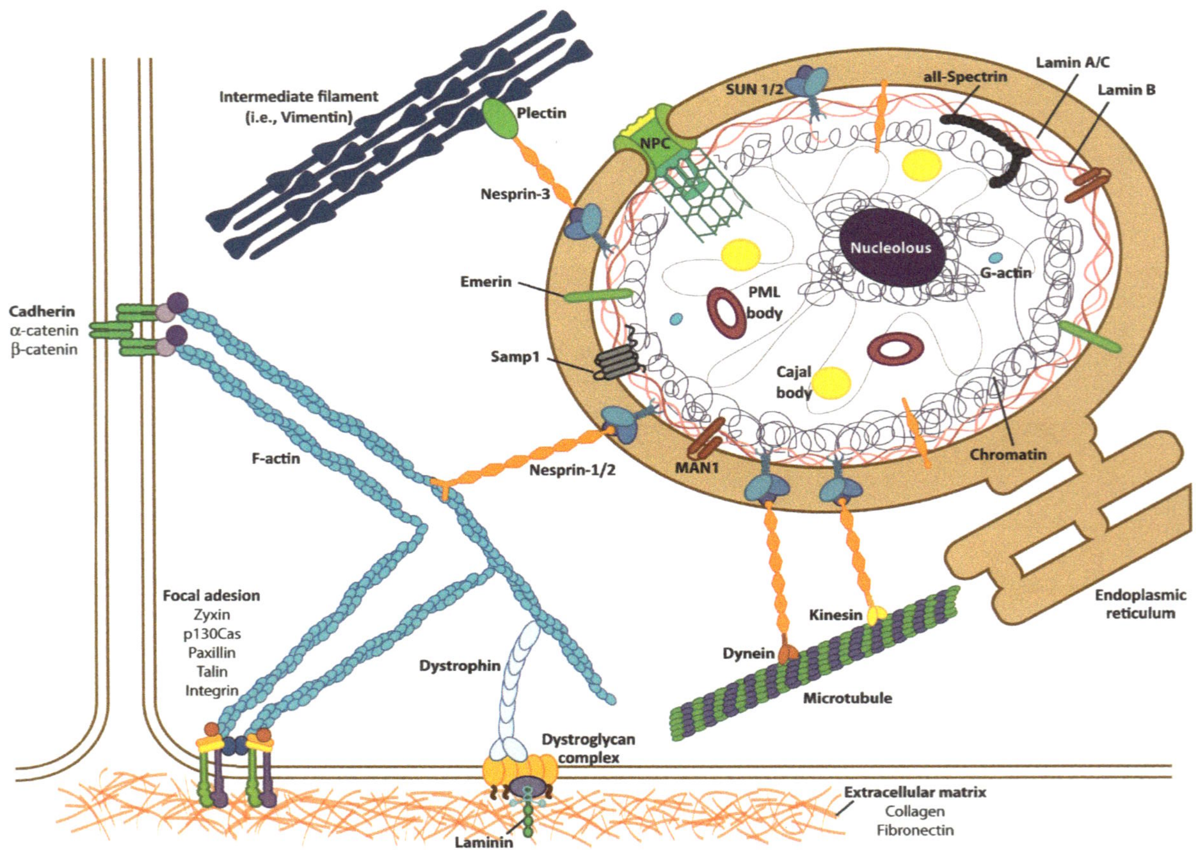

ASHLEY KAMINSKI *ET AL.*, FIGURE 7.1

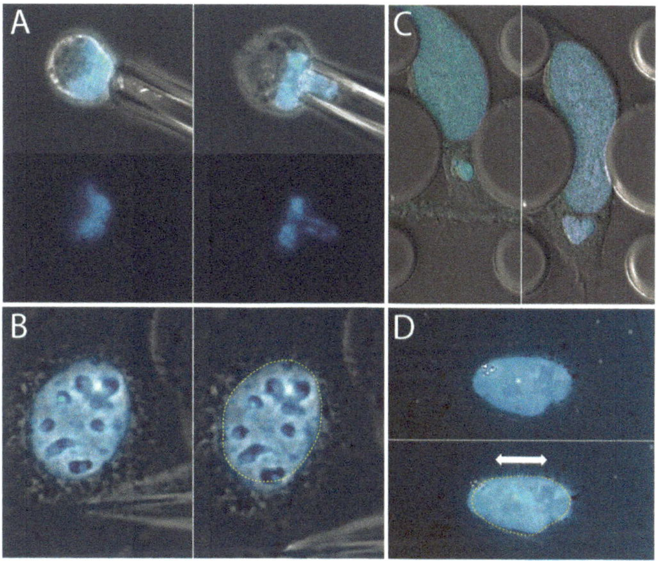

ASHLEY KAMINSKI *ET AL.*, FIGURE 7.3

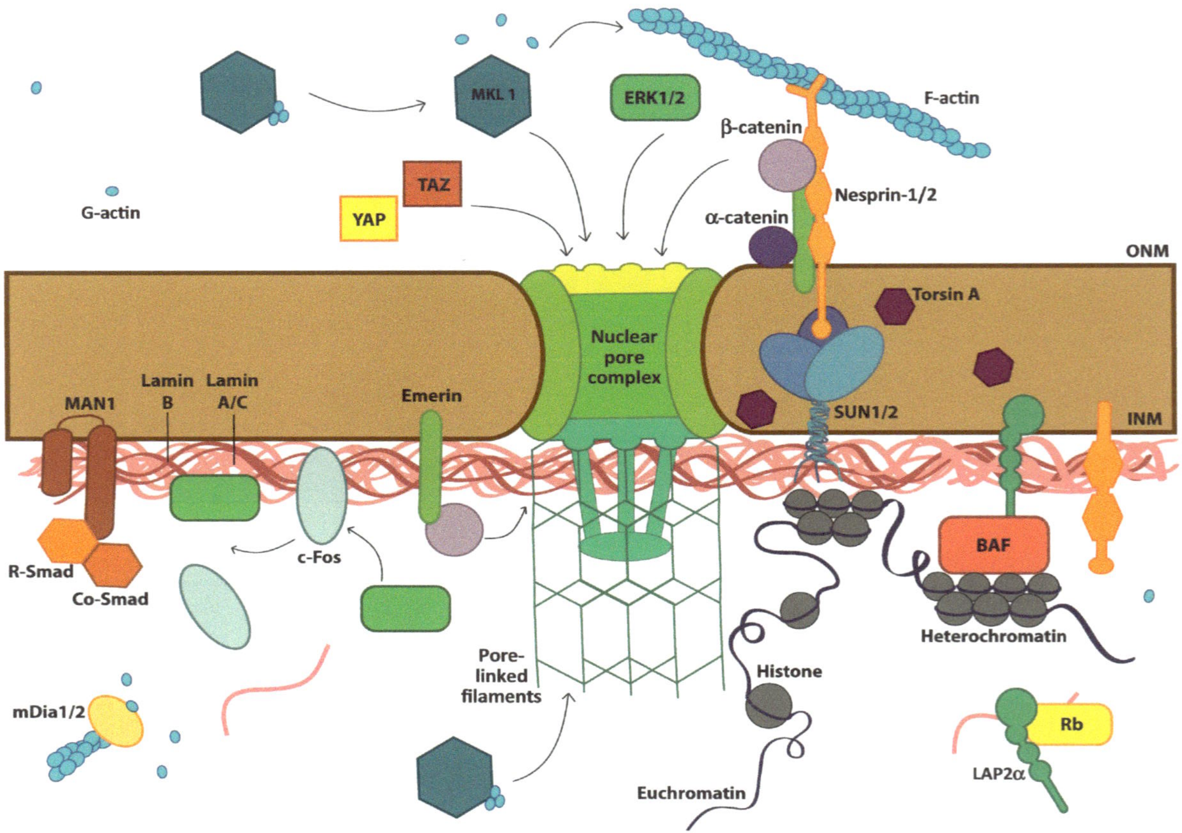

Ashley Kaminski *et al.*, Figure 7.4

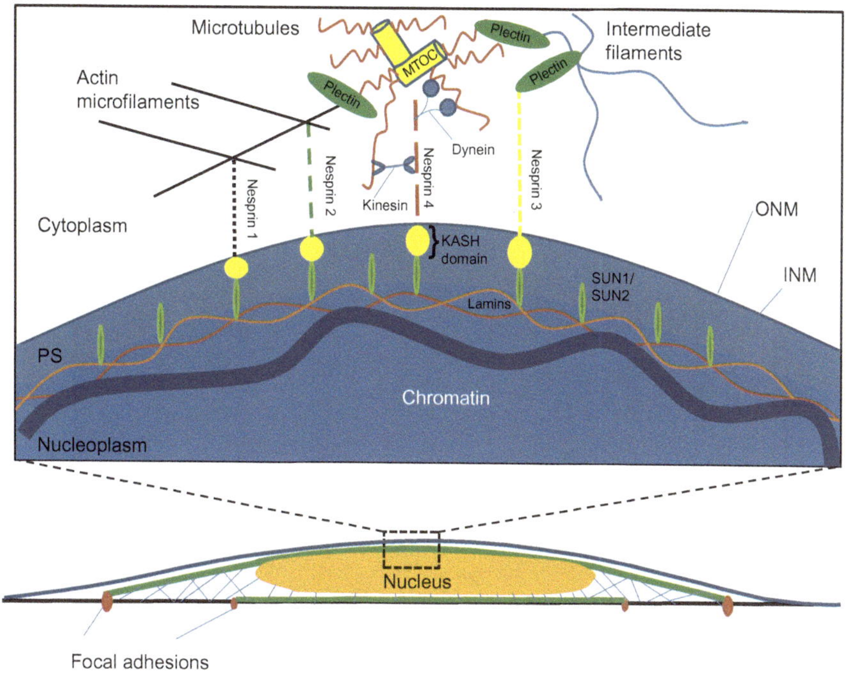

SAMER ALAM *ET AL.*, FIGURE 8.1

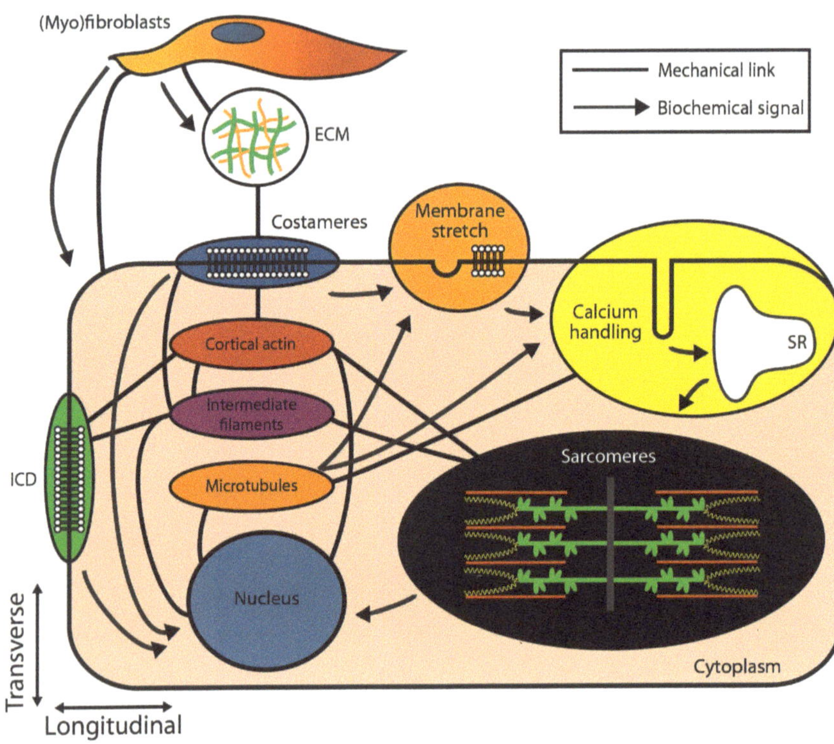

Gaurav Kaushik and Adam J. Engler, Figure 9.1

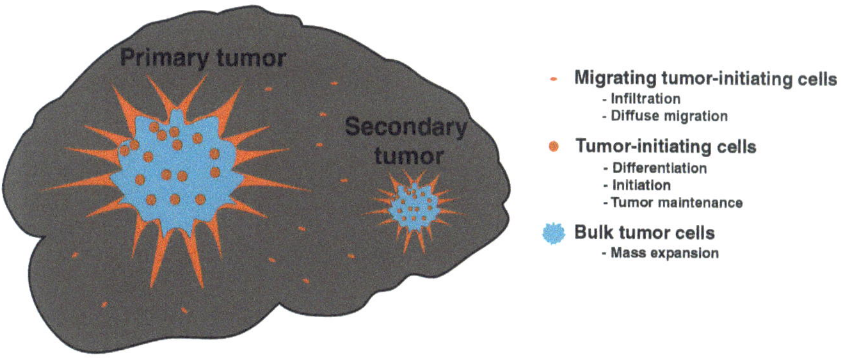

Sophie Y. Wong and Sanjay Kumar, Figure 10.1

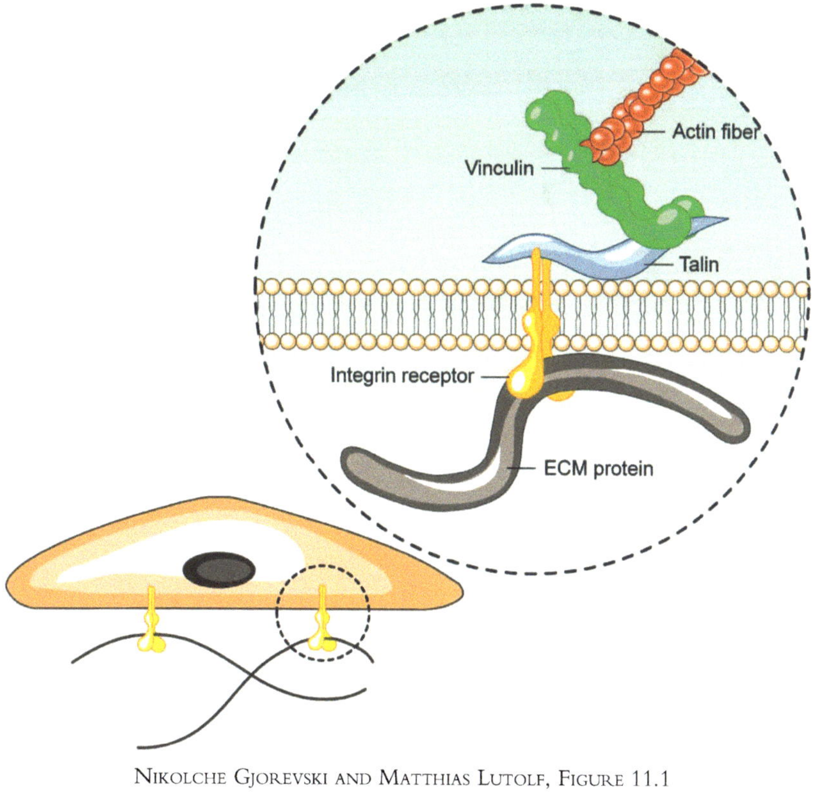

Nikolche Gjorevski and Matthias Lutolf, Figure 11.1

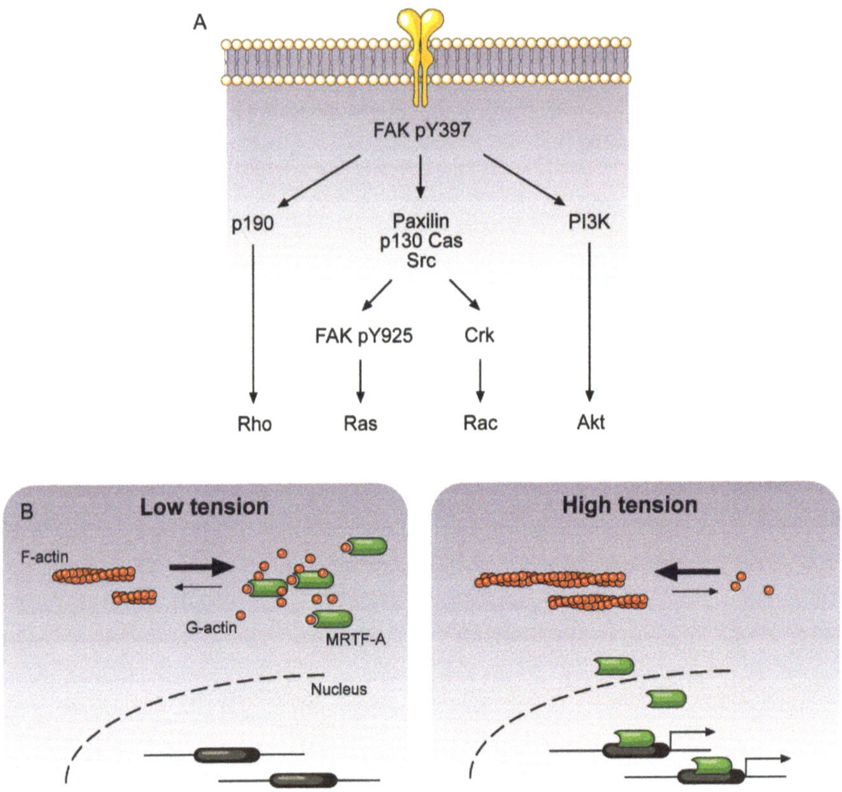

NIKOLCHE GJOREVSKI AND MATTHIAS LUTOLF, FIGURE 11.2

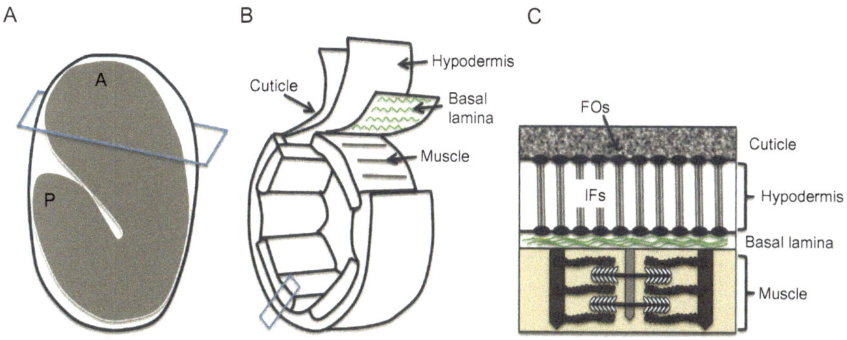

ERIN J. CRAM, FIGURE 12.3

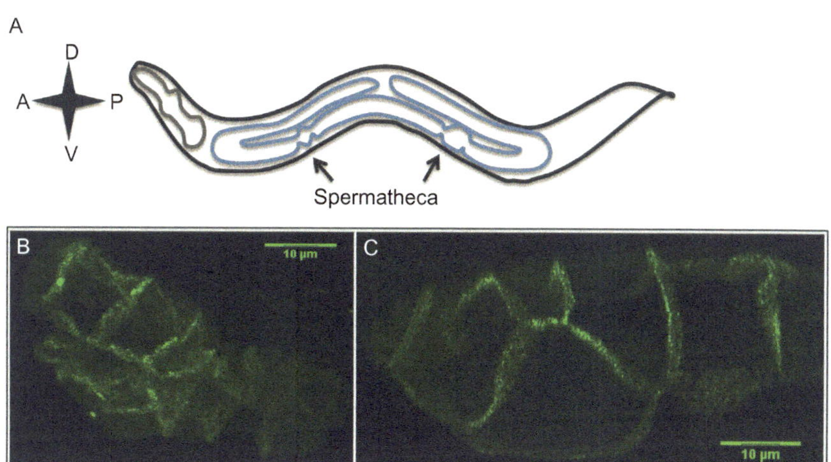

ERIN J. CRAM, FIGURE 12.5

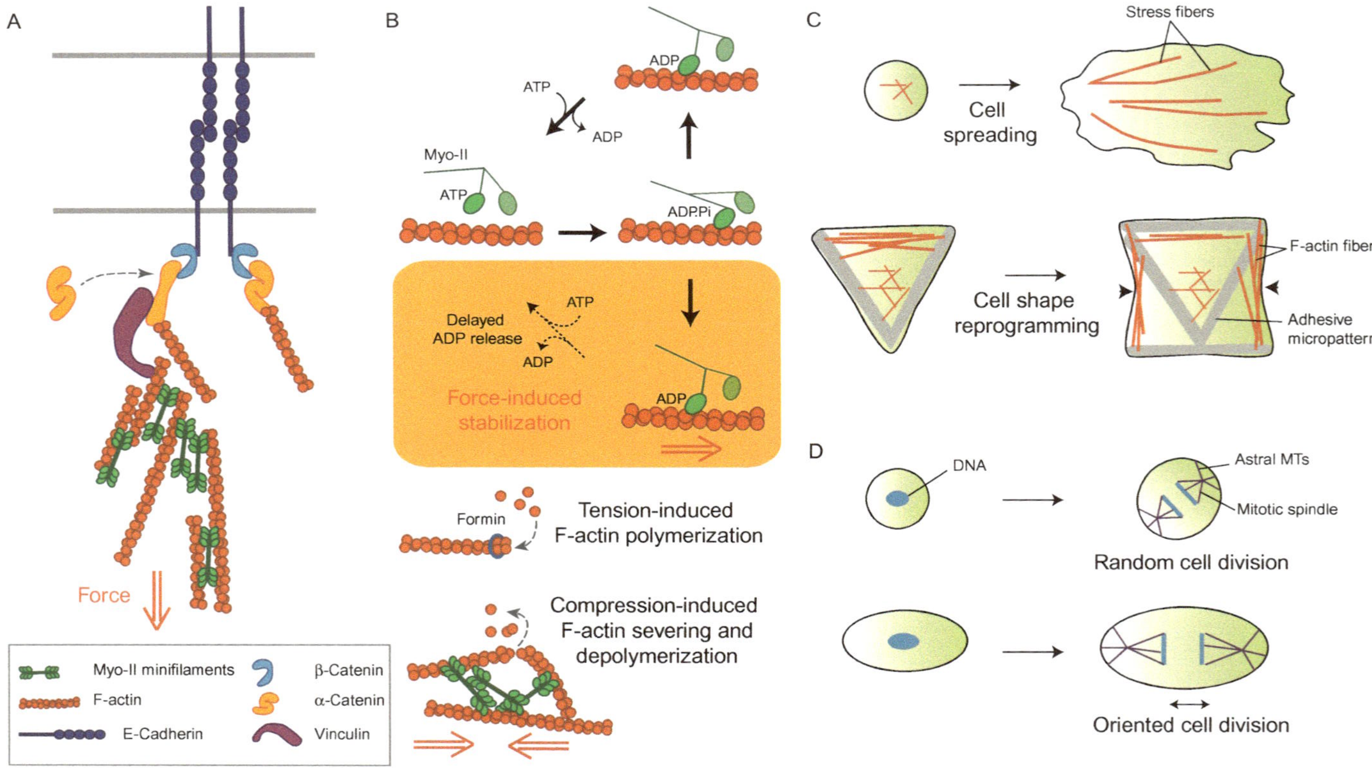

Soline Chanet and Adam C. Martin, Figure 13.1

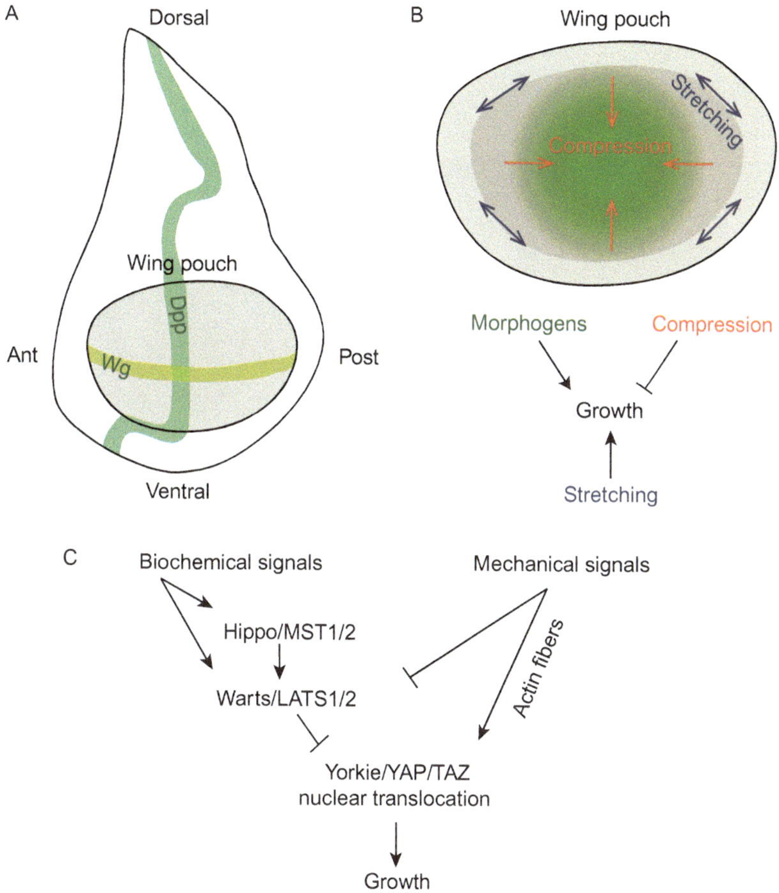

Soline Chanet and Adam C. Martin, Figure 13.2

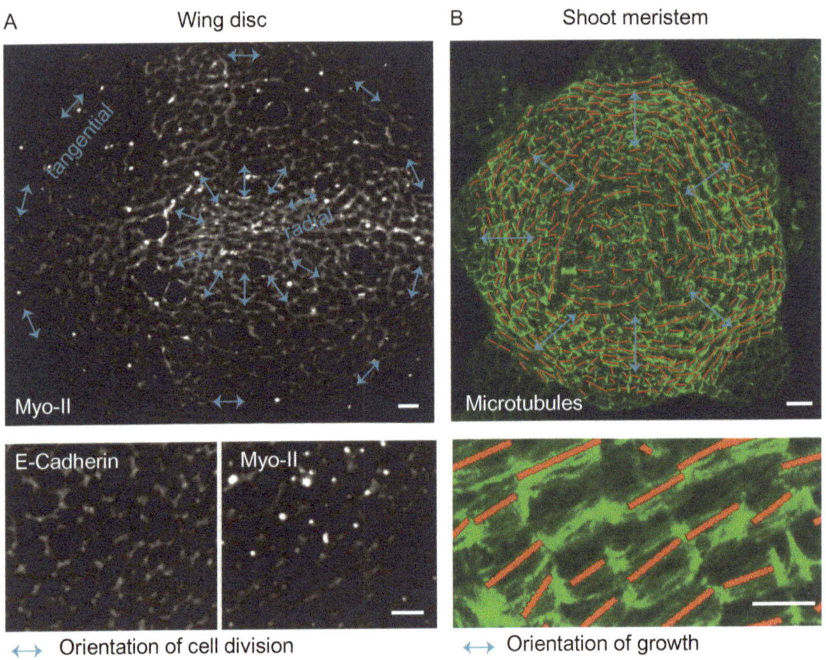

SOLINE CHANET AND ADAM C. MARTIN, FIGURE 13.3

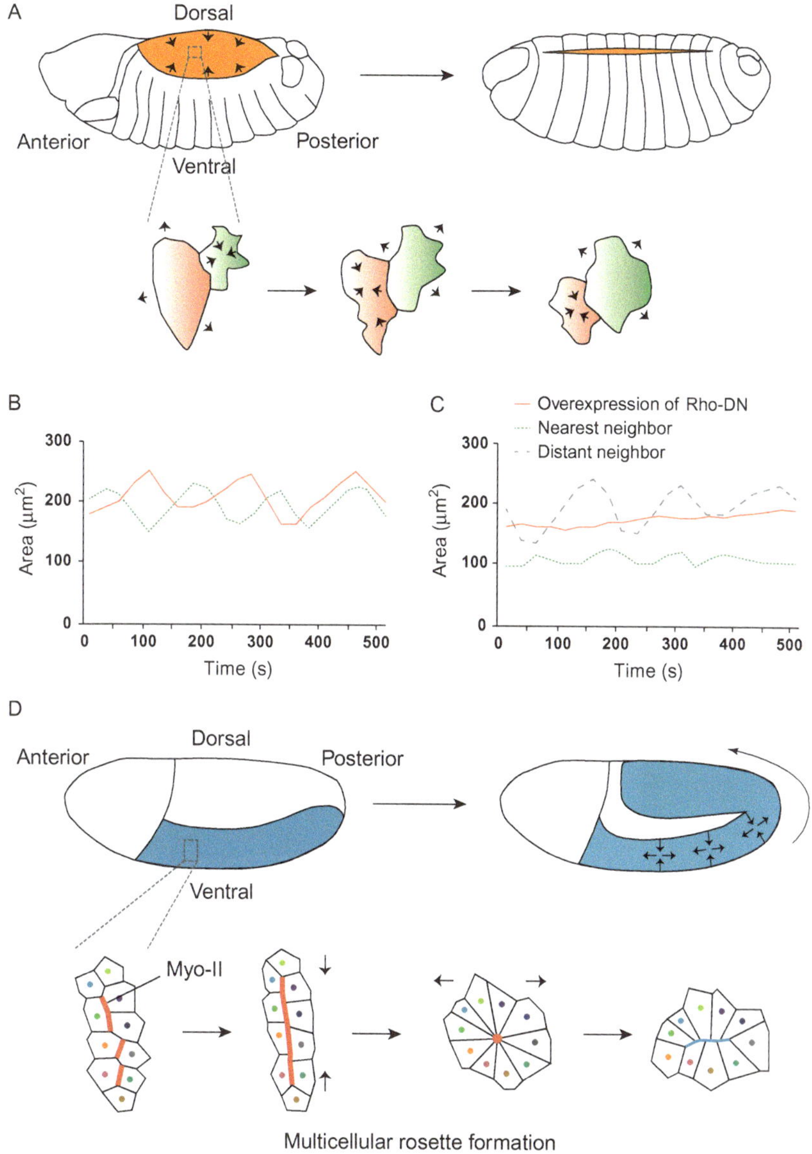

SOLINE CHANET AND ADAM C. MARTIN, FIGURE 13.4

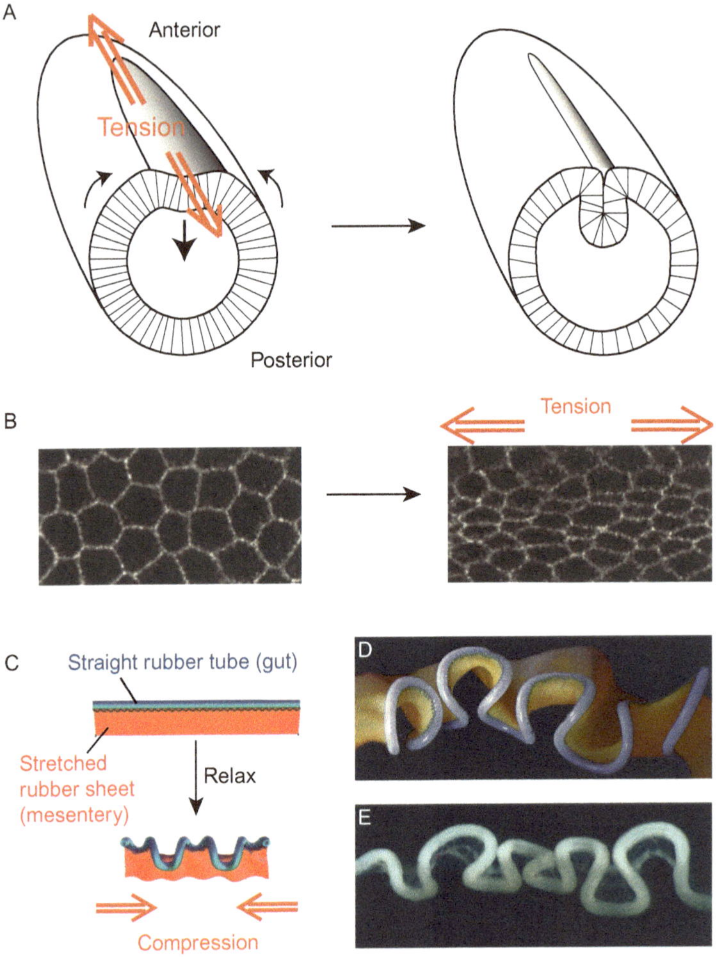

Soline Chanet and Adam C. Martin, Figure 13.5

Kolibri 360 GmbH
Ferdinand-Jühlke-Straße 7
99095 Erfurt, Deutschland
produktsicherheit@kolibri360.de